Stochastic Differential Equations

# Stochastic Differential Equations

An Introduction with Applications in Population
Dynamics Modeling

*Michael J. Panik*
*Department of Economics and Finance, Barney School of Business
and Public Administration*
*West Hartford, CT, USA*

This edition first published 2017

© 2017 John Wiley & Sons, Inc.

The right of Michael J Panik to be identified as the author of this work, has been asserted in accordance with law.

*Registered Office*
John Wiley & Sons, Inc., 111 River Street, Hoboken, NJ 07030, USA

*Editorial Office*
111 River Street, Hoboken, NJ 07030, USA

For details of our global editorial offices, customer services, and more information about Wiley products visit us at www.wiley.com.

Wiley also publishes its books in a variety of electronic formats and by print-on-demand. Some content that appears in standard print versions of this book may not be available in other formats.

*Library of Congress Cataloging-in-Publication Data*

Names: Panik, Michael J.
Title: Stochastic differential equations : an introduction with applications
   in population dynamics modeling / Michael J. Panik.
Description: 1st edition. | Hoboken, NJ : John Wiley & Sons, Inc., [2017] |
   Includes bibliographical references and index.
Identifiers: LCCN 2016056661 (print) | LCCN 2016057077 (ebook) | ISBN
   9781119377382 (cloth) | ISBN 9781119377412 (pdf) | ISBN 9781119377405 (epub)
Subjects: LCSH: Stochastic differential equations.
Classification: LCC QA274.23 .P36 2017 (print) | LCC QA274.23 (ebook) | DDC 519/.22–dc23
LC record available at https://lccn.loc.gov/2016056661

Cover image: © RKaulitzki/Getty Images
Cover design by Wiley

Set in 10/12pt Warnock by SPi Global, Pondicherry, India

Printed in the United States of America

10  9  8  7  6  5  4  3  2  1

# Contents

*In Memory of*
*John Snizik*

# Preface

The three common varieties of stochastic models that are typically used to study population dynamics are discrete-time Markov chain models, continuous-time Markov chain models, and stochastic differential equation (SDE) models. This book focuses on the third type—the use of (Itô) SDEs to determine a variety of population growth equations. While *deterministic* growth models have a rich history of applications in a multitude of fields, it should be obvious that such devices have their limitations in that they don't adequately account for, say, environmental noise (e.g., weather, natural disasters, and epidemics) as incorporated in and mirrored by a stochastic process. As will be seen later on, SDEs represent mathematical models that combine deterministic and stochastic components of dynamic behavior; their solutions are stochastic processes that depict diffusive dynamics.

A major difficulty associated with the derivation and application of SDEs is the up-front investment required of the reader in terms of preparation in the areas of mathematical statistics, real analysis and measure theory, stochastic calculus, and stochastic processes. Indeed, the scope of the requirements for studying SDEs can be rather daunting. To address any background deficiencies on the part of the reader, and to offer a comprehensive review for those who need a mathematical refresher, I have developed four detailed and very comprehensive chapters (Chapters 1–4) pertaining to the aforementioned areas of mathematics/statistics. These chapters are richer and much more extensive than what is normally included as a *review section* in virtually every book on SDEs. The perfunctory mathematical reviews typically offered are woefully inadequate for *practitioners* and *consumers* of SDE-based growth models. In addition, Appendices A–D are also offered for review purposes. These involve some basic calculus concepts (including the Riemann integral), the Lebesgue and Lebesgue–Stieltjes integrals, and ordinary differential equations.

Given that the book is written for beginners or novices to the area of SDEs, it is particularly well suited for advanced undergraduates and beginning graduate students in the areas of economics, population studies, environmental sciences, engineering, and biological sciences who have had a few courses in the calculus and statistics but have not been exposed to the full spectrum of mathematical study received by mathematics majors proper. It is also designed for practitioners in the aforementioned fields who need a gentle introduction to SDEs via a thorough review of the mathematical apparatus needed for studying this discipline. That said, the book will certainly be challenging to most readers. However, it does contain all that is needed for the successful study and application of (Itô-type) SDEs.

My sincere gratitude is extended to my colleagues in the Department of Economics and Finance at the University of Hartford. In particular, I have benefited considerably over the years from conversations with Bharat Kolluri, Farhad Rassekh, Rao Singamsetti, and Mahmoud Wahab. In addition, Alice Schoenrock accurately and steadfastly typed the entire manuscript. Her efforts are greatly appreciated.

I would also like to thank Kathleen Pagliaro, Assistant Editor, for helping to expedite the publication process. A special note of appreciation goes to Susanne Steitz-Filler—Editor (mathematics and statistics) at John Wiley & Sons—for her professionalism, vision, and encouragement.

# Symbols and Abbreviations

| | |
|---|---|
| ■ | denoting end of example |
| $2^{\Omega}$ | power set |
| $\Omega$ | the universal set or space or sample space |
| $(\Omega, A)$ | measurable space |
| $(\Omega, A, P)$ | probability space |
| $(\Omega, A, F, P)$ | filtered probability space |
| $(\Omega, A, \mu)$ | measure space |
| $(\Omega, A_c, \mu_c)$ | complete measure space |
| $(\Omega, T)$ | topological space, where $T$ is a topology on $\Omega$ |
| $\chi_A(x)$ | the indicator function of set $A$ |
| $\omega$ | outcome point in a sample space $\Omega$ |
| $\delta_{ij}$ | the Dirac delta function |
| $\delta_x(A)$ | the Dirac point mass function |
| $\mu$ | measure function |
| $\mu_c$ | the completion of $\mu$ |
| $\mu_r$ | the $r$th central moment |
| $\mu_r'$ | the $r$th moment about zero |
| $\mu^*$ | outer measure function |
| $\sigma(C)$ | the $\sigma$-algebra generated by $C$ |
| $\tau$ | stopping time |
| $\emptyset$ | null set |
| $\in$ | element inclusion |
| $\Gamma(\cdot)$ | the gamma function |
| $\sum$ | the $n$th-order variance–covariance matrix |
| $A$ | autonomous SDE |
| $A$ | $\sigma$-algebra |
| $A'$ | the complement of set $A$ |
| $\bar{A}$ | the closure of set $A$ |
| $A_1 \otimes A_2$ | the product $\sigma$-algebra |

| | |
|---|---|
| $\tilde{A}_t$ | the augmentation with respect to $P$ of $\left\{A_t^w\right\}_{t\geq0}$ |
| $\left\{A_t^w\right\}_{t\geq0}$ | the natural filtration generated by $\left\{W_t\right\}_{t\geq0}$ |
| $A - B$ | the difference between sets $A$, $B$ |
| $A\cup B$ | the union of sets $A$, $B$ |
| $A\cap B$ | the intersection of sets $A$, $B$ |
| $A\subset B$ | set $A$ is a proper subset of set $B$ |
| $A\subseteq B$ | set $A$ is a subset of set $B$ |
| $A\Delta B$ | the symmetric difference between sets $A$, $B$ |
| a.e. | almost everywhere |
| AL | autonomous linear SDE |
| ALH | autonomous linear homogeneous SDE |
| ALNS | autonomous linear narrow sense SDE |
| *arg max* | the argument of the maximum |
| a.s. | almost surely |
| $B$ | Banach space |
| $B$ | the Borel $\sigma$-algebra |
| BM | Brownian motion |
| CIR | Cox, Ingersol, and Ross SDE |
| CKLS | Chan, Karolyi, Longstaff, and Sanders SDE |
| $C(X)$ | the set of all bounded continuous functions on a metric space $X$ |
| DCT | dominated convergence theorem |
| $d(E, F)$ | the distance between sets $E$, $F$ |
| $d(x, y)$ | the distance between points $x$, $y$ |
| $e_i$ | the $i$th unit vector |
| EM | Euler–Maruyama method |
| $E(X)$ (or $\mu$) | the expectation of a random variable $X$ |
| $E\|X_t\|^p$ | $p$th-order moment |
| $E(X\|Y)$ | the conditional expectation of $X$ given $Y$, where $X$ and $Y$ are random variables |
| $F$ | filtration |
| $\|f\|$ | the norm of a function $f$ |
| $\|f\|_p$ | the $p$-norm of a function $f$ |
| $(f, g)$ | the inner product of functions $f$, $g$ |
| $f:X\to Y$ | $f$ is a point-to-point mapping from set $X$ to set $Y$ |
| $f^{-1}:Y\to X$ | $f^{-1}$ is the inverse (point-to-point) mapping from set $Y$ to set $X$ |
| $G_f$ | the graph of $f$ |
| $H$ | Hilbert space |
| $H_r$ | Hilbert space of random variables |
| $I(f)$ (or $I(t)$) | the Itô stochastic integral |
| $\mathbb{I}_n$ | the identity matrix of order $n$ |

| | |
|---|---|
| $I_q(\cdot)$ | the Bessel function of the first kind of order $q$ |
| $\mathbb{J}$ | the Jacobian determinant |
| $L$ | linear SDE |
| $L$ | likelihood function |
| $L^2(\mu)$ | the class of real-valued square integrable functions |
| $L^p(\mu)$ | the set of all measurable functions $f$ such that $|f|^p$ is integrable |
| LH | linear homogeneous SDE |
| $lim_{n\to\infty} \inf A_i$ | the limit inferior of a sequence $\{A_i\}$ of sets |
| $lim_{n\to\infty} \inf x_n$ | the limit inferior of a sequence $\{x_n\}$ of real numbers |
| $lim_{n\to\infty} \sup A_i$ | the limit superior of a sequence $\{A_i\}$ of sets |
| $lim_{n\to\infty} \sup x_n$ | the limit superior of a sequence $\{x_n\}$ of real numbers |
| $\ln L$ | log-likelihood function |
| LNS | linear narrow sense SDE |
| LS | Lebesgue–Stieltjes |
| $M$ | Lebesgue measurable sets |
| $M^2([\cdot,\cdot])$ | the space of all real-valued processes $f$ on the product space $\Omega \times [\cdot,\cdot]$ |
| MCT | monotone convergence theorem |
| ML | maximum likelihood |
| $N$ | the collection of $P$-null sets |
| $O$ | $f(x) = O(g(x))$ is the quantity whose ratio to $g(x)$ remains bounded as $x$ tends to a limit |
| ODE | ordinary differential equation |
| OU | Ornstein–Uhlenbeck |
| $|P|$ | mesh of partition $P$ |
| $P(A|B)$ | the conditional probability of event $A$ given event $B$ |
| $P_1 \otimes P_2$ | the product probability measure |
| PML | pseudo-maximum likelihood |
| $R$ | the set of real numbers |
| $R^*$ | the extended real numbers |
| $R^+$ | the non-negative real numbers |
| $R^n$ | the set of ordered $n$-tuples $(x_1, x_2, ..., x_n)$ or $n$-dimensional Euclidean space |
| RS | Riemann–Stieltjes |
| SBM | standard Brownian motion |
| SDE | stochastic differential equation |
| SO | Shoji–Ozaki method |
| $S(X)$(or $\sigma$) | the standard deviation of a random variable $X$ |
| $V(X)$(or $\sigma^2$) | the variance of a random variable $X$ |
| $\dot{W}(t)$ | white noise |
| $\{W_t\}_{t\geq 0}$ | Brownian motion or Wiener process |

| | |
|---|---|
| $\|x\|$ | the norm of an $n$-tuple $x = (x_1, x_2, \ldots, x_n)$ |
| $X^d$ | the derived set of all limit points of set $X$ |
| $X \sim N(0,1)$ | the random variable $X$ is standard normal |
| $X \sim N(\mu, \sigma)$ | the random variable $X$ is normally distributed with mean $\mu$ and standard deviation $\sigma$ |
| $\{X_t\}_{t \in T}$ | stochastic process |
| $X_t(w)$ | sample path of a stochastic process |
| $(x, x)$ | the inner product of an $n$-tuple $x = (x_1, x_2, \ldots, x_n)$ |
| $X \times Y$ | the product of sets $X$, $Y$ |

# 1

# Mathematical Foundations 1

Point-Set Concepts, Set and Measure Functions, Normed Linear Spaces, and Integration

## 1.1 Set Notation and Operations

### 1.1.1 Sets and Set Inclusion

We may generally think of a **set** as a collection or grouping of items without regard to structure or order. (Sets will be represented by capital letters, e.g., $A, B, C, \ldots$.) An **element** is an item within or a member of a set. (Elements are denoted by small case letters, e.g., $a, b, c, \ldots$.) A set of sets will be termed a **class** (script capital letters will denote a class of sets, e.g., $\mathcal{A}, \mathcal{B}, \mathcal{C}, \ldots$); and a set of classes will be called a **family**.

Let us define a **space** (denoted $\Omega$) as a type of *master* or *universal* set—it is the context in which discussions of sets occur. In this regard, an element of $\Omega$ is a point $\omega$. To define a set $X$, let us write $X = \{x \mid \text{the } x\text{'s possess some defining property}\}$, that is, this reads "$X$ is the set of all elements $x$ such that the $x$'s have some unique characteristic," where "such that" is written "$\mid$."

The set containing no elements is called the **empty set** (denoted $\phi$)—it is a member of every set. What about the size of a set? A set may be *finite* (it is either empty or consists of $n$ elements, $n$ a positive integer), *infinite* (e.g., the set of positive integers), or *countably infinite* (its elements can be put into one-to-one correspondence with the counting numbers).

We next look to inclusion symbols. Specifically, we first consider **element inclusion**. Element $x$ being a member of set $X$ is symbolized as $x \in X$. If $x$ is not a member of, say, set $Y$, we write $x \notin Y$. Next comes **set inclusion** (a subset notation). A set $A$ is termed a **subset** of set $B$ (denoted $A \subseteq B$) if $B$ contains the same elements that $A$ does and possibly additional elements that are not found in $A$. If $A$ is not a subset of $B$, we write $A \nsubseteq B$. Actually, two cases are subsumed in $A \subseteq B$: (1) either $A \subset B$ ($A$ is then called a **proper subset** of $B$, meaning that $B$ is a set that is larger than $A$; or (2) $A = B$ ($A$ and $B$ contain exactly the same

*Stochastic Differential Equations: An Introduction with Applications in Population Dynamics Modeling*, First Edition. Michael J. Panik.
© 2017 John Wiley & Sons, Inc. Published 2017 by John Wiley & Sons, Inc.

elements and thus are **equal**). More formally, $A = B$ if and only if $A \subseteq B$ and $B \subseteq A$. If equality between sets $A$ and $B$ does not hold, we write $A \neq B$.

### 1.1.2 Set Algebra

Given sets $A$ and $B$ within $\Omega$, their **union** (denoted $A \cup B$) is the set of elements that are in $A$, or in $B$, or in both $A$ and $B$. Here, we are employing the *inclusive or*. Symbolically, $A \cup B = \{x | x \in A \text{ or } x \in B\}$ (Figure 1.1a). The **intersection** of sets $A$ and $B$ (denoted $A \cap B$) is the set of elements common to both $A$ and $B$, that is, $A \cap B = \{x | x \in A \text{ and } x \in B\}$ (Figure 1.1b). The **complement** of a set $A$ is the set of elements within $\Omega$ that lie outside of $A$ (denoted $A'$). Here, $A' = \{x | x \notin A\}$ (Figure 1.1c).

If sets $A$ and $B$ do not intersect and thus have no elements in common, then $A$ and $B$ are said to be **disjoint** or **mutually exclusive** and we write $A \cap B = \emptyset$. The **difference** between sets $A$ and $B$ (denoted $A - B$) is the set of elements in $A$ but not in $B$ or $A - B = A \cap B'$. Thus, $A - B = \{x | x \in A \text{ and } x \notin B\}$ (Figure 1.1d). The **symmetric difference** between sets $A$ and $B$ (denoted $A \Delta B$) is the

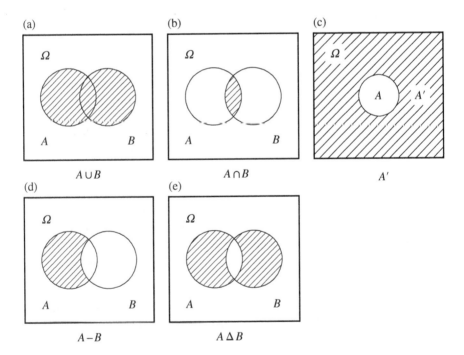

**Figure 1.1** (a) Union of $A$ and $B$, (b) intersection of $A$ and $B$, (c) complement of $A$, (d) difference of $A$ and $B$, and (e) symmetric difference of $A$ and $B$.

union of their differences in reverse order or $A \Delta B = (A - B) \cup (B - A) = (A \cap B') \cup (B \cap A')$ (Figure 1.1e).

A few essential properties of these set operations now follow. Specifically for sets $A$, $B$, and $C$ within $\Omega$:

## UNION

$A \cup A = A$, $A \cup \Omega = \Omega$, $A \cup \emptyset = A$
$A \cup B = B \cup A$ (commutative property)
$A \cup (B \cup C) = (A \cup B) \cup C$ (associative property)
$A \subseteq B$ if and only if $A \cup B = B$

## INTERSECTION

$A \cap A = A$, $A \cap \Omega = A$, $A \cap \emptyset = \emptyset$
$A \cap B = B \cap A$ (commutative property)
$A \cap (B \cap C) = (A \cap B) \cap C$ (associative property)
$A \subseteq B$ if and only if $A \cap B = A$

## COMPLEMENT

$(A')' = A$, $\Omega' = \emptyset$, $\emptyset' = \Omega$
$A \cup A' = \Omega$, $A \cap A' = \emptyset$
$$\left. \begin{array}{l} (A \cup B)' = A' \cap B' \\ (A \cap B)' = A' \cup B' \end{array} \right\} \text{De Morgan's laws}$$

## DIFFERENCE

$A - B = (A \cup B) - B = A - (A \cap B)$
$(A - B) - C = A - (B \cup C) = (A - B) \cap (A - C)$
$A - (B - C) = (A - B) \cup (A \cap C)$
$(A \cup B) - C = (A - C) \cup (B - C)$

## SYMMETRIC DIFFERENCE

$A \Delta A = \emptyset$, $A \Delta \emptyset = A$
$A \Delta B = B \Delta A$ (commutative property)
$A \Delta (B \Delta C) = (A \Delta B) \Delta C$ (associative property)
$A \cap (B \Delta C) = (A \cap B) \Delta (A \cap C)$

## DISTRIBUTIVE LAWS (connect the operations of union and intersection)

$A \cap (B \cup C) = (A \cap B) \cup (A \cap C)$
$A \cup (B \cap C) = (A \cup B) \cap (A \cup C)$

If $\{A_i, i = 1, ..., n\}$ is any arbitrary finite class of sets, then the extension of the union and intersection operations to this class can be written, respectively, as

$$\cup_{i=1}^{n} A_i \text{ and } \cap_{i=1}^{n} A_i.$$

Hence, the union of a class of sets is the collection of elements belonging to at least one of them; the intersection of a class of sets is the set of elements common to all of them. In fact, given these notions, De Morgan's laws may be extended to

$$\left(\cup_{i=1}^{n} A_i\right)' = \cap_{i=1}^{n} A_i' \text{ and } \left(\cap_{i=1}^{n} A_i\right)' = \cup_{i=1}^{n} A_i'.$$

Furthermore, if $\{A_i, i = 1, ..., n\}$ and $\{B_j, j = 1, ..., m\}$ are two finite classes of sets with $\{A_i\} \subseteq \{B_j\}$, then

$$\cup_{i=1}^{n} A_i \subseteq \cup_{j=1}^{m} B_j \text{ and } \cap_{j=1}^{m} B_j \subseteq \cap_{i=1}^{n} A_i.$$

In addition, if $\{A_i, i = 1, 2, ...\}$ represents a **sequence of sets**, then their union and intersection appears as

$$\cup_{i=1}^{\infty} A_i \text{ and } \cap_{i=1}^{\infty} A_i,$$

respectively.

## 1.2 Single-Valued Functions

Given two nonempty sets $X$ and $Y$ (which may or may not be equal), a **single-valued function** or **point-to-point mapping** $f: X \rightarrow Y$ is a rule or law of corres-pondence that associates with point $x \in X$ a unique point $y \in Y$. Here, $y = f(x)$ is the **image of $x$** under rule $f$. While set $X$ is called the **domain of $f$** (denoted $D_f$), the collection of those $y$'s that are the image of at least one $x \in X$ is called the **range of $f$** and denoted $R_f$. Clearly the range of $f$ is a subset of $Y$ (Figure 1.2a). If $R_f \subset Y$, then $f$ is an **into mapping**. In addition, if $R_f = Y$ (i.e., *every* $y \in Y$ is the image of at least one $x \in X$ or all the $y$'s are accounted for in the mapping process), then $f$ is termed an **onto** or **surjective mapping**. Moreover, $f$ is said to be **one-to-one** or **injective** if no $y \in Y$ is the image of more than one $x \in X$ (i.e., $x_1 \neq x_2$ implies $f(x_1) \neq f(x_2)$). Finally, $f$ is called **bijective** if it is both one-to-one and onto or both surjective and injective. If the range of $f$ consists of but a single element, then $f$ is termed a **constant function**.

Given a nonempty set $X$, if $Y$ consists entirely of real numbers or $Y = R$, then $f: X \rightarrow Y$ is termed a **real-valued function** or **mapping** of a point $x \in X$ into a unique real number $y \in R$.[1] Hence, the image of each point $x \in X$ is a real scalar $y = f(x) \in R$.

---

1 A discussion of real numbers is offered in Section 1.3.

(a)

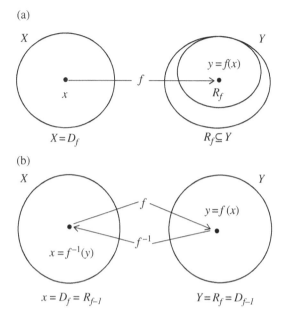

(b)

**Figure 1.2** (a) *f* is an into mapping and (b) *f* is one-to-one and onto.

For sets $X$ and $Y$ with set $A \subset X$, let $f_1: A \to Y$ be a point-to-point mapping of $A$ into $Y$ and $f_2: X \to Y$ be a point-to-point mapping of $X$ into $Y$. Then $f_1$ is said to be a **restriction** of $f_2$ and $f_2$ is termed an **extension** of $f_1$ if and only if for each $x \in A, f_1(x) = f_2(x)$.

Let $X_1, X_2, \ldots, X_n$ represent a class of nonempty sets. The **product set of** $X_1$, $X_2, \ldots, X_n$ (denoted $X_1 \times X_2 \times \cdots \times X_n$) is the set of all ordered $n$-tuples $(x_1, x_2, \ldots, x_n)$, where $x_i \in X_i$ for each $i = 1, \ldots, n$. Familiar particularizations of this definition are $R^1 = R$ (the real line); $R^2 = R \times R$ is the two-dimensional coordinate plane (made up of all ordered pairs $(x_1, x_2)$, where both $x_1 \in R$ and $x_2 \in R$); and $R^n = R \times R \times \cdots \times R$ (the product is taken $n$ times) depicts the collection of ordered $n$-tuples of real numbers. In this regard, for $f$ a point-to-point mapping of $X$ into $Y$, the subset $G_f = \{(x,y) | x \in X, y = f(x) \in Y\}$ of $X \times Y$ is called the **graph of** $f$.

If the point-to-point mapping $f$ is bijective ($f$ is one-to-one and onto), then its single-valued **inverse mapping** $f^{-1}: Y \to X$ exists. Thus to each point $y \in Y$, there corresponds a unique **inverse image** point $x \in X$ such that $x = f^{-1}(y) = f^{-1}(f(x))$ so that $x$ is termed the **inverse function** of $y$. Here, the domain $D_{f^{-1}}$ of $f^{-1}$ is $Y$, and its range $R_{f^{-1}}$ is $X$. Clearly, $f^{-1}$ must also be bijective (Figure 1.2b).

## 1.3   Real and Extended Real Numbers

We noted in Section 1.2 that a function $f$ is *real valued* if its range is the set of real numbers. Let us now explore some of the salient features of real numbers—properties that will be utilized later on.

The **real number system** may be characterized as a complete, ordered field, where a **field** is a set $F$ of elements together with the operations of addition and multiplication. Moreover, both addition and multiplication are associative and commutative, additive and multiplicative inverse and identity elements exist, and multiplication distributes over addition. Set $F$ is **ordered** if there is a binary order relation "<" in $F$ that satisfies the following conditions:

1. For any elements $x$, $y$ in $F$, either $x < y$, $y < x$, or $x = y$.
2. For any elements $x$, $y$, and $z$ in $F$, if $x < y$ and $y < z$, then $x < z$.

Now, if $F$ is an **ordered field**, then the order relation must be connected to the field operations according to the following conditions:

1. If $x < y$, then $x + z < y + z$.
2. If $x$, $y$, and $z$ is positive, then $zx < zy$.

Looking to the completeness property of the real number system, let us note first that a set $A$ ($\neq \varnothing$) of real numbers is **bounded above** if there is a real number $b$ (the **upper bound** for $A$) such that $a \leq b$ for every $a \in A$. The **least upper bound** or **supremum** of $A$ (denoted sup $A$) is a real number $b$ such that (1) $a \leq b$ for every $a \in A$; and (2) if $a \leq c$ for every $a \in A$, then $b \leq c$. So if $b$ is an upper bound for $A$ such that no smaller element of $A$ is also an upper bound for $A$, then $b$ is the least upper bound for $A$. In a similar vein, we can state that a set $A$ ($\neq \varnothing$) of real numbers is **bounded below** if there is a real number $b$ (the **lower bound** for $A$) such that $b \leq a$ for every $a \in A$. The **greatest lower bound** or **infimum** of $A$ (written $inf\ A$) is a real number $b$ such that (1) $b \leq a$ for every $a \in A$; and (2) if $c \leq a$ for every $a \in A$, then $c \leq b$. Hence, if $b$ is a lower bound for $A$ such that no larger element of $A$ is also a lower bound for $A$, then $b$ is the greatest lower bound for $A$. Clearly the supremum and infimum for $A$ must be unique.

Armed with these considerations, we can state the **completeness property** as every nonempty subset $A$ of the ordered field $F$ of real numbers which has an upper bound in $F$ has a least upper bound in $F$.

If we admit the elements $\{-\infty\}$ and $\{+\infty\}$ to our discussion of real numbers $R$, then the **extended real number system** (denoted $R^*$) consists of the set of real numbers $R$ together with $\pm\infty$, that is, $R^* = R \cup \{-\infty\} \cup \{+\infty\}$.

## 1.4   Metric Spaces

Given a space $\Omega$, a **metric** defined on $\Omega$ is an everywhere finite real-valued function $\mu$ of ordered pairs $(x, y)$ of points of $\Omega$ or $\mu(x,y) : \Omega \times \Omega \to [0, +\infty)$ satisfying the following conditions:

1. For $x \in \Omega$, $\mu(x, x) = 0$ (reflexitivity).
2. For $x, y \in \Omega$, $\mu(x, y) \geq 0$ and $\mu(x, y) = 0$ if and only if $x = y$.
3. For $x, y \in \Omega$, $\mu(x, y) = \mu(y, x)$ (symmetry).
4. For $x, y, z \in \Omega$, $\mu(x, y) \leq \mu(x, z) + (z, y)$ (triangle inequality).

Here, $\mu$ serves to define the **distance** between $x$ and $y$. A **metric space** consists of the space $\Omega$ and a metric $\mu$ defined on $\Omega$. Hence, a metric space will be denoted $(\Omega, \mu)$. For instance, if $\Omega = R$, then $R$ is a metric space if $\mu(x,y) = |x - y|$ (the distance between points $x$ and $y$ on the real line). In addition, if $\Omega = R^n$, then $R^n$ can be considered a metric space if

$$\mu(x,y) = \left[ \sum_{i=1}^{n} |x_i - y_i|^2 \right]^{\frac{1}{2}}, \tag{1.1}$$

where again $\mu(x, y)$ is interpreted as the distance between $x, y \in R^n$.[2]

Suppose $\Omega$ is a metric space with metric $\mu$ and $X (\neq \emptyset)$ is an arbitrary subset of $\Omega$. If $\mu$ is defined only for points in $X$, then $(X, \mu)$ is also a metric space. Then under this restriction on $\mu$, $X$ is termed a **subspace of $\Omega$.**

The importance of a metric space is that it incorporates a concept of distance ($\mu$) that is applicable to the points within $\Omega$. In addition, this distance function will enable us to tackle issues concerning the convergence of sequences in $\Omega$ and continuous functions defined on $\Omega$.

---

2  Equation (1.1) is actually a generalization of the absolute value function $|x - y|$. To see this, let us define on $R^n$ a **norm** (denoted $\| \cdot \|$)—a function $\| \cdot \| : R^n \to [0, +\infty)$ which assigns to each $x \in R^n$ some number $\|x\|$ such that

a. $\|x\| \geq 0$ and $\|x\| = 0$ if and only if $x = 0$;
b. $\|x + y\| \leq \|x\| + \|y\|$ (triangle inequality);
c. for a scalar $c, \|cx\| = |c| \|x\|$ (homogeneity); and
d. $\left| \sum_{i=1}^{n} x_i y_i \right| \leq \|x\| \|y\|$ (Cauchy–Schwarz inequality).

Then the distance between points $x, y \in R^n$ induced by the norm "$\| \cdot \|$" on $R^n$ is

$$\|x - y\| = \left[ \sum_{i=1}^{n} |x_i - y_i|^2 \right]^{\frac{1}{2}} \tag{1.2}$$

or Equation (1.1). So if $\Omega = R^n$ and $\mu$ is given by (1.1), then $R^n$ is a metric space with metric (1.1).

## 1.5   Limits of Sequences

Let $X$ be a subset of $R^n$. A **sequence of points** in $X$ is a function whose domain is the set of all positive integers $I$ and whose range appears in $X$. If the value of the function at $n \in I$ is $x_n \in X$, then the range of the sequence will be denoted by $\{x_n\} = \{x_1, x_2,...\}$ and interpreted as "the sequence of points $x_1, x_2, ...$ in $X$." (Note that the sequence of points $\{x_n\}$ mapped into $X$ *is not* a subset of $X$.) By deleting certain elements of the sequence $\{x_n\}$, we obtain the **subsequence** $\{x_n\}_{n \in J}$, where $J$ is a subset of the positive integers.

A sequence $\{x_n\}$ in $R^n$ **converges to a limit** $\bar{x}$ if and only if $lim_{n \to \infty}$ $\mu(x_n, \bar{x}) = lim_{n \to \infty} \|x_n - \bar{x}\| = 0$. (This is alternatively expressed as $lim_{n \to \infty}$ $x_n = \bar{x}$ or $x_n \to \bar{x}$ as $n \to \infty$.) That is, $\bar{x}$ is the limit of $\{x_n\}$ if for each $\varepsilon > 0$ there exists an index value $\bar{n}_\varepsilon$ such that $n > \bar{n}_\varepsilon$ implies $\|x_n - \bar{x}\| < \varepsilon$. If we think of the condition $\|x_n - \bar{x}\| < \varepsilon$ as defining an open sphere of radius $\varepsilon$ about $\bar{x}$, then we can say that $\{x_n\}$ converges to $\bar{x}$ if for each open sphere of radius $\varepsilon > 0$ centered on $\bar{x}$, there exists an $\bar{n}_\varepsilon$ such that $x_n$ is within this open sphere for all $n > \bar{n}_\varepsilon$. Hence, the said sphere contains all points of $\{x_n\}$ from $x_{\bar{n}_\varepsilon}$ on, that is, $\bar{x}$ is the limit of the sequence $\{x_n\}$ in $R^n$ if, given $\varepsilon > 0$, all but a finite number of terms of the sequence are within $\varepsilon$ of $\bar{x}$.

A point $\hat{x} \in R^n$ is a **limit (cluster) point** of an infinite sequence $\{x_k\}$ if and only if there exists an infinite subsequence $\{x_k\}_{k \in K}$ of $\{x_k\}$ that converges to $\hat{x}$, that is, there exists an infinite subsequence $\{x_k\}$ such that $lim_{j \to \infty} \|x_{k_j} - \hat{x}\| = 0$ or $x_{k_j} \to \hat{x}$ as $j \to \infty$. Stated alternatively, $\hat{x}$ is a limit point of $\{x_k\}$ if, given a $\delta > 0$ and an index value $\bar{k}$, there exists some $k > \bar{k}$ such that $\|x_k - \hat{x}\| < \delta$ for infinitely many terms of $\{x_k\}$.

What is the distinction between the limit of a sequence and a limit point of a sequence? To answer this question, we state the following:

a. $\bar{x}$ is a limit of a sequence $\{x_k\}$ in $R^n$ if, given a small and positive $\varepsilon \in R$, all but a finite number of terms of the sequence are within $\varepsilon$ of $\bar{x}$.

b. $\hat{x}$ is a limit point of $\{x_k\}$ in $R^n$ if, given a real scalar $\varepsilon > 0$ and given $\bar{k}$, infinitely many terms of the sequence are within $\varepsilon$ of $\hat{x}$.

Thus, a sequence $\{x_k\}$ in $R^n$ may have a limit but no limit point. However, if a convergent sequence $\{x_k\}$ in $R^n$ has infinitely many distinct points, then its limit is a limit point of $\{x_k\}$. Likewise, $\{x_k\}$ may possess a limit point but no limit. In fact, if the sequence $\{x_k\}$ in $R^n$ has a limit point $\hat{x}$, then there is a subsequence $\{x_k\}_{k \in K}$ of $\{x_k\}$ that has $\hat{x}$ as a limit; but this does not necessarily mean that the entire sequence $\{x_k\}$ converges to $\hat{x}$.[3]

---

3 If $x_k = n =$ constant for all $k$, then $\{x_k\}$ converges to the limit $n$. But since the range of this sequence contains only a single point, it is evident that the sequence has no limit point. If $x_k = 1/k$, then the sequence $\{x_k\}$ converges to a limit of zero, which is also a limit point. In addition, if $x_k = (-1)^k$, then the sequence $\{x_k\}$ has limit points at $\pm 1$, but has no limit.

A sufficient condition that at least one limit point of an infinite sequence $\{x_k\}$ in $R^n$ exists is that $\{x_k\}$ is **bounded**, that is, there exists a scalar $M \in R$ such that $\|x_k\| \leq M$ for all $k$. In this regard, if an infinite sequence of points $\{x_k\}$ in $R^n$ is bounded and it has only one limit point, then the sequence converges and has as its limit that single limit point.

The preceding definition of the limit of the sequence $\{x_n\}$ explicitly incorporated the actual limit $\bar{x}$. If one does not know the actual value of $\bar{x}$, then the following theorem enables us to prove that a sequence converges even if its actual limit is unknown. To this end, we state first that a sequence is a **Cauchy sequence** if for each $\varepsilon > 0$ there exists an index value $N_{\varepsilon/2}$ such that $m, n > N_{\varepsilon/2}$ implies $d(x_m, x_n) = \|x_m - x_n\| < \varepsilon.$[4] Second, $R^n$ is said to be **complete** in that to every Cauchy sequence $\{x_n\}$ defined on $R^n$ there corresponds a point $\bar{x}$ such that $lim_{n \to \infty} x_n = \bar{x}$. Given these concepts, we may now state the

> **Cauchy Convergence Criterion**: Given that $R^n$ is complete, a sequence $\{x_n\}$ in $R^n$ converges to a limit $\bar{x}$ if and only if it is a Cauchy sequence, that is, a necessary and sufficient condition for $\{x_n\}$ to be convergent in $R^n$ is that $d(x_m, x_n) \to 0$ as $m, n \to \infty$.

Hence, every convergent sequence on $R^n$ is a Cauchy sequence. The implication of this statement is that if the terms of a sequence approach a limit, then, beyond some point, the distance between pairs of terms diminishes.

It should be evident from the preceding discussion that a **complete metric space** is a metric space in which every Cauchy sequence converges, that is, the space contains a point $\bar{x}$ to which the sequence converges or $lim_{n \to \infty} x_n = \bar{x}$. In this regard, it should also be evident that the real line $R$ is a complete metric space as is $R^n$.

We next define the **limit superior** and **limit inferior** of a sequence $\{x_n\}$ of real numbers as, respectively,

a. $\displaystyle \lim_{n \to \infty} \sup x_n = \lim_{n \to \infty} \left( \sup_{m \geq n} x_m \right)$ and

b. $\displaystyle \lim_{n \to \infty} \inf x_n = \lim_{n \to \infty} \left( \inf_{m \geq n} x_m \right)$. $\qquad (1.3)$

Hence, the limit superior of the sequence $\{x_n\}$ is the largest number $\bar{x}$ such that there is a subsequence of $\{x_n\}$ that converges to $\bar{x}$—and no subsequence converges to a higher value. Similarly, the limit inferior is the smallest limit attainable for some convergent subsequence of $\{x_n\}$—and no subsequence converges to a lower value. Looked at in another fashion, for, say, Equation (1.3a), a

---

4  That is, for $\varepsilon > 0$ there exists a positive integer $N_{\varepsilon/2}$ such that $m \geq N_{\varepsilon/2}$ implies $d(x_m, \bar{x}) < \varepsilon/2$; and $n \geq N_{\varepsilon/2}$ implies $d(x_n, \bar{x}) < \varepsilon/2$. Hence, both $m, n > N_{\varepsilon/2}$ imply, via the triangle inequality, that

$d(x_m, x_n) \leq d(x_m, \bar{x}) + d(x_n, \bar{x}) < \frac{\varepsilon}{2} + \frac{\varepsilon}{2} = \varepsilon.$

number $\bar{x}$ is the limit superior of a sequence $\{x_n\}$ if (1) for every $x < \bar{x}$, we have $x < x_n$ for infinitely many $n$'s; and (2) for every $x > \bar{x}$, we have $x < x_n$ for only finitely many $n$'s. Generally speaking, when there are multiple points around which the terms of a sequence tend to "pile up," the limit superior and limit inferior select the largest and smallest of these points, respectively.

We noted earlier that a sequence defined on a subset $X$ of $R^n$ is a function whose range is $\{x_n\}$. If this function is **bounded**, then its range $\{x_n\}$ is bounded from both above and below. In fact, if $\{x_n\}$ is a bounded sequence of real numbers, then the limit superior and limit inferior both exist. It is also important to note that $lim_{n\to\infty} x_n$ exists if and only if the limit superior and limit inferior are equal. We end this discussion of limits by mentioning that since any set of extended real numbers has both a supremum and an infimum, it follows that every sequence of extended real numbers has both a limit superior and a limit inferior.

## 1.6 Point-Set Theory

Let $\delta$ be any positive scalar. A **$\delta$-neighborhood of a point** $x_0 \in R^n$ or **sphere of radius $\delta$ about $x_0$** is the set $\delta(x_0) = \{x | \|x - x_0\| < \delta, \delta > 0\}$. A point $\bar{x}$ is an **interior point** of a set $X$ in $R^n$ if there exists a $\delta$-neighborhood about $\bar{x}$ that contains only points of $X$.

A set $X$ in $R^n$ is said to be **open** if, given any point $x_0 \in X$, there exists a positive scalar $\delta$ such that $\delta(x_0) \subseteq X$. Hence, $X$ is open if it contains only interior points. Moreover,

a. $\varnothing$, $\delta(x_0)$, and $R^n$ are all open sets.
b. Any union of open sets in $R^n$ is open; and any finite intersection of open sets in $R^n$ is open.

Let $X$ be a set in $R^n$. The **complementary set of X**, denoted $X'$, is the collection of all points of $R^n$ lying outside of $X$. A point $\bar{x} \in X'$ is an exterior point of $X$ in $R^n$ if there exists a $\delta$-neighborhood of $\bar{x}$ that contains only points of $X'$. A point $\bar{x}$ is a **boundary point** of a set $X$ in $R^n$ if every $\delta$-neighborhood of $\bar{x}$ encompasses points in $X$ and in $X'$.

A set $X$ in $R^n$ is **bounded** if there exists a scalar $M \in R$ such that $\|x\| \leq M$ for all $x \in X$. Stated alternatively, $X$ is bounded if it has a finite **diameter** $d(X) = sup\{\|x - y\| | x, y \in X\}$.

A set $X$ in $R^n$ has an **open cover** if there exist a collection $\{G_i\}$ of open subsets from $R^n$ such that $X \subseteq \cup_i G$. The open cover $\{G_i\}$ of $X$ in $R^n$ is said to contain a **finite subcover** if there are finitely many indices $i_1, ..., i_m$ for which $X \subseteq \cup_{j=1}^{m} G_{i_j}$.

A point $\bar{x}$ is termed a **point of closure** of a set $X$ in $R^n$ if every $\delta$-neighborhood of $\bar{x}$ contains at least one point of $X$, that is, $\delta(\bar{x}) \cap X \neq \phi$. It is important to note that a point of closure of $X$ need not be a member of $X$; however, every element within $X$ is also a point of closure of $X$. A subset $X$ of $R^n$ is **closed** if every point of closure of $X$ is contained in $X$. The **closure** of a set $X$ in $R^n$, denoted $\bar{X}$, is the set of points of closure of $X$. Clearly, a set $X$ in $R^n$ is closed if and only if $X = \bar{X}$. A set $X$ in $R^n$ has a **closed cover** if there exists a collection $\{G_i\}$ of closed subsets from $R^n$ such that $X \subseteq \cup_i G$.

Closely related to the concept of a point of closure of $X$ is the notion of a **limit (cluster) point** of a set $X$ in $R^n$. Specifically, $\bar{x}$ is a limit point of $X$ if each $\delta$-neighborhood about $\bar{x}$ contains at least one point of $X$ different from $\bar{x}$, that is, points of $X$ different from $\bar{x}$ tend to "pile up" at $\bar{x}$. So if $\bar{x}$ is a limit point of a set $X$ in $R^n$, then $X \cap \delta(\bar{x})$ is an infinite set—every $\delta$-neighborhood of $\bar{x}$ contains infinitely many points of $X$. Moreover,

a. If $X$ is a finite set in $R^n$, then it has no limit point.
b. The limit point of $X$ need not be an element of $X$.
c. The collection of all limit points of $X$ in $R^n$ is called the **derived set** and will be denoted $X^d$.

Based on the preceding discussion, we can alternatively characterize a set $X$ in $R^n$ as **closed** if it contains each of its limit points or if $X^d \subseteq X$. In addition, we can equivalently state that the **closure** of a set $X$ in $R^n$ is $X$ together with its collection of limit points or $\bar{X} = X \cup X^d$. Furthermore,

a. $\varnothing$, a single point, and $R^n$ are all closed sets.
b. Any finite union of closed sets in $R^n$ is closed; any intersection of closed sets in $R^n$ is closed.
c. The closure of any set $X$ in $R^n$ is the smallest closed set containing $X$.
d. A subset $X$ in $R^n$ is closed if and only if its complementary set $X'$ is open.
e. A subset $X$ in $R^n$ is closed if and only if $X$ contains its boundary.

Let's now briefly relate the concepts of a limit and a limit point of a sequence in $R^n$ to some of the preceding point-set notions that we just developed. In particular, we shall take another look at the *point of closure* concept. To this end, a limit point (as well as a limit) of a sequence $\{x_k\}$ in $R^n$ is a **point of closure** of a set $X$ in $R^n$ if $X$ contains $\{x_k\}$. Conversely, if $\hat{x}$ is a point of closure of a set $X$ in $R^n$, then there exists a sequence $\{x_k\}$ in $X$ (and hence also a subsequence $\{x_k\}_{k \in K}$ in $X$) such that $\hat{x}$ is a limit point of $\{x_k\}$ (and thus a limit of $\{x_k\}_{k \in K}$). Hence, the **closure** of $X, \bar{X}$, consists of all limit points of convergent sequences $\{x_k\}$ from $X$.

Similarly, we note that a subset $X$ in $R^n$ is **closed** if and only if every convergent sequence of points $\{x_k\}$ from $X$ has a limit in $X$, that is, $X$ is closed if for $\{x_k\}$ in $X$, $lim_{k \to \infty} x_k = \hat{x} \in X$. Also, a set $X$ in $R^n$ is **bounded** if every sequence of points $\{x_k\}$ formed from $X$ is bounded. In addition, if a set $X$ in $R^n$ is both closed and bounded, then it is termed **compact**. (Equivalently, a set $X$ in $R^n$ is compact if it has the **finite intersection property**: every finite subclass has a nonempty intersection.) We mention briefly the following:

a. A closed subset of a compact set $X$ in $R^n$ is compact.
b. The union of a finite number of compact sets in $R^n$ is compact; the intersection of any number of compact sets in $R^n$ is compact.
c. A set $X$ in $R^n$ is compact if and only if it is closed and bounded.
d. Any finite set of points in $R^n$ is compact.
e. If $X$ in $R^n$ is a set consisting of a convergent sequence $\{x_k\}$ and its limit $\bar{x} = lim_{k \to \infty} x_k$, then $X$ is compact. Conversely, if $X$ in $R^n$ is compact, every sequence $\{x_k\}$ has a convergent subsequence $\{x_k\}_{k \in K}$ whose limit belongs to $X$.

A set $X$ in $R^n$ is **locally compact** if each of its points has a $\delta$-neighborhood with compact closure, that is, for each $x \in X$, $\overline{\delta(x)}$ is compact. In this regard, any compact space is locally compact but not conversely, for example, $R^n$ is locally compact but not compact.

## 1.7 Continuous Functions

For metric spaces $X$ and $Y$ with metrics $d_1$ and $d_2$, respectively, let $f: X \to Y$ be a point-to-point mapping of $X$ into $Y$. $f$ is said to be **continuous at a point** $x_0 \in X$ if either

a. for any $\varepsilon > 0$ there exists a $\delta_\varepsilon > 0$ such that $d_1(x, x_0) < \delta_\varepsilon$ implies $d_2(f(x), f(x_0)) < \varepsilon$. (Note that the subscript on $\delta$ means that "$\delta$ depends upon the $\varepsilon$ chosen."); or
b. for each $\varepsilon$-neighborhood of $f(x_0)$, $\varepsilon(f(x_0))$, there exists a $\delta_\varepsilon$-neighborhood about $x_0$, $\delta_\varepsilon(x_0)$, such that $f(\delta_\varepsilon(x_0)) \subseteq \varepsilon(f(x_0))$, that is, points "near" $x_0$ are mapped by $f$ into points "near" $f(x_0)$.

In general, the point-to-point mapping $f: X \to Y$ is **continuous on $X$** if it is continuous at each point of $X$.

Theorems 1.7.1 and 1.7.2 provide us with a set of necessary and sufficient conditions for the continuity of a point-to-point mapping at a specific point $x_0 \in X$ and at any arbitrary $x \in X$, respectively. Specifically, we start with Theorem 1.7.1.

**Theorem 1.7.1** (continuity in terms of convergent sequences). For metric spaces $X$ and $Y$, the point-to-point mapping $f$ of $X$ into $Y$ is continuous at $x_0 \in X$ if and only if $x_k \to x_0$ implies $f(x_k) \to f(x_0)$ for every subsequence $\{x_k\}$ in $X$.

Hence, $f$ is a continuous mapping of $X$ into $Y$ if it "sends convergent sequences in $X$ into convergent sequences in $Y$." Next comes Theorem 1.7.2.

**Theorem 1.7.2** (continuity in terms of open (resp. closed) sets). For metric spaces $X$ and $Y$, let $f$ be a point-to-point mapping of $X$ into $Y$. Then, (a) $f$ is continuous if and only if $f^{-1}(A)$ is open in $X$ whenever set $A$ is open in $Y$; and (b) $f$ is continuous if and only if $f^{-1}(A)$ is closed in $X$ whenever $A$ is closed in $Y$.

Thus, $f$ is continuous if it "pulls open (resp. closed) sets back to open (resp. closed) sets," that is, the inverse images of open (resp. closed) sets are open (resp. closed).

We next consider Theorem 1.7.3 which states that continuous mappings preserve compactness. That is,

**Theorem 1.7.3** For metric spaces $X$ and $Y$, let $f$ be a continuous point-to-point mapping from $X$ into $Y$. If $A$ is a compact subset of $X$, then so is its range $f(A)$.

Next, let $X$ be a subset of $R^n$. A continuous point-to-point mapping $g : R^n \to X$ is termed a **retraction mapping** on $R^n$ if $g(x) = x$ for all $x \in X$. Here, $X$ is called a **retraction** of $R^n$. If $X$ is contained within an arbitrary subset $A$ of $R^n$, then $g : A \to X$ is a retraction of $A$ onto $X$ if $g(x) = x$ for all $x \in X$.

## 1.8 Operations on Sequences of Sets

Let $\{A_i\}, i = 1, 2, \ldots$, represent a sequence of sets in a metric space $X$. If $\{A_i\}$ is such that $A_i \subseteq A_{i+1}, i = 1, 2, \ldots$, then $\{A_i\}$ is said to be a **nondecreasing sequence** (if $A_i \subset A_{i+1}$, then $\{A_i\}$ is said to be an **expanding sequence**). In addition, if $\{A_i\}$ is such that $A_i \supseteq A_{i+1}, i = 1, 2, \ldots$, then $\{A_i\}$ is called a **nonincreasing sequence** (if $A_i \supset A_{i+1}, i = 1, 2, \ldots$, then $\{A_i\}$ is termed a **contracting sequence**). A **monotone sequence** of sets is one which is either an expanding or contracting sequence.

If the sequence $\{A_i\}, i = 1, 2, \ldots$, in $X$ is nondecreasing or nonincreasing, then its limit exists and we have the following:

$$\lim_{i \to \infty} A_i = \cup_{i=1}^{\infty} A_i \quad (\text{if } \{A_i\} \text{ is nondecreasing});$$

$$\lim_{n \to \infty} A_i = \cap_{i=1}^{\infty} A_i \quad (\text{if } \{A_i\} \text{ is nonincreasing}).$$

In addition, for any sequence of sets $\{A_i\}, i = 1, 2, \ldots$, in $X$,

$$sup\{A_i\} = \cup_{i=1}^{\infty} A_i, \; inf\{A_i\} = \cap_{i=1}^{\infty} A_i;$$

with

$$sup\left(\cup_{i=1}^{\infty} A_i\right) = sup\{sup\, A_i, i = 1, 2, \ldots\},$$
$$inf\left(\cup_{i=1}^{\infty} A_i\right) = inf\{inf\, A_i, i = 1, 2, \ldots\}; \text{and}$$

$$sup\left(\cap_{i=1}^{\infty}A_i\right) \le inf\left\{sup\,A_i, i=1,2,...\right\},$$
$$inf\left(\cap_{i=1}^{\infty}A_i\right) \ge sup\left\{inf\,A_i, i=1,2,...\right\}.$$

Let $\{A_i\}, i=1,2,...$, again depict a sequence of sets in $X$.

Then there are subsets $E_i \subset A_i$ of disjoint sets, with $E_j \cap E_k = \phi$ for $j \ne k$, such that

$$\cup_{i=1}^{\infty}E_i = \cup_{i=1}^{\infty}A_i.$$

We next consider the concepts of the limit superior and limit inferior of a sequence of sets $\{A_i\}, i=1,2,...$, in a metric space $X$. To this end, the **limit superior of a sequence** $\{A_i\}$ is defined as

$$\lim_{i\to\infty} sup\,A_i = \cap_{i=1}^{\infty}\left(\cup_{k\ge i}A_k\right)$$

$$= (A_1 \cup A_2 \cup \cdots) \cap (A_2 \cup A_3 \cup \cdots) \cap \cdots$$

$$= \{x \in X | x \in A_i \text{ for infinitely many } i\}.$$

Hence, *lim sup* $A_i$ is the set $S$ of points such that, for every positive integer $i$, there exists a positive integer $k \ge i$ such that $S \in A_i$; thus $S$ consists of those points that belong to $A_i$ for an infinite number of $i$ values. Looked at in another fashion, if $x \in S$, then $x$ is in all of $\cup_{k\ge i}A_k$. Hence, no matter how large of an $i$ value is chosen, you can find a $k \ge i$ for which $x$ is a member of $A_i$.

Similarly, the **limit inferior of a sequence** $\{A_i\}$ is

$$\lim_{i\to\infty} inf\,A_i = \cup_{i=1}^{\infty}\left(\cap_{k\ge i}A_k\right)$$

$$= (A_1 \cap A_2 \cap \cdots) \cup (A_2 \cap A_3 \cap \cdots) \cup \cdots$$

$$= \{x \in X | x \in A_i \text{ for all but finitely many } i\}.$$

Thus, *lim inf* $A_i$ is the set $I$ of points such that, for some positive integer $i$, $I \in A_i$ for all positive integers $k \ge i$; hence, $I$ consists of those points that belong to $A_i$ for all except a finite number of $i$ values. Stated alternatively, if $x \in I$, then $x$ is an element of $\cap_{k\ge i}A_k$ so that $x \in A_i$ for $k \ge i$—$x$ must be in $I$ with only finitely many exceptions, that is, for $x \in I$, there is an index value such that $x$ is in every $A_i$ in the remaining portion of the limit.[5]

---

5 Alternative definitions of the limit superior and limit inferior of a sequence of sets are the following. Again, let $\{A_i\}$ be a sequence of sets in a metric space $X$. Then

$$\lim_{i\to\infty} sup\,A_i = \left\{x \in X | \lim_{i\to\infty} inf\, d(x,A_i) = 0\right\};$$
$$\lim_{i\to\infty} inf\,A_i = \{x \in X | d(x,A_i) = 0\},$$

where $d(x, A_i)$ is the distance from $x$ to $A_i$.

We note briefly that if $\{A_i\}, i = 1, 2, \ldots,$ is any sequence of sets in a metric space $X$, then *lim inf* $A_i \subset$ *lim sup* $A_i$. A sequence of sets $\{A_i\}$ is **convergent** (or a subset $A$ of $X$ is the **limit** of $\{A_i\}$) if

$$\lim_{i \to \infty} \sup A_i = \lim_{i \to \infty} \inf A_i = \lim_{i \to \infty} A_i = A.$$

Here, $A$ is termed the **limit set**. In this vein, any monotone sequence of sets $\{A_i\}, i = 1, 2, \ldots,$ is convergent.

## 1.9 Classes of Subsets of $\Omega$

### 1.9.1 Topological Space

We previously defined a metric space $(\Omega, \mu)$ as consisting of the space $\Omega$ and a metric $\mu$ defined on $\Omega$. Let $\mathcal{A}$ denote the class of open sets in the metric space. Then $\mathcal{A}$ satisfies the following conditions:

i. $\varnothing, \Omega \in \mathcal{A}$.
ii. If $A_1, A_2, \ldots, A_n \in \mathcal{A}$, then $\cap_{i=1}^{n} A_i \in \mathcal{A}$ (the intersection of every *finite* class of sets in $\mathcal{A}$ is itself a set in $\mathcal{A}$).
iii. If $A_\alpha \in \mathcal{A}$ for $\alpha \in I$, then $\cup_{\alpha \in I} A_\alpha \in \mathcal{A}$ (the union of every arbitrary class of sets in $\mathcal{A}$ is itself a set in $\mathcal{A}$).

Armed with properties (i)–(iii), let us generalize a metric space to that of a topological space. That is, given a nonempty space $\Omega$ and a given class $\mathcal{A}$ of subsets of $\Omega$ consisting of the "open sets" in $\Omega$, a class $\mathcal{J}$ of subsets of $\Omega$ is called a **topology** on $\Omega$ if (i)–(iii) hold. (Thus, the class of open sets $\mathcal{A}$ determines the topology in $\Omega$.) Hence, a **topological space** consists of $\Omega$ and a topology $\mathcal{J}$ on $\Omega$ and is denoted $(\Omega, \cdot)$.

A subset $A$ of a topological space $(\Omega, \mathcal{J})$ is said to be (everywhere) **dense** if its closure $\bar{A}$ equals $(\Omega, \cdot)$. Hence, $A$ is dense if and only if (a) $A$ intersects every nonempty set; or (b) the only open set disjoint from $A$ is $\varnothing$.

### 1.9.2 $\sigma$-Algebra of Sets and the Borel $\sigma$-Algebra

A **ring** $\mathcal{R}$ is a nonempty class of subsets that contains $\varnothing$ and is closed under the operations of union, intersection, and difference. A $\sigma$-**ring** is a ring $\mathcal{R}$ that is closed under countable unions and intersections, that is, if $A_i \in \mathcal{R}$, $i = 1, 2, \ldots,$ then $A = \cup_{i=1}^{\infty} A_i \in \mathcal{R}$ and $\cap_{i=1}^{\infty} A_i = \left( A - \cup_{i=1}^{\infty} (A - A_i) \right) \in \mathcal{R}$.

Next, we can define a $\sigma$-**algebra** as a class of sets $\mathcal{F}$ that contains $\Omega$ and is a $\sigma$-ring. More formally, for $\Omega$ a given space, a $\sigma$-algebra on $\Omega$ is a family $\mathcal{F}$ of subsets of $\Omega$ that satisfies the following conditions:

i. $\Omega \in \mathcal{F}$.
ii. If a set $A \in \mathcal{F}$, then its complement $A' \in \mathcal{F}$, where $A' = \Omega - A$.

iii. If $\{A_i\}_{i \geq 1} \in \mathcal{I}$, then $\cup_{i=1}^{\infty} A_i \in \mathcal{I}$, that is, countable unions of sets in $\mathcal{I}$ are also in $\mathcal{I}$.

Note that since $\Omega \in \mathcal{I}$, we must have $\Omega' = \emptyset \in \mathcal{I}$; with $\left(\cup_{i=1}^{\infty} A_i\right)' = \cap_{i=1}^{\infty} A_i'$, it follows that $\mathcal{I}$ is closed under countable intersections as well. Note also that if $\{A_i\}_{i \geq 1} \in \mathcal{I}$, then $\lim_{i \to \infty} A_i \in \mathcal{I}$, $\lim_{i \to \infty} \sup A_i \in \mathcal{I}$, and $\lim_{i \to \infty} \inf A_i \in \mathcal{I}$. The pair $(\Omega, \mathcal{I})$ is called a **measurable space** and the sets in $\mathcal{I}$ are termed ($\mathcal{I}$-) **measurable sets**.

Given a family $\mathcal{C}$ of subsets of $\Omega$, there exists a smallest $\sigma$-algebra $\sigma(\mathcal{C})$ on $\Omega$ that contains $\mathcal{C}$, is contained in every $\sigma$-algebra that contains $\mathcal{C}$, and is unique. Here, $\sigma(\mathcal{C})$ is termed the $\sigma$-**algebra generated by** $\mathcal{C}$ and is specified as

$$\sigma(\mathcal{C}) = \cap \{\mathcal{H}_j | \mathcal{H}_j \text{ a } \sigma\text{-algebra on } \Omega, \mathcal{C} \subset \mathcal{H}_j\}.$$

(For instance, if $\mathcal{C} = \{E\}$, $E \subset \Omega$, then $\sigma(\mathcal{C}) = \{\emptyset, E, E', \Omega\}$.) Now, if $\Omega = R^n$ and $\mathcal{C}$ is a family of open sets in $R^n$, then $\mathcal{B}^n = \sigma(\mathcal{C})$ is called the **Borel $\sigma$-algebra on** $\Omega$ and an element $B \in \mathcal{B}^n$ is called a **Borel set**. Hence, the Borel $\sigma$-algebra on $\Omega$ is the smallest $\sigma$-algebra generated by all the open subsets of $R^n$; and the **class of Borel sets** $\mathcal{B}^n$ in $R^n$ is the $\sigma$-algebra generated by the open sets in $R^n$. In fact, the class of half-open intervals in $R^n$ generates the $\sigma$-algebra $\mathcal{B}^n$ of Borel sets in $R^n$. Borel sets also include all open and closed sets, all countable unions of closed sets, among others.

Since $\sigma$-algebras will be of paramount importance in our subsequent analysis (especially in our review of the essentials of probability theory), let us consider Example 1.1.

**Example 1.1** To keep the analysis manageable, suppose $\Omega = \{1,2,3,4\}$. Then possibly, $\mathcal{I} = \{\emptyset, \{1,2\}, \{3,4\}, \Omega\}$. Does $\mathcal{I}$ satisfy (i)–(iii) given earlier? If so, then $\mathcal{I}$ is a legitimate $\sigma$-algebra. Specifically,

1. As constructed, $\emptyset \in \mathcal{I}$ and $\Omega \in \mathcal{I}$. Hence, (i) holds.
2. $\emptyset' = \Omega, \Omega' = \emptyset$ $\{1,2\}' = \{3,4\}, \{3,4\}' = \{1,2\}$. Clearly each of these subsets is a member of $\mathcal{I}$ and thus (ii) is valid.
3. Since $\mathcal{I}$ contains four disjoint subsets, partition the index set $I = \{1,2,3,4\}$ into four disjoint subsets according to

$$I_1 = \{i | A_i = \emptyset\} \neq \emptyset \quad I_2 = \{i | A_i = \{1,2\}\} \neq \emptyset$$
$$I_3 = \{i | A_i = \{3,4\}\} \neq \emptyset, \text{ and } I_4 = \{i | A_i = \Omega\} \neq \emptyset.$$

Then

$$\cup_{i=1}^{\infty} A_i = \bigcup_{i \in I} A_i$$

$$= \left(\bigcup_{i \in I_1} A_i\right) \cup \left(\bigcup_{i \in I_2} A_i\right) \cup \left(\bigcup_{i \in I_3} A_i\right) \cup \left(\bigcup_{i \in I_4} A_i\right)$$

$$= \emptyset \cup \{1,2\} \cup \{3,4\} \cup \Omega \in \mathcal{I}$$

and thus $\mathcal{I}$ is a $\sigma$-algebra on $\Omega$. ∎

## 1.10 Set and Measure Functions

### 1.10.1 Set Functions

We previously defined the concept of a point-to-point function or mapping as a rule $f$ that associates with a point $x$ from a nonempty set $X$ a unique point $y = f(x)$ in a nonempty set $Y$, where $X$, *a set of points*, was called the domain of the function. Now, let's consider a real-valued function whose domain is a *class of sets*, that is, we have a function of sets rather than a function of points. In this regard, consider a function $\mu: \mathcal{C} \to R^*$, where $\mathcal{C}$ is a nonempty class of sets and $R^*$ denotes the set of extended real numbers. Thus $\mu$ is a rule that associates with each set $E \in \mathcal{C}$ a unique element $\mu(E)$, which is either a real number or $\pm\infty$. Some important types of set functions follow.

First, a set function $\mu: \mathcal{C} \to R^*$ is said to be (finitely) **additive** if

  i. $\mu(\varnothing) = 0$; and
  ii. for every finite collection $E_1, E_2, ..., E_n$ of disjoint sets $(E_j \cap E_k = \varnothing, j \neq k)$ in $\mathcal{C}$ such that $\cup_{i=1}^n E_i \in \mathcal{C}$, we have $\mu\left(\cup_{i=1}^n E_i\right) = \sum_{i=1}^n \mu(E_i) \in R^*$.

Remember that the domain $\mathcal{C}$ of $\mu$ is a finitely additive class of sets $\{E_i, i = 1, ..., n\}$ and $\sum_{i=1}^n \mu(E_i)$ is defined in $R^*$. (If $\Omega = R$, $\mathcal{C}$ is the class of all *finite* intervals of $R$, and if $E$ is taken to be $(a, b)$ or $(a, b]$ or $[a, b)$ or $[a, b]$, then $\mu(E) = b - a$.)

It should be evident that a suitable domain of definition of an additive set function $\mu$ is a ring $R$ since, if $E_i \in R, i = 1, ..., n$, then $\cup_{i=1}^n E_i \in R$. So if $\mathcal{C}$ is a ring, then the set function $\mu: \to R^*$ is additive if and only if $\mu(\varnothing) = 0$ and, if $E_j$ and $E_k$ are disjoint sets in $\mathcal{C}$, then $\mu\left(E_j \cup E_k\right) = \mu\left(E_j\right) + \mu(E_k)$. In this regard, suppose $\mu: \mathcal{C} \to R^*$ is an additive set function defined as a ring $\mathcal{C}$ with sets $E_j, E_k \in \mathcal{C}$. Then

  i. if $E_j \subset E_k$ and $\mu(E_j)$ is finite, then $\mu(E_k - E_j) = \mu(E_k) - \mu(E_j) \geq 0$;
  ii. if $E_j \subset E_k$ and $\mu(E_j)$ is infinite, then $\mu(E_j) = \mu(E_k)$;
  iii. if $E_j \subset E_k$ and $\mu(E_k)$ is finite, then $\mu(E_j)$ is finite; and
  iv. if $\mu(E_k) = +\infty$, then $\mu(E_j) \neq -\infty$.

Next, a set function $\mu: \mathcal{C} \to R^*$ is termed $\sigma$-**additive** (or **countably** or **completely additive**) provided

  i. the domain of $\mu$ is a $\sigma$-ring of sets $\mathcal{C}$;
  ii. $\mu(\varnothing) = 0$; and
  iii. for any disjoint sequence $E_1, E_2, ...$ of sets in $\mathcal{C}$ such that $\cup_{i=1}^\infty E_i \in \mathcal{C}$, we have

$$\mu\left(\cup_{i=1}^\infty E_i\right) = \sum_{i=1}^\infty \mu(E_i) \in R^*.$$

Here, the domain $\mathcal{C}$ of $\mu$ is a *countably additive* class of sets $\{E_i, i = 1, 2, ...\}$ and $\sum_{i=1}^\infty \mu(E_i) \in R^*$ is defined in the extended real numbers. Clearly, a $\sigma$-additive set function is also (finitely) additive, though the converse is not generally true. However, if $\mathcal{C}$ is a finite class of sets, then the additivity of $\mu\colon \mathcal{C} \to R^*$ implies $\sigma$-additivity.

A set function $\mu\colon \to R^*$ is said to be **$\sigma$-finite** if, for each set $E \in \mathcal{C}$, there is a sequence of sets $E_i \in \mathcal{C}$, $i = 1, 2, \ldots$, such that $E = \cup_{i=1}^{\infty} E_i$ and $\mu(E_i) < +\infty$ for all $i$. As this definition reveals, additivity is not a property of $\sigma$-finite set functions. For instance, consider the Borel $\sigma$-algebra in $R^n$ that is generated by the collection of all "cubes" of the form $C = (a_1, b_1) \times (a_2, b_2) \times \cdots \times (a_n, b_n)$, with $b_i > a_i, i = 1, \ldots, n$. Then, $\mu(C) = \prod_{i=1}^{n}(b_i - a_i)$. Here, $\mu$ is $\sigma$-finite since $R^n = \cup_{i=1}^{\infty}(-i, i)^n$.

A set function $\mu$ defined on $\mathcal{C}$ is **nondecreasing** if $\mu(E_k) \geq \mu(E_j)$ whenever $E_j \subset E_k$; it is **nonincreasing** if $\mu(E_k) \leq \mu(E_j)$ when $E_j \subset E_k$; and it is said to be **monotone** if it is either nondecreasing or nonincreasing. Now, if $\mu$ is additive and nondecreasing (resp. nonincreasing), then it is everywhere non-negative (resp. nonpositive). In fact, the reverse implication holds, that is, if $\mu$ is additive and everywhere non-negative (resp. nonpositive), then it is also nondecreasing (resp. nonincreasing).

### 1.10.2 Measure Functions

Let $R^+$ denote the set of non-negative real numbers together with $+\infty$, that is, $R^+ = \{x \in R^* \,|\, x \geq 0\}$. A **measure function** on a $\sigma$-ring $\mathcal{C}$ is any non-negative $\sigma$-additive set function $\mu\colon \mathcal{C} \to R^+$. (For any subset $A \in \mathcal{C}$ we assume that $-\infty < \mu(A) < +\infty$.) Note that since a measure function $\mu$ on $\mathcal{C}$ is non-negative, it must also be nondecreasing. A **Borel measure** is a measure function $\mu$ on the $\sigma$-algebra $\mathcal{B}$ of Borel subsets of a given topological space $(\Omega, \mathcal{J})$, that is,

$$\mu : \mathcal{B} \to [0, +\infty).$$

A couple of important characteristics of measure functions are as follows:

i. If $\mu$ is a measure function on $\mathcal{C}$ and if $\{E_i, i = 1, 2, \ldots\}$ is any sequence of sets from $\mathcal{C}$, then $\mu$ is **countably subadditive** or $\mu\left(\cup_{i=1}^{\infty} E_i\right) \leq \sum_{i=1}^{\infty} \mu(E_i)$.

ii. If $\mu$ is a measure function on $\mathcal{C}$ and if $\{E_i, i = 1, 2, \ldots\}$ is any sequence of sets from $\mathcal{C}$ with $\mu\left(\cup_{i=1}^{\infty} E_i\right) < +\infty$, then $\mu(lim_{i \to \infty} E_i) = lim_{i \to \infty} \mu(E_i)$.

We next examine the continuity of set functions. To this end, suppose $\mathcal{R}$ is a ring and the set function $\mu\colon \mathcal{R} \to R^*$ is additive with $\mu(A) > -\infty$ for all sets $A \in \mathcal{R}$:

i. $\mu$ is **continuous from below** at $A$ if $lim_{i \to \infty} \mu(E_i) = \mu(A)$ for every monotone increasing sequence $\{E_i\}$ in $\mathcal{R}$ that converges to $A$.

ii. $\mu$ is **continuous from above** at $A$ if $lim_{i \to \infty} \mu(E_i) = \mu(A)$ for every monotone decreasing sequence $\{E_i\}$ in $\mathcal{R}$ for which $\mu(E_i) < +\infty$ for some $i$.

iii. $\mu$ is continuous at $A \in \mathcal{R}$ if it is continuous at $A$ from both above and below. Moreover, under the aforementioned assumptions, if

iv. $\mu$ is $\sigma$-additive (and thus additive) on $\mathcal{R}$, then $\mu$ is continuous at $A$ for all sets $A \in \mathcal{R}$.

Given two classes $\mathcal{C}$ and $\mathcal{D}$ of subsets of $\Omega$, with $\mathcal{C} \subset \mathcal{D}$, and set functions $\mu$: $\mathcal{C} \rightarrow R^*$ and $\tau$: $\mathcal{D} \rightarrow R^*$ respectively, $\tau$ is termed an **extension** of $\mu$ if, for all $A \in \mathcal{C}$, $\tau(A) = \mu(A)$; and $\mu$ is called a **restriction** of $\tau$ to $\mathcal{C}$.

In later chapters, we shall be concerned with issues pertaining to the convergence of sequences of random variables. To adequately address these issues, we need to be able to define measures on countable unions and intersections of "measurable sets." To accomplish this task, we need to assume that the collection of measurable sets is a $\sigma$-algebra $\mathcal{I}$, that is, $\mathcal{I}$ contains $\Omega$ and is a $\sigma$-ring. This requirement enables us to confine our analysis, for the most part, to "measure spaces," where a **measure space** is a triple $(\Omega, \mathcal{I}, \mu)$ consisting of a space $\Omega$, a $\sigma$-algebra $\mathcal{I}$ on $\Omega$ (a collection subsets of $\Omega$), and $\mu$: $\mathcal{I} \rightarrow R^+$ is a measure on $\mathcal{I}$.

### 1.10.3  Outer Measure Functions

Suppose $\mathcal{C}$ is the class of all subsets of a space $\Omega$. Then $\mu^*$: $\mathcal{C} \rightarrow R^+$ is an **outer measure function** on $\Omega$ if

i. $\mu^*(\varnothing) = 0$;
ii. $\mu^*$ is nondecreasing (i.e., for subsets $E_j \subset E_k$, $\mu(E_j) \leq \mu(E_k)$); and
iii. $\mu^*$ is countably subadditive—i.e., for any sequence $\{E_i,\ i = 1,\ 2,\ ...\}$ of subsets of $\Omega$,

$$\mu^*\left(\cup_{i=1}^{\infty} E_i\right) \leq \sum_{i=1}^{\infty} \mu^*(E_i). \tag{1.4}$$

In sum, $\mu^*$ is said to be non-negative, monotone, and countably subadditive. We note that every measure on the class of all subsets of $\Omega$ is an outer measure on $\Omega$; and, in defining an outer measure on $\Omega$, no "additivity" requirement was in effect.

Given that $\mu^*$ is an outer measure on $\Omega$, a subset $E$ is said to be **measurable** with respect to $\mu^*$, or simply $\mu^*$-measurable, if for every set $A \subset \Omega$,

$$\mu^*(A) = \mu^*(A \cap E) + \mu^*(A \cap E') \tag{1.5}$$

(given that $A = (A \cap E) \cup (A \cap E')$). Thus, a subset $E$ of $\Omega$ is $\mu^*$-measurable if it partitions a set $A \subset \Omega$ into two subsets, $A \cap E$ and $A \cap E'$, on which $\mu^*$ is additive. As this definition reveals, a set $E$ is not innately measurable—its measurability depends upon the outer measure employed. That is, to define the measurability of a set $E$, we start with an arbitrary set $A$ and we examine the effect of $E$ on the outer measure of $A$, $\mu^*(A)$. If $E$ is measurable, then it is sufficiently "well-behaved" in that it does not partition $A$ in a way that compromises the additivity of $\mu^*$, that is, if we partition $A$ into $A \cap E$ and $A \cap E'$, then the outer measures of $A \cap E$ and $A \cap E'$ add up correctly to $\mu^*(A)$.

Since $\mu^*$ is countably subadditive, we have, from (1.4), $\mu^*(A) = \mu^*(A \cap E) + \mu^*(A \cap E')$ for all sets $A$, $E \in \Omega$. Hence, $E$ is $\mu^*$-measurable if and only if

$$\mu^*(A) \geq \mu^*(A \cap E) + \mu^*(A \cap E') \tag{1.6}$$

for every set $A \in \Omega$. Since this inequality holds for any set $A$ for which $\mu^*(A) = +\infty$, it follows that a necessary and sufficient condition for $E$ to be $\mu^*$-measurable is that $\mu^*(A) < +\infty$ for every $A \in \Omega$.

Some important properties of outer measures are the following:

i. If $E \in \Omega$ is $\mu^*$-measurable, then $E'$ is also $\mu^*$-measurable.
ii. If $\mu^*(E) = 0$, then $E \in \Omega$ is measurable.
iii. Any finite union of $\mu^*$-measurable sets in $\Omega$ is $\mu^*$-measurable.
iv. If $\{E_i, i = 1, 2, ...\}$ is a sequence of disjoint $\mu^*$-measurable sets in $\Omega$ and if $\mathcal{G} = \cup_{j=1}^{\infty} E_j$, then for any set $A \in \Omega$, $\mu^*(A \cap \mathcal{G}) = \sum_{j=1}^{\infty} \mu^*(A \cap E_j)$.
v. Any countable union of $\mu^*$-measurable sets in $\Omega$ is $\mu^*$-measurable.
vi. Any countable union of disjoint $\mu^*$-measurable sets in $\Omega$ is $\mu^*$-measurable.
vii. If $\{E_i, i = 1, 2, ...\}$ is a sequence of disjoint $\mu^*$-measurable sets in $\Omega$ and if, for each $n$, $\mathcal{G}_n = \cup_{j=1}^{n} E_j$, then, for each set $A \in \Omega$, $\mu^*(A \cap \mathcal{G}_n) = \sum_{j=1}^{n} \mu^*(A \cap E_j)$.

Why are outer measures important? Simply because they are useful for constructing measure functions. That is, given that the outer measure $\mu^*$ has as its domain the class of all subsets of the space $\Omega$, a *restriction* of $\mu^*$ to a "smaller" domain always generates a measure function. In this regard, suppose $\mu^*$ is an outer measure function on $\Omega$ and let $\mathcal{E}$ be the class of $\mu^*$-measurable sets. Then $\mathcal{E}$ is a completely additive class (a $\sigma$-algebra) and the restriction of $\mu^*$ to $\mathcal{E}$ is a measure function $\mu$.

An outer measure $\mu^*$ is said to be **regular** if, for every subset $A \in \Omega$, there is a $\mu^*$-measurable set $E \supset A$ such that $\mu^*(E) = \mu^*(A)$. (Here, $E$ is said to be a **measurable cover** for $A$.) Thus, an outer measure is regular if it effectuates measurable sets in a manner that guarantees that every set $A \in \Omega$ has a measurable cover $E$.

Key properties of regular outer measures are the following:

i. If $\mu^*$ is a regular outer measure on $\Omega$ and $\{E_i, i = 1, 2, ...\}$ is an increasing sequence of sets, then $\mu^*(lim_{i \to \infty} E_i) = lim_{i \to \infty} \mu^*(E_i)$.
ii. If $\mu^*$ is a regular outer measure on $\Omega$ for which $\mu^*(\Omega) < +\infty$, then a subset $E \in \Omega$ is measurable if and only if $\mu^*(\Omega) = \mu^*(\Omega \cap E) + \mu^*(\Omega \cap E') = \mu^*(E) + \mu^*(E')$. (This result follows from Equation (1.5) with $A = \Omega$, since (1.5) must hold for *any* set $A$.)

Next, let $\Omega$ be a metric space. An outer measure $\mu^*$ on $\Omega$ is a **metric outer measure** if $\mu^*(\emptyset) = 0$; $\mu^*$ is nondecreasing and countably subadditive; and $\mu^*$

is additive on separated sets (i.e., for subsets $E$ and $F$ in $\Omega$ with $d(E, F) > 0$, $\mu^*(E \cup F) = \mu^*E + \mu^*(F)$).[6] We note briefly the following:

i. If $\mu^*$ is a metric outer measure, then any closed set is measurable.
ii. If $\mu^*$ is a metric outer measure, then every Borel set is measurable (since the class $\mathcal{E}$ of $\mu^*$-measurable sets contains the open sets, and thus contains $\mathcal{B}$, the class of Borel sets).

### 1.10.4 Complete Measure Functions

Given a measure function $\mu \colon \mathcal{C} \to R^+$, the class $\mathcal{C}$ of subsets of $\Omega$ is **complete with respect to $\mu$** if $E \subset F$, $F \in \mathcal{C}$, and $\mu(F) = 0$ implies $E \in \mathcal{C}$. Now, if $\mu \colon \mathcal{C} \to R^+$ is such that $\mathcal{C}$ is complete with respect to $\mu$, then $\mu$ is said to be **complete**. Hence, $\mu$ is complete if its domain contains all subsets of sets of measure zero, that is, every subset of a set of measure zero is measurable.

For a measure space $(\Omega, \mathcal{F}, \mu)$, the **completion of $\mathcal{F}$**, denoted $\mathcal{F}_c$, with respect to a measure $\mu$ on $\mathcal{F}$ involves all subsets $A \in \Omega$ such that there exist sets $E, F \in \mathcal{F}$, with $E \subset A \subset F$, and $\mu(F - E) = 0$. The **completion of $\mu$**, $\mu_c$, is defined on $\mathcal{F}_c$ as $\mu_c(A) = \mu_c\ (E) = \mu_c(F)$; it is the unique extension of $\mu$ to $\mathcal{F}_c$. For $A \in \mathcal{F}_c, \mu_c(A) = inf\{\mu(F)|F \in \mathcal{F}, A \subset F\} = sup\{\mu(E)|E \in \mathcal{F}, E \subset A\}$.

The **complete measure space** $(\Omega, \mathcal{F}_c, \mu_c)$ is thus the completion of $(\Omega, \mathcal{F}, \mu)$. In fact, $(\Omega, \mathcal{F}_c, \mu_c)$ is the smallest complete measure space that contains $(\Omega, \mathcal{F}, \mu)$.

If a measure $\mu$ is obtained by restricting an outer measure $\mu^*$ to $\mathcal{E}$, the class of sets of $\Omega$ that are $\mu^*$-measurable, then $\mu$ is a complete measure. In fact, any measure generated by an outer measure is complete.

### 1.10.5 Lebesgue Measure

In what follows, our discussion will focus in large part on a class $\mathcal{M}$ of open sets (containing $\varnothing$) in $\Omega = R$. This will then facilitate our development of the Lebesgue integral.

Let us express the **length of a bounded interval** $I$ (which may be open, closed, or half-open) with endpoints $a$ and $b$, $a < b$, as $l(I) = b - a$. Our objective herein is to extend this "length" concept to arbitrary subsets of $R$, for example, for a subset $E \subset R$, the notion of the "length of $E$" is simply its measure $\mu(E)$. In particular, we need to explore the concept of *Lebesgue measure* of a set $E$, $\mu(E)$, and specify the family of *Lebesgue measurable sets*. Our starting point is the concept of Lebesgue outer measure.

---

6 For sets $E, F \in R^n$, the **distance between sets** $E, F$ is $d(E,F) = inf\{\|x - y\| | x \in E, y \in F\}$. If $E \cap F \neq \varnothing, d(E,F) = 0$.

For each subset $E \subset R$, the **Lebesgue outer measure** $\mu^*(E)$ is defined as

$$\mu^*(E) = inf \left\{ \sum_{i=1}^{n} l(I_i) \,|\, \{I_i\} \text{ is a sequence of open intervals with } E \subset \cup_{i=1}^{n} I_i \right\}.$$

What is the significance of this expression? Suppose $E$ can be covered by multiple sets of open intervals, where the union of each particular set of open intervals contains $E$. Since the total length of any set of intervals can overestimate the measure of $E$ (it may contain points not in $E$), we need to take the greatest lower bound of the lengths of the interval sets in order to isolate the covering set whose length fits $E$ *as closely as possible* and whose constituent intervals *do not overlap*.

Given the discussion on outer measures in Section 1.10.3, it follows that the key properties of Lebesgue outer measures are the following:

i. For every set $E \subset R, 0 \le \mu^*(E) \le +\infty$.
ii. $\mu^*$ is nondecreasing.
iii. $\mu^*$ is countably subadditive, that is, for any sequence $\{E_i, i = 1, 2, ...\}$ of subsets of $R$,

$$\mu^*\left(\cup_{i=1}^{\infty} E_i\right) \le \sum_{i=1}^{n} \mu^*(E_i).$$

iv. $\mu^*$ generalizes or extends the concept of "length" in that $\mu^*(I) = l(I)$.

How does the concept of *Lebesgue outer measure* translate to the notion of *Lebesgue measure* itself? In order to transition from $\mu^*(E)$ to $\mu(E)$, we need an additional condition on $E$. Specifically, a set $E \subset R$ is **Lebesgue measurable** if for every set $A \subset R$,

$$\mu^*(A) = \mu^*(A \cap E) + \mu^*(A \cap E'). \tag{1.5}$$

This requirement is not new (see the discussion underlying Equation (1.5) of Section 1.10.3). As explained therein, if for every $A$ the partition of $A$ induced by $E$ (the sets $A \cap E$ and $A \cap E'$) has outer measures that correctly add up to the outer measure of $A$ itself, then set $E$ is "well-behaved" in that $E$ does not adversely impact or distort the outer measure of $A$ when $E$ is used to partition $A$. The upshot of all this is that, under (1.5), $\mu^*(E)$ yields $\mu(E)$. That is, if $E$ is Lebesgue measurable, then the **Lebesgue measure** of $E$ is defined to be its outer measure $\mu^*(E)$ and simply written as $\mu(E)$.

As far as the properties of Lebesgue measure $\mu(E)$ are concerned, they mirror those of $\mu^*(E)$ (see properties (i)–(iv)), but with one key exception—property (iii) involving *countable subadditivity* is replaced by *countable additivity*: if $\{E_i, i = 1, 2, ...\}$ is a sequence of disjoint subsets of $R$, then

$$\text{(iii)}' \mu\left(\cup_{i=1}^{\infty} E_i\right) = \sum_{i=1}^{\infty} \mu(E_i).$$

How should the family of Lebesgue measurable sets (denoted $\mathcal{M}$) be defined? Clearly, we need to specify the *largest* family $\mathcal{M}$ of subsets of $R$ for which $\mu\colon \mathcal{M} \to R^+$ and properties (i), (ii), (iii)′, and (iv) hold. Hence, the family **of Lebesgue measurable sets** $\mathcal{M}$ encompasses the collection of all open intervals as well as all finite unions of intervals on the real line. Then for $E \in \mathcal{M}$, $\mu(E)$, the Lebesgue measure of $E$, is the total length of $E$ when $E$ is decomposed into the union of a finite number of disjoint intervals.

We note in passing that $\emptyset$ and $R$ are Lebesgue measurable with $\mu(\emptyset) = 0$ and $\mu(R) = +\infty$, respectively; open and closed intervals of real numbers are Lebesgue measurable; every open set and every closed set is Lebesgue measurable; every Borel set (which includes countable sets, open and closed intervals, all open sets and all closed sets) is Lebesgue measurable; any countable set of real numbers has Lebesgue measure equal to zero; if $E$ is Lebesgue measurable, then so is $E'$; and if $\{E_i, i = 1, 2, \ldots\}$ is a sequence of Lebesgue measurable sets, then $\cup_{i=1}^{\infty} E_i$ and $\cap_{i=1}^{\infty} E_i$ are Lebesgue measurable sets.

**Example 1.2** Let $E = [a, b] \subset \mathcal{M}$ with $\{x_1, x_2, x_3\} \subset [a, b], a < x_1 < x_2 < x_3 < b$. Consider the set $A = E\{x_1, x_2, x_3\}$. For the measure function $\mu\colon \mathcal{M} \to R^+$,

$$\mu(A) = \mu\{[a, x_1) \cup (x_1, x_2) \cup (x_2, x_3) \cup (x_3, b]\}$$
$$= \mu[a, x_1) + \mu(x_1, x_2) + \mu(x_2, x_3) + \mu(x_3, b]$$
$$= (x_1 - a) + (x_2 - x_1) + (x_3 - x_2) + (b - x_3) = b - a = \mu(E). \blacksquare$$

### 1.10.6  Measurable Functions

Let $(X, \mathcal{D})$ and $(Y, \mathcal{G})$ be measurable spaces, where $\mathcal{D}$ is a $\sigma$-algebra on $X$ and is a $\sigma$-algebra on $Y$, respectively. A **measurable function** is a mapping $f\colon X \to Y$ such that $f^{-1}(G) \in \mathcal{D}$ for every set $G \in \mathcal{G}$. Clearly, the measurability of $f$ depends upon $\mathcal{D}$ and $\mathcal{G}$ and not on the particular measures defined on these $\sigma$-algebras. As this definition indicates, measurable functions are defined in terms of inverse images of sets. (Thus, measurable functions are mappings that occur between measurable spaces in much the same way that continuous functions are mappings between topological spaces.) To elaborate on this notion, if $f\colon \Omega \to \mathcal{G}$ and $A \in \cdot$, let $f^{-1}(A) = \{x \in \Omega | f(x) \in A\}$ and call $f^{-1}(A)$ the **inverse image** of set $A$ under rule $f$. (Note: $f^{-1}(A)$ contains all of the points in the domain $\Omega$ of $f$ mapped by $f$ into $A$; it does *not* denote the inverse function of $f$.) Key properties of the inverse image of $A$ are the following:

   i. $f^{-1}(A') = (f^{-1}(A))'$ for all $A \in \mathcal{G}$.
  ii. If $A, B \in \mathcal{G}$, then $f^{-1}(A \cap B) = f^{-1}(A) \cap f^{-1}(B)$.
 iii. If $\{A_k\} \subset \mathcal{G}$, then $f^{-1}\left(\cup_{k=1}^{\infty} A_k\right) = \cup_{k=1}^{\infty} f^{-1}(A_k)$.

In addition, if $\mathcal{C}$ is a collection of subsets of $\mathcal{G}$, let $f^{-1}(\mathcal{C}) = \{f^{-1}(A)|A \in \mathcal{C}\}$. In this regard, if $f: \Omega \to \mathcal{G}$ and $\mathcal{C}$ is a collection of subsets of $\mathcal{G}$, then (a) for $\mathcal{C}$ a $\sigma$-algebra on $\mathcal{G}$, $f^{-1}(\mathcal{C})$ is a $\sigma$-algebra on $\Omega$; and (b) $f^{-1}(\mathcal{G}(\mathcal{C})) = \mathcal{G}(f^{-1}(\mathcal{C}))$, $(\Omega, \mathcal{G})$ a measurable space.

A measurable mapping $g: X \to Y$ on a measure space $(\Omega, \mathcal{G}, \mu)$ is **measure preserving** if $\mu(g^{-1}(A)) = \mu(A)$ for all measurable sets $A$.

We previously termed the $\sigma$-algebra generated by intervals (open, closed, half-open) in $R$ the **Borel $\sigma$-algebra** $\mathcal{B}$. In this regard, if $(\Omega, \mathcal{G})$ is a measurable space, the mapping $f: \Omega \to R$ is $\mathcal{G}$-**measurable** if $f^{-1}(B) \in \mathcal{G}$ for every Borel set $B \in R$. In fact, the collection of sets $f^{-1}(B)$, where $B$ is contained within the Borel subsets of $R$, is a $\sigma$-algebra on $\Omega$. In addition, if the collection $\mathcal{C}$ of Borel subsets of $R$ generates the Borel $\sigma$-algebra, then $f: \Omega \to R$ is $\mathcal{G}$-measurable if and only if $f^{-1}(\mathcal{C}) \subset \mathcal{G}$. Equivalently, the mapping $f: \Omega \to R$ is $\mathcal{G}$-measurable if and only if the set $\{x \in \Omega | f(x) \le a\}$ is measurable (i.e., it is a member of $\mathcal{G}$) for all $a \in R$. (Note: "$\le$" can be replaced by "$<, \ge, >$.")

The **indicator** or **characteristic function** of a set $A \in \Omega$ is defined as

$$\chi_A(x) = \begin{cases} 1, x \in A; \\ 0, x \notin A. \end{cases} \tag{1.7}$$

If $A, B$ are two subsets of $\Omega$, then

$$\chi_{A \cap B} = min\{\chi_A, \chi_B\} = \chi_A \cdot \chi_B;$$

$$\chi_{A \cup B} = max\{\chi_A, \chi_B\} = \chi_A + \chi_B - \chi_A \cdot \chi_B; \text{ and}$$

$$\chi_{A'} = 1 - \chi_A.$$

Moreover, if $A_i, i = 1, \dots, n$, and $B_j, j = 1, \dots, m$, are subsets of $\Omega$ and $X = \sum_{i=1}^{n} x_i \chi_{A_i}$ and $Y = \sum_{j=1}^{m} y_j \chi_{B_j}$, then

$$X \cdot Y = \sum_{i=1}^{n} \sum_{j=1}^{m} x_i y_j \chi_{A_i \cap B_j}.$$

In addition, if $\{A_i\}_{i=1}^{n}$ and $\{B_j\}_{j=1}^{m}$ are partitions of $\Omega$, then $\{A_i \cap B_j\}_{\text{all } i,j}$ is also a partition of $\Omega$, and thus

$$X + Y = \sum_{i=1}^{n} \sum_{j=1}^{m} (x_i + y_j) \chi_{A_i \cap B_j}.$$

If $\mathcal{G}$ is a $\sigma$-algebra on $\Omega$, then $\emptyset$ and $\Omega$ are members of $\mathcal{G}$. In addition, with set $A \in \mathcal{G}$, it follows that $A' \in \mathcal{G}$. Hence, $\mathcal{G}_{\chi_A} = \{\emptyset, A, A', \Omega\}$, and thus $\chi_A$ is $\mathcal{G}$-**measurable** if and only if $A \in \mathcal{G}$. Note also that if $X$ and $Y$ are $\mathcal{G}$-measurable functions on $\Omega$, then $X + Y, X - Y, X \cdot Y$, and $cX$ ($c$ a real scalar) are all $\mathcal{G}$-measurable. Suppose $Y(x) \ne 0$ for all $x \in \Omega$. Then $X/Y$ is also $\mathcal{G}$-measurable.

Suppose $f : \Omega \to R^*$ is a measurable function with $A = \{x | f(x) \geq 0\}$ and $B = \{x | f(x) \leq 0\}$. If $f^+ = f \cdot \chi_A$ and $f^- = -f \cdot \chi_B$, then the **positive part of $f$** is defined as

$$f^+ = max\{f(x), 0\} = \begin{cases} f(x), f(x) \geq 0; \\ 0, f(x) < 0; \end{cases}$$

and the **negative part of $f$** is defined as

$$f^- = max\{-f(x), 0\} = \begin{cases} -f(x), f(x) \leq 0; \\ 0, f(x) \geq 0, \end{cases}$$

where $f^+$ and $f^-$ are both positive functions on $\Omega$. With $\mathfrak{I}$ a $\sigma$-algebra on $\Omega$ and $f$ is measurable, sets $G = f^{-1}(\{x | x \geq 0\})$ and $H = f^{-1}(\{x | x \leq 0\})$ are in the $\sigma$-algebra generated by $\mathfrak{I}$ (denoted $\mathfrak{I}_f$). Hence, $f^+, f^-$ and $|f|$ are all $\mathfrak{I}_f$-measurable. The upshot of this discussion is that an arbitrary measurable function $f$ can be written in a canonical way as the difference between two positive measurable functions as $f = f^+ - f^-$. In addition, $|f| = f^+ + f^-$.

A function $\Phi : \Omega \to R$ defined on a measurable space $(\Omega, \mathfrak{I})$ is a **simple function** if there are disjoint measurable sets $A_1, \ldots, A_n$ and real scalars $c_1, \ldots, c_n$ such that

$$\Phi = \sum_{i=1}^{n} c_i \chi_{A_i}. \tag{1.8}$$

Clearly, $\Phi$ takes on finitely many, finite values $c_i, i = 1, \ldots, n$. Since simple functions are measurable, any measurable function may be approximated by simple functions. In fact, for $f : \Omega \to R^+$ a non-negative measurable function, there is a monotone increasing sequence $\{\Phi_i\}$ of simple functions that converges pointwise to $f$.

Given a measure space $(\Omega, \mathfrak{I}, \mu)$, if $\Omega = \cup_{i=1}^{n} A_i$ and the sets $A_i$ are disjoint, then these sets are said to form a (finite) **dissection** of $\Omega$. They are said to form an $\mathfrak{I}$-**dissection** if $A_i \in \mathfrak{I}, i = 1, \ldots, n$. A function $f : \Omega \to R$ is termed $\mathfrak{I}$-**simple** if it can be expressed as $f(x) = \sum_{i=1}^{n} c_i \chi_{A_i}$, where the $A_i$'s, $i = 1, \ldots, n$, form an $\mathfrak{I}$-dissection of $\Omega$. Thus, $f(x)$ takes on a constant value $c_i$ on the set $A_i$, given that the $A_i$'s are disjoint subsets of $\mathfrak{I}$.

A sequence of measurable functions $\{f_n\}$ from a measure space $(\Omega, \mathfrak{I}, \mu)$ to $R*$ **converges pointwise** to a function $f : \Omega \to R^*$ if $lim_{n \to \infty} f_n(x) = f(x)$ for every $x \in \Omega$. Moreover, $f$ itself is measurable. A sequence $\{f_n\}$ **converges pointwise a.e.**[7] to $f$ if it converges pointwise to $f$ except on a set $M$ of measure zero.

---

7 A **set of measure zero** is a measurable set $M$ such that $\mu(M) = 0$. A property or condition that holds for all $x \in \Omega - M$, where $M$ is a set of measure zero, is said to hold **almost everywhere** (abbreviated "a.e.") or "except on a set of measure zero." Note: a subset of a set of measure zero need not be measurable; but if it is measurable, then it must have measure zero.

If $(\Omega, \mathcal{F}, \mu)$ is a complete measure space and $\{f_n\}$ converges pointwise a.e. to $f$, then $f$ is measurable.

We note briefly that if the measurable space is $(R^n, \mathcal{B}^n)$, a $\mathcal{B}^n$-measurable function is termed a **Borel-measurable function**.

### 1.10.7 Lebesgue Measurable Functions

Suppose $(X, \mathcal{D})$ and $(Y, \mathcal{G})$ are measurable spaces, with $X$ and $Y$ equipped with the $\sigma$-algebras $\mathcal{D}$ and $\mathcal{G}$, respectively. Then, as indicated in Section 1.10.6, the function $f : X \to Y$ is **measurable** if the anti-image of $E$ under $f$ is in $\mathcal{D}$ for every $E \subset \mathcal{G}$, that is, $f^{-1}(E) = \{x \in X | f(x) \in E\} \in \mathcal{D}$ for all $E \subset \mathcal{G}$.

Let us now get a bit more specific. Suppose $(R, \mathcal{L})$ and $(R, \mathcal{B})$ are measurable spaces, with $\mathcal{L}$ the $\sigma$-algebra of Lebesgue measurable sets and $\mathcal{B}$ the Borel $\sigma$-algebra on $R$. (Remember that $\mathcal{B}$ is the smallest $\sigma$-algebra containing all the open sets.) The function $f : R \to R$ is **Lebesgue measurable** if the anti-image of $B$ under $f$ is a Lebesgue measurable subset of $R$ for every Borel subset $B$ of $R$, that is, $f^{-1}(B) = \{x \in R | f(x) \in B\} \in \mathcal{L}$ for all $B \in \mathcal{B}$. (Clearly, the domain and range of $f$ involve different $\sigma$-algebras defined on the same set $R$.) In very basic terms, for a bounded interval $I$, a function $f : I \to R$ is Lebesgue measurable if, for every open set $B \subseteq R$, the anti-image $f^{-1}(B)$ is measurable in $I$.

An important alternative way of specifying a function that is Lebesgue measurable is the following. If $(X, \mathcal{L})$ is a measurable space, then $f : X \to R$ is **Lebesgue measurable** if and only if $f^{-1}([a, +\infty]) = \{x \in X | f(x) > a\} \in \mathcal{L}$ for all $a \in R$ (Figure 1.3). (Note: equivalent statements involve ">" being replaced by "$\geq$" or "<" or "$\leq$.")

To summarize: A function $f$ between measurable spaces is measurable if the anti-image of each measurable set is *measurable*. A function $f$ is *Lebesgue*

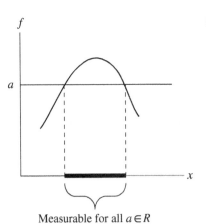

**Figure 1.3** A measurable subset of $\mathcal{L}$.

Measurable for all $a \in R$

*measurable* if and only if the anti-image of each of the sets $[a, +\infty]$ is a Lebesgue measurable set.

We end this discussion by commenting that continuous functions, monotone functions, step functions, and Riemann integrable functions are all Lebesgue measurable. Moreover, if $f,g : R \to R$ are Lebesgue measurable functions and $c \in R$, then $cf, f + g, fg, |f|$ and $|g|$, and $max\{f,g\}$ are all Lebesgue measurable.

## 1.11 Normed Linear Spaces

### 1.11.1 Space of Bounded Real-Valued Functions

Given a nonempty set $A$, suppose that each pair of elements $x, y \in A$ can be operated on, by a process called **addition** ("+"), to yield a new element $x + y = z \in A$, where the operator "+" satisfies the following:

a. $x + y = y + x$ (commutative law).
b. $x + (y + z) = (x + y) + z$ (associative law).
c. $0 \in A$ such that $x + 0 = x$ for every $x \in A$ (zero is the **additive identity**).
d. $-x \in A$ such that $x + (-x) = 0$ ($-x$ is the **additive inverse**).

Suppose also that, for each real scalar $\alpha$ and for each element $x \in A$, $\alpha$ and $x$ can be operated on, by a process called **scalar multiplication** ("·"), to yield a new element $\alpha x = y \in A$, where the operator "·" satisfies the following:

a. $\alpha(x + y) = \alpha x + \alpha y$
b. $(\alpha + \beta)x = \alpha x + \beta x$ $\Big\}$ distributive laws.

c. $(\alpha\beta)x = \alpha(\beta x)$ (associative law).

d. $1 \cdot x = x$ ($1$ is the **multiplicative identity**).

Now, if $A$ is closed under the operations of addition and scalar multiplication, then $A$ (which can be viewed as an algebraic system) is termed a **linear space** (or **vector space**). A nonempty subset $C$ of a linear space $A$ constitutes a **linear subspace** of $A$ if $x + y$ is in $C$ when $x, y \in C$; and $\alpha x \in C$ ($\alpha$ a real scalar) when $x \in C$. As was the case with $A$ itself, $0$ and $-x$ are elements of $C$ whenever $x \in C$.

In Section 1.4, we introduced the concept of a norm—a function that assigns to each $x$ within a space a real number $\|x\|$ such that the properties of non-negativity, homogeneity, and the triangle inequality hold. If the norm "$\|\cdot\|$" is defined on a linear space $A$, then $A$ becomes a **normed linear space**. We also noted in Section 1.4 that a normed linear space is a metric space with respect to the metric $d(x, y) = \|x - y\|$ induced by the norm (Equation (1.2)).

We next consider the notion of a **function space**—a linear space whose elements are functions defined on some nonempty set $X$, with pointwise addition and scalar multiplication satisfying the following:

    a. $(f + g)(x) = f(x) + g(x)$.

    b. $(\alpha f)(x) = \alpha f(x)$.

    c. the zero element $0 \in X$ is the constant function "0"        (1.9)

       whose only value is the scalar $0$.

    d. $(-f)(x) = -f(x)$.

Consider now the set of all real-valued functions defined on $X$. Clearly, $X$ is a *real linear space* whose elements satisfy (1.9). If we separate out from $X$ the subset $B$ of all bounded real-valued functions ($f \in B$ is **bounded** if there exists a real scalar $M$ such that $|f(x)| \le M$), then $B$ is itself a linear space. In addition, if we define on the elements of $B$ the norm

$$\|f\| = sup|f(x)|, \qquad (1.10)$$

then $B$ is a metric space.

Suppose $\{f_n\}$ is a sequence of real-valued functions defined on $B$ and that, for each $x \in B$, $\{f_n(x)\}$ is a Cauchy sequence (to review, $\{f_n(x)\}$ is a Cauchy sequence if for all $\varepsilon > 0$ there exists an $N_{\varepsilon/2} > 0$ such that for all $m$, $n >$ $N_{\varepsilon/2}, \|f_m(x) - f_n(x)\| < \varepsilon$). Thus, $B$ is complete in that, for each $x \in B$ and every Cauchy sequence $\{f_n\}$ defined on $B$, there exists a well-defined continuous **limit function** $f(x) = lim_{n \to \infty} f_n(x)$ so that $\{f_n(x)\}$ **converges pointwise** to $f(x)$. In fact, any normed space with the property that every Cauchy sequence defined on it is convergent is complete.

The preceding discussion enables us to conclude that $B$ constitutes an important type of function space, namely, a **Banach space**—a complete normed linear (metric) space. We next turn to another type of function space that is also a Banach space.

### 1.11.2 Space of Bounded Continuous Real-Valued Functions

A key property that the elements of a function space can possess is *continuity*. To explore this characteristic, let's assume at the outset that the set of all real-valued functions is defined on a metric space $X$. Furthermore, given $B$ above (as defined earlier, $B$ is the subset of $X$ containing all bounded real-valued functions), let $C(X) \subset B$ denote the set of all bounded continuous functions defined on $X$.

It should thus be evident that

    a. if $f, g$ are continuous real-valued functions defined on $X$, then, pointwise, $f + g$ and $\alpha f$ ($\alpha$ a real scalar) are also continuous;

b. $C(X)$ is a linear subspace of the linear (metric) space $B$; and

c. $C(X)$ is a closed[8] subset of the linear (metric) space $B$.

But remember that $B$ is a Banach space and, since a closed linear subspace of a Banach space is also a Banach space, it follows that $C(X)$, the set of all bounded continuous real-valued functions defined on a metric space $X$ with norm (1.10) is a Banach space.

### 1.11.3 Some Classical Banach Spaces

1. Let $R^n$ denote the set of all **vectors** or ordered $n$-tuples $x = (x_1, x_2, ..., x_n)$ of real numbers. For elements $x = (x_1, x_2, ..., x_n)$ and $y = (y_1, y_2, ..., y_n)$ in $R^n$, let us define, coordinatewise, addition and scalar multiplication as

$$x + y = (x_1 + y_1, x_2 + y_2, ..., x_n + y_n),$$
$$\alpha x = (\alpha x_1, \alpha x_2, ..., \alpha x_n), \alpha \text{ a real scalar,}$$

respectively. In addition,, the zero (or null) element $0 = (0, 0, ..., 0)$ (also an $n$-tuple) and $-x = (-x_1, -x_2, ..., -x_n)$ are elements in $R^n$. For any element $x \in R^n$, let us define the norm of $x$, $\|x\|$, by

$$\|x\| = \left( \sum_{i=1}^{n} |x_i|^2 \right)^{\frac{1}{2}}. \tag{1.11}$$

(See Section 1.4 for a discussion of the properties of this norm.) Given (1.11), it is evident that $R^n$ can be characterized as a normed linear (metric) space (also called $n$-dimensional **Euclidean space**—since (1.11) is the **Euclidean norm**). Moreover, it is complete with respect to the metric (1.2) and thus amounts to a complete normed linear space or Banach space.

2. Let $L^p(\mu)$ denote the set of all measurable functions $f$ defined on a measure space $(\Omega, \Im, \mu)$ and having the property that $|f(x)|^p$ is integrable, with **p-norm**

$$\|f\|_p = \left( \int |f(x)|^p d\mu(x) \right)^{\frac{1}{p}}, 1 \le p < +\infty. \tag{1.12}$$

For $p \in [1, +\infty), L^p(\mu)$ is complete with respect to (1.12) and thus constitutes a Banach space.

---

8 Suppose $f \in B$ with $f \in \overline{C(X)}$ (the closure of $C(X)$). Let $d$ be the metric on $X$, with $\varepsilon > 0$ given. Since $f$ is in $\overline{C(X)}$, there exists a function $f_0$ in $C(X)$ such that $\|f - f_0\| < \varepsilon'$ implies $|f(x) - f(x_0)| < \varepsilon'$ for each $x \in X$, where $\varepsilon'$ is proportional to $\varepsilon$. With $f_0$ continuous at $x_0$, there exists a $\delta_{\varepsilon'} > 0$ such that $d(x, x_0) = \|x - x_0\| < \delta_{\varepsilon'}$ implies $|f(x_0) - f_0(x_0)| < \varepsilon'$. Since $\|x - x_0\| < \delta_\varepsilon$ implies that $|f(x) - f(x_0)| < \varepsilon$, we see that $f$ is continuous at $x_0$ ($x_0$ arbitrary). Hence, $f \in C(X) = \overline{C(X)}$ so that $C(X)$ must be closed (Simmons, 1963, p. 83).

3. Let $H$ be an $n$-dimensional linear (vector) space with the **inner product norm** defined by

$$\|x\| = (x,x)^{\frac{1}{2}},$$ (1.13)

where $(x, x)$ is the **inner product**[9] defined by

$$(x,x) = \sum_{i=1}^{n} |x_i|^2, x \in H.$$

(A linear space equipped with an inner product is called an **inner product space**.) Clearly, $H$ is a normed linear space and is complete with respect to the norm given by (1.13); it will be called a **Hilbert space**—a complete normed inner product space. Although $H$ is always a Banach space whose norm is determined by an inner product $(\cdot, \cdot)$ (i.e., $\|f\| = (f,f)^{1/2}$ for all $f$ in the space), the converse does not generally hold. So what is the essential difference between a Banach space and a Hilbert space? The difference is in the source of the norm, that is, for a Banach space, the norm is defined directly as $\|\cdot\| : B \to [0, +\infty)$ for all points $x$, $y$ (and scalar $c$) satisfying the properties outlined earlier in footnote 2; and for a Hilbert space, the norm is defined by an inner product (Equation (1.13)). The inner product is not defined on a Banach space.

The ordered $n$-tuples (vectors) $x, y \in R^n$ are said to be **orthogonal** if $(x,y) = \sum_{i=1}^{n} x_i y_i = 0$. Elements $x$ and $y$ within a Hilbert space ($H$) are orthogonal if $(x, y) = 0$ and **orthonormal** if, in addition, $\|x\| = \|y\| = 1$. An **orthonormal set** in $H$ is a nonempty subset of $H$ that consists of mutually orthogonal unit vectors $e_i, i = 1, 2, \dots$. (A **unit vector** $e_i$ has a "1" as its $i$th component and "0's" elsewhere.) That is, an orthonormal set is a non-empty subset $\{e_i, i = 1, 2, \dots\}$ of $H$ with the properties (1) $(e_i, e_j) = 0, i \neq j$; and (2) $\|e_i\| = 1$ for all $i$.

An orthonormal sequence $\{e_i, i = 1, 2, \dots\}$ in $H$ is **complete** if the only member of $H$ that is orthogonal to every $e_i$ is the **null vector** 0 (which contains all zero components). (Stated alternatively, we cannot find a vector $e$ such that $\{\{e_i\}, e\}$ is an orthonormal set that properly contains $\{e_i, i = 1, 2, \dots\}$.)

Suppose $\{e_1, e_2, \dots, e_n\}$ is a finite orthonormal set in $H$. If $x$ is any vector in $H$, then

1. $\sum_{i=1}^{n} |(x, e_i)|^2 \leq \|x\|^2$;

---

9 The inner product $(x, x)$ satisfies the following conditions:

a. For $x \in H$, $(x, x) \geq 0$ and $(x, x) = 0$ if and only if $x = 0$ (positive semidefiniteness).
b. For $x, y \in H$, $(x,y) = (y,x)$ (symmetry).
c. For $x, y \in H$, with $a, b$ real scalars, $((ax_1 + bx_2), y) = (ax_1, y) + (bx_2, y)$ (linear in its first argument).
d. For $x, y \in H$, $|(x,y)| \leq \|x\| \|y\|$ (**Cauchy–Schwarz inequality**).

2. $x = \sum_{i=1}^{n} (x, e_i) e_i$; and

3. $\left( x - \sum_{i=1}^{n} (x, e_i) e_i, e_j \right) = 0$ for each $j$.

An **orthonormal basis** for a Hilbert space ($H$) is a basis[10] consisting of non-zero orthonormal vectors. Such vectors are linearly independent and span $H$ in the sense that every element in $H$ can be written as a linear combination of the basis vectors. In fact, every Hilbert space contains a maximal orthonormal set that serves as a basis.

Suppose in (1.12), we set $p = 2$. Then the class of real-valued square-integrable functions

$$L^2(\mu) = \|f\|_2 = \left( \int |f|^2 d\mu(x) \right)^{\frac{1}{2}} < +\infty \qquad (1.12.1)$$

is a Hilbert space. In addition, if ($\Omega$, $\mathcal{F}$, $\mu$) is a measure space and the functions $f, g \in L^2(\mu)$, then the inner product of $f$ and $g$ is

$$(f, g) = \int f g \, d\mu, \qquad (1.14)$$

where $|(f \cdot g)| \leq \|f\| \|g\|$.

The space $L^2(a, b)$, the collection of Borel measurable real-valued square integrable functions $f$ on $(a, b)$ (i.e., $\int_a^b |f(t)|^2 dt < +\infty$), is a Hilbert space. For this space, the inner product is $(f, g) = \int_a^b f(t) g(t) dt$, and the associated norm and metric are, respectively, $\|f\|_2 = \left( \int_a^b |f(t)|^2 dt \right)^{1/2}$ and $d(f, g) = \|f - g\| = \left( \int_a^b |f(t) - g(t)|^2 dt \right)^{1/2}$. (Here, the functions $f$, $g$ are considered equal if they differ on $(a, b)$ only on a set of measure zero.)

## 1.12 Integration

Our approach in this section is to first define the *integral of a non-negative simple function*. We then define the *integral of a non-negative measurable*

---

10 To review, a vector $x \in R^n$ is a **linear combination** of the vectors $x_j \in R^n, j = 1, \ldots, m$, if there exists scalars $\lambda_j, j = 1, \ldots, m$, such that $x = \sum_{j=1}^{m} \lambda_j x_j$. A set of vectors is **linearly independent** if the trivial combination $0x_1 + \cdots + 0x_n$ is the only linear combination of the $x_j$ which equals the null vector. (The set of vectors $\{x_j, j = 1, \ldots, m\}$ is said to be **linearly dependent** if there exists scalars $\lambda_j, j = 1, \ldots, m$, not all zero such that $\sum_{j=1}^{m} \lambda_j x_j = 0$.) The vectors $x_j, j = 1, \ldots, m$, **span** $R^n$ if every element of $R^n$ can be written as a linear combination of the $x_j$'s. Hence, the $x_j$'s constitute a **spanning set** for $R^n$. A **basis** for $R^n$ is a linearly independent set of vectors from $R^n$ which spans $R^n$. Thus, every vector in $R^n$ can be expressed as a linear combination of the basis vectors.

*function via an approximation by simple functions.* Next comes the definition of the *integral of a measurable function*, followed by the specification of the *integral of a measurable function on a measurable set*. In what follows, $(\Omega, \mathcal{I}, \mu)$ is taken to be a measure space. However, if $\Omega = R$ admits the Borel $\sigma$-algebra, $\mathcal{I}$ is the $\sigma$-algebra of Lebesgue measurable sets in $R$, and the measure $\mu : \mathcal{I} \rightarrow [0, +\infty]$ is given by $\mu(E) = \mu^*(E), E \in \mathcal{I}$, then the integrals defined below are also *Lebesgue integrals*. (Readers not familiar with the *Lebesgue integral* are encouraged to read Appendix B to this chapter along with Taylor (1973) before tackling this section.)

### 1.12.1  Integral of a Non-negative Simple Function

Recall (Section 1.10.5) that a non-negative simple function has the form

$$\emptyset(x) = \sum_{i=1}^{n} c_i \chi_{E_i}(x), c_i \geq 0, i = 1, ..., n, \tag{1.15}$$

where the indicator function $\chi_{E_i}$ is defined as

$$\chi_{E_i}(x) = \begin{cases} 1, x \in E_i; \\ 0, x \in E_i'. \end{cases}$$

The **integral of a non-negative simple function** with respect to $\mu$ is defined in terms of the integral operator "$\int$" as

$$\int_{\Omega} \emptyset \, d\mu = \sum_{i=1}^{n} c_i \mu(E_i), \tag{1.16}$$

where $E_i = \{x | \emptyset(x) = c_i\}$, $\int_{\Omega} \chi_{E_i} d\mu = \int_{E_i} d\mu = \mu(E_i) < +\infty$ and the sum on the right-hand side of (1.16) is well defined since each of its terms is non-negative. (It is important to note that since the specification of a simple function in terms of indicator functions is not unique, this definition of the integral is independent of the actual specification used.) For the simple function given in Equation (1.15), suppose set $A \in \mathcal{I}$ is measurable. Then the **integral of a non-negative simple function over a set $A$** is defined as

$$\int_{A} \emptyset \, d\mu = \sum_{i=1}^{n} c_i \mu(E_i \cap A). \tag{1.16.1}$$

As far as the essential properties of the integral operator "$\int$" are concerned, it is *linear* as well as *order preserving* on the class of non-negative simple functions. That is, given two non-negative simple functions $\emptyset = \sum_{i=1}^{n} c_i \chi_{E_i}$ and $\psi = \sum_{j=1}^{m} d_j \chi_{F_j}$, the simple function

$\emptyset + \psi = \sum_{i=1}^{n} \sum_{j=1}^{m} (c_i + d_j) \chi_{E_i \cap F_j}$, and thus

$$\int_{\Omega} (\emptyset + \psi) d\mu = \int_{\Omega} \emptyset \, d\mu + \int_{\Omega} \psi d\mu; \left( \text{``} \int \text{'' is \textbf{linear}} \right) \tag{1.17}$$

while, for $\emptyset \geq \psi$,

$$\int_{\Omega} \emptyset \, d\mu \geq \int_{\Omega} \psi d\mu. \left( \text{``} \int \text{'' is \textbf{order preserving or monotonic}} \right) \tag{1.18}$$

### 1.12.2 Integral of a Non-negative Measurable Function Using Simple Functions

Suppose the non-negative function $f: \Omega \to R^+$ is measurable. It was noted in Section 1.10.5 that there exists a monotone increasing sequence $\{f_n\}$ of simple functions that converge pointwise to $f$. Given that $\int_{\Omega} f_n d\mu$ is defined for all $n$, and the said sequence is monotonic, it follows that the limit of $\int_{\Omega} f_n d\mu$ is an element of $R^+$. Hence, we may define the operation of **integration for non-negative measurable functions** as

$$\int_{\Omega} f d\mu = \lim_{n \to \infty} \int_{\Omega} f_n d\mu. \tag{1.19}$$

Since $\int_{\Omega} f d\mu$ may be finite or infinite in $R^+$, we may conclude that a non-negative measurable function $f$ is integrable with respect to a measure $\mu$ if the limit in (1.19) is finite. In addition, if $f \geq 0$ is measurable, then **integration for non-negative measurable functions over a set $A$** is defined as

$$\int_{A} f d\mu = \sup_{\emptyset} \int_{A} \emptyset \, d\mu < +\infty. \tag{1.19.1}$$

where the supremum is taken over all simple functions $\emptyset$ with $0 \leq \emptyset \leq f$.

### 1.12.3 Integral of a Measurable Function

Suppose the function $f: \Omega \to R^+$ is measurable. Then, as indicated earlier in Section 1.10.5, so are $f_+$ and $f_-$. If $f_+$ and $f_-$ are integrable with respect to $\mu$, then $f = f_+ - f_-$ itself is integrable with respect to $\mu$, and thus

$$\int_{\Omega} f d\mu = \int_{\Omega} f_+ d\mu - \int_{\Omega} f_- d\mu \tag{1.20}$$

so that this expression defines **integration for the class of integrable measurable functions**. Also, for set $A \in \mathcal{I}$, $f$ is integrable over a set $A$ if

$$\int_A |f| \, d\mu = \int_A f_+ \, d\mu + \int_A f_- \, d\mu < +\infty. \tag{1.20.1}$$

### 1.12.4  Integral of a Measurable Function on a Measurable Set

Let set $A \in \mathcal{I}$. Suppose $\int f\chi_A \, d\mu$ is defined (e.g., either $f\chi_A$ is non-negative and measurable, or $f\chi_A$ is measurable and integrable). Then

$$\int_A f \, d\mu = \int_A f\chi_A \, d\mu. \tag{1.21}$$

Thus, $f$ **is integrable over a set** $A$ **if** $f\chi_A$ **is integrable**. (Note that if $A \in \mathcal{I}$ and $\mu(A) = 0$, then $f \colon \Omega \to R^*$ is integrable over $A$ with $\int_A f \, d\mu = 0$.)

For a measure space $(\Omega, \mathcal{I}, \mu)$ and $f \colon \Omega \to R^*$ an integrable function with respect to $\mu$ over $\Omega$, some additional properties of the integral operator "$\int$" are the following:

  i. For $A$ and $B$ disjoint sets in $\mathcal{I}$,

$$\int_{A \cup B} f \, d\mu = \int_A f \, d\mu + \int_B f \, d\mu.$$

  ii. $|f|$ is integrable and $\left| \int_\Omega f \, d\mu \right| = \int_\Omega |f| \, d\mu.$

  iii. For a constant $c \in R$, $cf$ is integrable and $\int_\Omega cf \, d\mu = c \int_\Omega f \, d\mu.$

  iv. If $f \geq 0$, then $\int_\Omega f \, d\mu \geq 0$; but if $f \geq 0$ and $\int f \, d\mu = 0$, then $f = 0$ a.e.

  v. If $g \colon \Omega \to R^*$ is integrable with respect to $\mu$ over $\Omega$, then if $f = g$ a.e., it follows that

$$\int_\Omega f \, d\mu = \int_\Omega g \, d\mu.$$

  vi. If sets $A, B \in \mathcal{I}$ with $A \subset B$ and $f \geq 0$, then $\int_A f \, d\mu \leq \int_B f \, d\mu.$

  vii. Let $\mu$ be the counting measure[11] on $\Omega = \{1, 2, 3, \ldots\}$ and define the measurable function $f \colon \Omega \to R$ as $f(j) = a_j, j \in \Omega$. Then

---

11 Let $(\Omega, \mathcal{I}, \nu)$ be a measure space. The **counting measure** $\nu$ on $\Omega$ is defined as $\nu(A) =$ number of elements in $A \in \mathcal{I}$. This measure is finite if $\Omega$ is a finite set; it is $\sigma$-finite if $\Omega$ is countable.

$$\int_{\Omega} f(j)\, d\mu(j) = \sum_{j=1}^{\infty} a_j.$$

This integral is well-defined if $f \geq 0$, or if the sum on the right-hand converges absolutely. (If $\sum_{j=1}^{\infty} |a_j|$ is convergent, then $\sum_{j=1}^{\infty} |a_j|$ is termed **absolutely convergent**.) In either instance we say that $f$ is integrable with respect to $\mu$.

viii. A measurable function $f$ is integrable on $A \in \mathcal{I}$ if and only if $|f|$ is integrable on $A$.

ix. If $f$ is integrable on set $A \in \mathcal{I}$, if $g$ is measurable, and if $|g| \leq f$ a.e. on $A$, then $g$ is integrable on $A$ and $\int_A g\, d\mu \leq \int_A f\, d\mu$.

x. If $f$ is *any* function and for set $A \in \mathcal{I}$, if $\mu(A) = 0$, then $\int_A f\, d\mu = 0$.

### 1.12.5 Convergence of Sequences of Functions

Let $(\Omega, \mathcal{I}, \mu)$ be a measure space. A sequence of functions $\{f_n\}$, where $f_n : \Omega \to R^+$, **converges pointwise** to a function $f : \Omega \to R^+$ if $\lim_{n\to\infty} f_n(x) = f(x)$ for every $x \in \Omega$. Here $f$ is termed a **limiting function**. The sequence $\{f_n\}$ **converges pointwise a.e.** to $f$ if it converges pointwise to $f$ on $\Omega - A$, where $A \in \mathcal{I}$ is a set of measure zero.

In this regard, let $\{f_n\}$ be a sequence of functions that converges pointwise to a limiting function $f$. When can we legitimately conclude that $\int_{\Omega} f_n d\mu$ converges to $\int_{\Omega} f d\mu$? Two conditions that guarantee the convergence of the integrals $\int_{\Omega} f_n d\mu$ are (1) the monotone convergence of the sequence $\{f_n\}$; and (2) a uniform bound on $\{f_n\}$ by an integrable function.

To set the stage for a discussion of the first condition, let us define a sequence of functions $\{f_n\}$, where $f_n : \Omega \to R^+$, as **monotone increasing** if $f_1(x) \leq \cdots \leq f_n(x) \leq \ldots$ for every $x \in \Omega$. We then have Theorem 1.12.1.

**Theorem 1.12.1**  (Lebesgue) Monotone Convergence Theorem (MCT)
Let $\{f_n\}$ be a monotone increasing sequence of non-negative measurable functions $f_n : \Omega \to [0, +\infty]$ on a measure space $(\Omega, \mathcal{I}, \mu)$ and let $f : \Omega \to [0, +\infty]$ be the pointwise limit of $\{f_n\}$ or $f(x) = \lim_{n\to\infty} f_n(x)$. Then

$$\lim_{n\to\infty} \int_{\Omega} f_n d\mu = \int_{\Omega} f d\mu.$$

(Note that if $f$ is integrable on $\Omega$ $\left(\lim_{n\to\infty} \int_{\Omega} f_n d\mu < +\infty\right)$, this theorem posits the convergence of the integrals $\int_{\Omega} f_n d\mu$ to $\int_{\Omega} f d\mu$. If $f$ is not integrable on $\Omega$, then possibly $f_n$ is integrable for all $n$ and $\int_\Omega f_n d\mu \to +\infty$ as $n \to +\infty$.) A consequence of the MCT is Corollary 1.12.1.

**Corollary 1.12.1** (corollary to the MCT). Let $\{f_n\}$, $f : \Omega \to [0, +\infty]$, be a sequence of non-negative measurable functions and set $f = \sum_{n=1}^{\infty} f_n$. Then

$$\int_\Omega f d\mu = \sum_{n=1}^{\infty} \int_\Omega f_n d\mu.$$

A generalization of the MCT is provided by Lemma 1.12.1.

**Lemma 1.12.1** Fatou's lemma
Let $\{f_n\}$ be a sequence of non-negative measurable functions $f_n : \Omega \to [0, \infty]$ on a measure space $(\Omega, \vartheta, \mu)$. Then

$$\int_\Omega \left( \lim_{n \to \infty} \inf f_n \right) d\mu \le \lim_{n \to \infty} \inf \int_\Omega f_n d\mu.$$

As this lemma indicates, the limit of the integrals on the right-hand side of this inequality is always at least as large as the integral of the limit function on the left-hand side.

The second aforementioned condition is incorporated in Theorem 1.12.2.

**Theorem 1.12.2** (Lebesgue) Dominated Convergence Theorem (DCT)
Let $\{f_n\}$ be a sequence of integrable functions, where $f_n : \Omega \to R^*$, on a measure space $(\Omega, \vartheta, \mu)$ that converge pointwise to a limit function $f : \Omega \to R^*$. If there is an integrable function $g : \Omega \to [0, \infty]$ such that $|f_n(x)| \le g(x)$ for all $x \in \Omega$ (and independent of $n$), then $f$ is integrable and

$$\lim_{n \to \infty} \int_\Omega f_n d\mu = \int_\Omega f d\mu.$$

We close this section with an additional useful Theorem 1.12.3.

**Theorem 1.12.3** Let $\{A_n\}$ be a sequence of disjoint measurable sets with $A = \cup_{n=1}^{\infty} A_n$ and let $f$ be a non-negative measurable function that is integrable on each set $A_n$. Then $f$ is integrable on $A$ if and only if $\sum_{n=1}^{\infty} \int_{A_n} f d\mu < +\infty$, so that

$$\int_A f d\mu = \sum_{n=1}^{\infty} \int_{A_n} f d\mu.$$

# 2

# Mathematical Foundations 2

Probability, Random Variables, and Convergence of Random Variables

## 2.1 Probability Spaces

Let us define a **random experiment** as a class of occurrences that can happen repeatedly, for an unlimited number of times, under essentially unchanged conditions. A **random phenomenon** is a happening such that, on any given trial of a random experiment, the outcome is unpredictable. The outcome of a random experiment is called an **event**.

For instance, consider the following game of chance: we roll a fair pair of six-sided dice. (Clearly, this is a process that has random outcomes.) Consider the event "the sum of the faces showing is nine." Let's denote this event as $A$. What is the probability that event $A$ occurs? To answer this question, we need to examine the *set of all possible outcomes* that can obtain on any roll of a pair of dice. This set is called the **sample space** ($\Omega$) and has the form $\Omega = \left\{ \omega_{ij} = (i,j) \mid i,j = 1,\ldots,6 \right\}$ (see Figure 2.1). Clearly, this set has as its elements 36 **simple events** (points or ordered pairs). Are there other events in $\Omega$ besides $A$? (Since $A$ is composed of more than one simple event, it will be termed a **compound event**.) Obviously, the answer is "yes." How many possible subsets of $\Omega$ can exist? The answer to this question is provided by the **power set** of $\Omega$, $2^{\Omega}$, which is the set of all subsets of $\Omega$.[1]

---

1 For each set $A \subseteq \Omega$, let the **indicator function of $A$** be denoted as $\chi_A : \Omega \to \{0,1\}$ or

$$\chi_A(\omega) = \begin{cases} 1, \omega \in A; \\ 0, \omega \notin A. \end{cases}$$

Thus, $2^{\Omega}$ is actually a mapping from $\Omega$ to $\{0,1\}$. However, since the set $\{0,1\}$ has but two elements, this mapping is written more succinctly as $2^{\Omega}$. $2^{\Omega}$ has the following basic properties: (a) $\Omega \in 2^{\Omega}$; (b) if $A \in 2^{\Omega}$ then $A' \in 2^{\Omega}$; and (c) if $A,B \in 2^{\Omega}$, then $A \cup B \in 2^{\Omega}$. Classes of subsets of $\Omega$ with these properties are called "algebras."

*Stochastic Differential Equations: An Introduction with Applications in Population Dynamics Modeling*, First Edition. Michael J. Panik.
© 2017 John Wiley & Sons, Inc. Published 2017 by John Wiley & Sons, Inc.

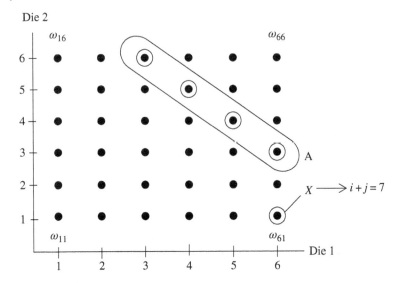

**Figure 2.1** Event $A = \{\omega_{36}, \omega_{45}, \omega_{54}, \omega_{63}\}$.

We previously specified event $A$ as "the sum of the faces showing is nine." The number 9 is actually the value of a **random variable** defined on $\Omega$, that is, a random variable $X$ is a real-valued function defined on the elements of a sample space $\Omega$. Here, $X$ (the sum of the faces) takes a point $\omega_{ij} \in \Omega (i,j = 1,...,6)$ and maps it into a real number according to the rule $i + j$ (Figure 2.1). Hence, the role of a random variable is to define the possible outcomes of a random experiment—it tells us how to get some real numbers out of our random experiment. In this case, $X$ assumes the 11 values $\{2, 3, ..., 12\}$.

To find the probability of event $A$, $P(A)$, we need the **probability measure** $P : 2^{\Omega} \rightarrow [0,1]$, a set function that ascribes to each event $A \subseteq \Omega$ its probability value $P(A)$. To assign some number $P(A)$ to each $A \in 2^{\Omega}$, let us employ the concept of **relative frequency** or $f_n(A) = n_A/n$, where $n_A$ is the number of simple events in $A$ and $n$ is the total number of simple events within $\Omega$. Since $A = \{\omega_{i,j} = (i,j) | i + j = 9\} = \{\omega_{36}, \omega_{45}, \omega_{54}, \omega_{63}\}$ (see Figure 2.1), it is readily determined that $f_n(A) = 4/36 = P(A)$. (Note that the relative frequency concept is applicable because the $n = 36$ simple events in $\Omega$ are *equally likely*.) It should be intuitively clear that $P$ must satisfy two conditions: (1) $P(\Omega) = 1$ ($\Omega$ is called the **certain event**); and (2) if $A,B \in 2^{\Omega}$ are disjoint events ($A \cap B = \emptyset$), then $P$ must be **finitely additive** or $P(A \cup B) = P(A) + P(B)$. (More on all this shortly.)

Given that $\Omega$ contains a finite number of simple events, the triple ($\Omega, 2^{\Omega}$, $P$) completely describes the given random experiment. But what about the circumstance in which a random experiment generates a countably infinite or uncountable number of outcomes? In these circumstances, we need to devise a more comprehensive analytical representation of a random experiment. To this end, a general description of a random process is incorporated in a particular type of measure space called a **probability space** $(\Omega, \mathcal{A}, P)$, where $(\Omega, \mathcal{A})$ is a measurable space; $\Omega$ is the sample space of the random experiment; $\mathcal{A}$ is a $\sigma$-algebra of event subsets of $\Omega$; and $P: \mathcal{A} \to [0,1]$ is a **probability measure** on $\mathcal{A}$. In this regard, the $\sigma$-algebra $\mathcal{A}$ must satisfy the following conditions:

1. $\Omega \in \mathcal{A}$.
2. If event $A \in \mathcal{A}$, then $A' \in \mathcal{A}$ ($A'$ denotes the non-occurrence of $A$)
3. If events $A_i \in \mathcal{A}$, then $\cup_i A_i \in \mathcal{A}$.

In addition, the probability measure must satisfy the following conditions:

a. For every event $A \in \mathcal{A}$, $P(A) \geq 0$.
b. $P(\Omega) = 1$ ($\Omega$ is treated as the "certain event")
c. If $\{A_i, i = 1, 2, \ldots\}$ is a countably infinite sequence of pairwise disjoint events in $\mathcal{A}(A_j \cap A_k = \varnothing, j \neq k)$, then $P(\cup_i A_i) = \Sigma_i P(A_i)$, that is, $P$ is $\sigma$-**additive**. (If $\{A_i, i = i, \ldots, n\}$ is a finite sequence of such events in $\mathcal{A}$, then $P$ is termed **finitely additive**.)

As corollaries to properties (a)–(c) of $P: P(\varnothing) = 0$ (think of $\varnothing$ as the "impossible event"); $P(A') = 1 - P(A)$ (the probability that event $A$ does not occur is 1 minus the probability that it does occur). In addition, if events $A, B \in \mathcal{A}$, then $A \subseteq B$ implies

$$P(A) \leq P(B).$$

What about the random variable $X$ defined on the probability space $(\Omega, \mathcal{A}, P)$? For its specification, let's first focus on the following generalized mapping between measurable spaces. Let $(\Omega_1, \mathcal{A}_1)$ and $(\Omega_2, \cdot_2)$ be measurable spaces. A mapping $X: \Omega_1 \to \Omega_2$ that assigns to every element $\omega \in \Omega_1$ the image $\omega' = X(\omega) \in \Omega_2$ is said to be $(\mathcal{A}_1 - \mathcal{A}_2)$-**measurable**. This mapping is called a $\Omega_2$-**valued random variable on** $(\Omega_1, \mathcal{A}_1)$ if the pre-images (or anti-images) of measurable sets in $\Omega_2$ are measurable sets in $\Omega_1$, that is, for $B \in \mathcal{A}_2$, $\{\omega | X(\omega) \in B\} = X^{-1}(B) \in \mathcal{A}_1$. The set $\mathcal{A}_1(X)$ of pre-images of measurable sets in $\Omega_2$ is itself a $\sigma$-algebra in $\Omega_1$ and is the smallest $\sigma$-algebra with respect to which $X$ is measurable—it is termed the $\sigma$-**algebra generated by** $X$ **in** $\Omega_1$.

Now, back to our original question. Let $(\Omega, \mathcal{A}, P)$ be a probability space and let $(R, \mathcal{U})$ be a measurable space. A **random variable** $X$ on $(\Omega, \mathcal{A})$ is a real-valued function from $\Omega$ to $R$ such that, for all $B \in \mathcal{U}$, $X^{-1}(B) \subseteq \mathcal{A}$. (Clearly, $X$ is a $(\mathcal{A} - \mathcal{U})$-measurable map.) Thus, as required, pre-images of measurable sets in the range of $X$ are measurable sets in $\mathcal{A}$.

How are probabilities defined on $(\Omega, \mathcal{A})$? Given that the probability measure $P$ is a set function with domain $\Omega$ and $A$ is an event within the range of $X$, let $X \in A$ denote the subset $X^{-1}(A) = \{\omega | X(\omega) \in A\}$ of $\Omega$. For $P(X \in A)$ to exist, $X^{-1}(A)$ must lie within the domain $\Omega$ of $P$, that is, $X^{-1}(A) \in \mathcal{A}$.

A few additional points are in order. First, $(\Omega, \mathcal{A}, P)$ is a **complete probability space** if $\mathcal{A}$ contains all subsets $E$ of $\Omega$ having $P$-outer measure $(P*(\cdot))$ zero, where

$$P*(E) = inf\{P(F) | E \subset F, F \in \mathcal{A}\}.$$

In fact, any probability space can be made complete simply by augmenting $\mathcal{A}$ by all sets of measure zero and extending $P$ accordingly.[2]

Next, for $\mathcal{C}$ a class of subsets of $\Omega$, the smallest $\sigma$-algebra containing $\mathcal{C}$, $\sigma(\mathcal{C})$, is termed the $\sigma$-**algebra generated by** $\mathcal{C}$—it is formed as the intersection of all $\sigma$-algebras containing $\mathcal{C}$. Now, let $\mathcal{C}$ be an algebra of subsets of $\Omega$ with $P : \mathcal{C} \rightarrow [0,1]$ a probability measure on $(\Omega, \mathcal{C})$. If $P$ has the following properties,

1. $P(\Omega) = 1$.
2. For pairwise disjoint events $A_i \in \mathcal{C}$, with $\cup_i A_i \in \mathcal{C}$, $P$ is $\sigma$-additive,

then $P$ is *uniquely extended* to a probability measure on $(\Omega, \sigma(\mathcal{C}))$.

Suppose $(\Omega, \mathcal{B})$ is a topological space. The **Borel** $\sigma$-**algebra** $\mathcal{B}$ is the $\sigma$-algebra generated by the collection of open sets in $\Omega$. As indicated in Section 1.9.2, $\mathcal{B}$ is the smallest $\sigma$-algebra containing all open sets on $\Omega$. An element $B \in \mathcal{B}$ is termed a **Borel set**. For instance, if a set $A \in R$ is open, then $A \in \mathcal{B}$ ($A$ is Borel). If the set $C \in R$ is closed, then $R - C$ is open, and thus $R - C \in \mathcal{B}$ so that $\mathcal{C} \in \mathcal{B}$. Thus, every closed set is a Borel set. In addition, countable intersections of open sets and countable unions of closed sets are Borel.

Let $(\Omega, \cdot, P)$ be a probability space and $(R, \mathcal{B})$ a topological space. A mapping $X : \Omega \rightarrow R$ which is $(\mathcal{A} - \mathcal{B})$-measurable (i.e., for an event $B \in \mathcal{B}$, $X^{-1}(B) = \{\omega | X(\omega) \in B\} \subseteq \mathcal{A}$) is a real-valued random variable. Then for any such $X, P(X \in B) = P(\omega | X(\omega) \in B) = P(X^{-1}(B))$, $B \in \mathcal{B}$, defines a probability measure on $(R, \mathcal{B})$.

Suppose $(\Omega_1, \mathcal{B})$ is a topological space, where $\mathcal{B}$ is a Borel $\sigma$-algebra. If $f : (\Omega_1, \mathcal{B}) \rightarrow (\Omega_2, \mathcal{C})$ and $f^{-1}(\omega) \in \mathcal{B}$ for $\omega \in \mathcal{C}$, then the mapping $f$ is said to be **Borel measurable**. For $(\Omega, \mathcal{A}, P)$ a probability space, a function $f : \Omega \rightarrow R$ is called $\mathcal{A}$-**measurable** if $f^{-1}(B) = \{\omega | f(\omega) \in B\} \in \mathcal{A}$ for all open (Borel) sets $B \in R$.

Let's consider a few of example problems.

---

2 For $P$ a probability measure on $(\Omega, \mathcal{A})$, **the completion of** $\mathcal{A}$, $\mathcal{A}_c = \{A \in \Omega|$, there exist events $B_1$, $B_2 \in \mathcal{A}$ such that $B_1 \subset A \subset B_2$ and $P(B_1 - B_2) = 0\}$ is a $\sigma$-algebra on $\Omega$ containing $\mathcal{A}$. Once $P$ is extended to $\mathcal{A}_c$ (the **completion of** $P$, $P_c$, is defined on $\mathcal{A}_c$ as $P_c(A) = P_c(B_1) = P_c(B_2)$), we say that the complete probability space $(\Omega, \mathcal{A}_c, P_c)$ is the **completion of** $(\Omega, P)$.

**Figure 2.2** The sample space $\Omega$.

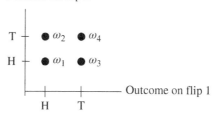

**Example 2.1** (A finite number of outcomes)

Suppose a random experiment consists of flipping a fair coin twice (equivalently, we can simultaneously flip two coins once). The sample space is $\Omega = \{\omega_1, \omega_2, \omega_3, \omega_4\}$ (Figure 2.2). There are many possible sets $\mathcal{A}$ that can satisfy the properties of a $\sigma$-algebra. The smallest is $\mathcal{A} = \{\emptyset, \Omega\}$. (Clearly, $\emptyset, \Omega \in \mathcal{A}$; $\emptyset' = \Omega \in \mathcal{A}$; $\Omega' = \emptyset \in \mathcal{A}$; and $\emptyset \cup \Omega \in \mathcal{A}$.) If it is required that the simple events $\{\omega_1\}, \{\omega_2\} \in \mathcal{A}$, then the smallest $\sigma$-algebra containing $\{\omega_1\}, \{\omega_2\}$ is

$$\mathcal{A} = \{\emptyset \{\omega_1\}, \{\omega_2\}, \{\omega_1, \omega_2\}, \{\omega_3, \omega_4\}, \{\omega_1, \omega_3, \omega_4\}, \{\omega_2, \omega_3, \omega_4\}, \Omega\},$$

the $\sigma$-algebra generated by $\{\omega_1\}, \{\omega_2\}$. The largest $\sigma$-algebra containing all four simple events in $\Omega$ is

$$\mathcal{A} = \{\emptyset, \{\omega_1\}, \{\omega_2\}, \{\omega_3\}, \{\omega_4\}, \{\omega_1, \omega_2\}, \{\omega_1, \omega_3\}, \{\omega_1, \omega_4\} \{\omega_2, \omega_3\} \{\omega_2, \omega_4\} \{\omega_3, \omega_4\},$$

$$\{\omega_1, \omega_2, \omega_3\}, \{\omega_1, \omega_2, \omega_4\}, \{\omega_1, \omega_3, \omega_4\}, \{\omega_2, \omega_3, \omega_4\}, \Omega\}.$$

For the probability measure defined on $\Omega$, take $P(\omega_i) = 1/4$, $i = 1, 2, 3, 4$ (since the $\omega_i$ are equally likely). Then, for instance, $P(\omega_1') = 1 - P(\omega_1) = 3/4$, $P(\omega_1 \cup \omega_2) = P(\omega_1) + P(\omega_2) = 1/2$, and $P(\cup_{i=1}^{4} \omega_i) = \sum_{i=1}^{4} P(\omega_i) = 1$. Thus, the triple $(\Omega, \mathcal{A}, P)$ constitutes a probability space. ∎

**Example 2.2** (A continuous set of outcomes)

Let a random experiment consist of selecting a real number $x$ between 0 and 1 inclusive. Hence, $\Omega = \{x | 0 \le x \le 1\} = [0,1]$. Suppose we define $A$ as $x \in (a, b] \subset [0,1]$. Then the $\sigma$-algebra $\mathcal{A}$ is the set generated by all half-open intervals $(a, b]$; it consists of all intervals of the form $(a, b]$, all unions of the same, and the complements of all the effected sets. This $\sigma$-algebra is the Borel $\sigma$-algebra on $\Omega$. For our probability measure $P$ on $\mathcal{A}$, let's define $P(A)$ as the probability that $x \in [0,1]$ belongs to $A \in \mathcal{A}$. Then $P(A) = b - a$. As usual, the probability space is $(\Omega, \mathcal{A}, P)$. ∎

**Example 2.3** (An infinite number of outcomes)

Suppose our random experiment involves tossing a single six-sided (fair) die repeatedly until a 4 shows for the first time. Here, $\Omega = \{\omega_1, \omega_2, ...\}$, where $\omega_i$ is the outcome where the first $i - 1$ tosses show 1 or 2 or 3 or 5 or 6 and the $i$th toss shows 4. Then $\mathcal{A} = \{\emptyset, \{\omega_1\}, \{\omega_2\}, ... \{\omega_1, \omega_2\}, \{\omega_1, \omega_3\}, ...,\}$ is a $\sigma$-algebra of event subsets of $\Omega$ so that $\omega_i \in \mathcal{A}, i = 1, 2, ....$ If our probability measure on $\mathcal{A}$ is defined as $P(\{\omega_i\}) = (1/6)^i, i = 1, 2, ...$, then the triple $(\Omega, \mathcal{A}, P)$ represents the probability space. ∎

## 2.2  Probability Distributions

Suppose $(\Omega, \mathcal{A}, P)$ is a probability space and $(R, \mathcal{U})$ is a measurable space. As was noted in Section 2.1, a random variable $X$ is a real-valued function defined on the sample space $\Omega$, that is, $X : \Omega \to R$. For each outcome, $\omega \in \Omega, X$ assigns the value $X(\omega) \in R$. Consider the (cumulative) **distribution function** $F(x) = P(\{\omega \in \Omega | X(\omega) \leq x\})$ of the random variable $X$. Since $X$ is taken to be $(\Omega - \mathcal{U})$-measurable, the set $A(x) = \{\omega \in \Omega | X(\omega) \leq x\}$ must be a subset of $\mathcal{A}$ for each $x \in R$. Clearly, $F(x) = P(A(x))$.

Let's assume that the random variable $X$ is **discrete**—it assumes either a finite or a countably infinite number of values, that is, $X(\omega) \in \{x_1, x_2, ...\} \subset R, \omega \in \Omega$. To display the possible values of a discrete random variable $X$, we utilize its **probability mass function** $p : \{x_1, x_2, ...\} \to [0,1]$, where $p(x_j) = P(X = x_j)$. Here, we require that $p(x_j) \geq 0$ and $\Sigma_j p(x_j) = 1, j = 1, 2, ....$ In addition, the distribution function can now be written as

$$F(x) = P(X \leq x) = \sum_{x_j < x} p(x_j). \tag{2.1}$$

As to the properties of $F(x)$,

a. $F(x)$ is monotone nondecreasing (i.e., if $x \leq y$, then $F(x) \leq F(y)$).

b. $F(-\infty) = \lim_{x \to -\infty} F(x) = \lim_{x \to -\infty} P(X \leq x) = 0$ and

$F(+\infty) = \lim_{x \to +\infty} F(x) = P(X < +\infty) = 1.$

c. $F(x)$ is continuous to the right or $\lim_{y \to x+} F(y) = F(x + 0) = F(x)$, where $y \to x +$ means that $y$ approaches $x$ from the right.

Since $X$ is a discrete random variable, $F$ increases by "jumps" that occur at points of discontinuity of $F$. Moreover, if $F(x+) = \lim_{y \to x+} F(y)$ and $F(x-) = \lim_{y \to x-} F(y)$ denote, respectively, right and left limits of $F$ at $x$, then $F(x-) \leq F(x) \leq F(x+)$.

Now, suppose $F$ is given. Then we can recover $p$ from the difference

$$p(x) = F(x+) - F(x-). \tag{2.2}$$

**Example 2.4** Let our random experiment consist of flipping a (fair) coin once. The associated probability space is $(\Omega, \mathcal{A}, P)$, where $\Omega = \{H, T\}$, $\mathcal{A} = \{\emptyset, \{H\}, \{T\}, \Omega\}$, and $P(\{H\}) = P(\{T\}) = 1/2$. In addition, with $X$ discrete on $\Omega$, define $X(H) = 0$ and $X(T) = 1$ so that, as required, $X(\omega) = \{x_1 = 0, x_2 = 1\}$ and $P(\{0\}) = P(\{1\}) = 1/2$. Then $X$'s probability mass and distribution functions are provided in Table 2.1 and illustrated in Figure 2.3. Note also that

$$p(0) = F(0+) - F(0-) = \frac{1}{2} - 0 = \frac{1}{2};$$

$$p(1) = F(1+) - F(1-) = 1 - \frac{1}{2} = \frac{1}{2}. \quad \blacksquare$$

**Table 2.1** The probability mass and distribution function for $X$.

| $X$ | $p(x)$ | $F(x)$ |
|---|---|---|
| 0 | 1/2 | 1/2 |
| 1 | 1/2 | 1 |
| | 1 | |

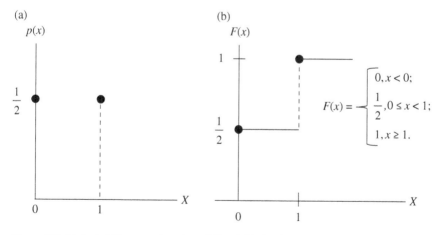

(a) $p(x)$     (b) $F(x)$

$$F(x) = \begin{cases} 0, x < 0; \\ \dfrac{1}{2}, 0 \le x < 1; \\ 1, x \ge 1. \end{cases}$$

**Figure 2.3** (a) Probability mass function of $X$ and (b) distribution function for $X$.

**Table 2.2** The probability mass and distribution function for X.

| X | p(x) | F(x) |
|---|------|------|
| 1 | 1/6 | 1/6 |
| 2 | 1/6 | 2/6 |
| 3 | 1/6 | 3/6 |
| 4 | 1/6 | 4/6 |
| 5 | 1/6 | 5/6 |
| 6 | 1/6 | 1 |
|   | 1   |   |

**Example 2.5** Suppose our random experiment consists of rolling a single (fair) six-sided die. Here, $\Omega = \{\omega_1, \omega_2, ..., \omega_6\}$, where $\omega_i$ is the face showing, $i = 1, ..., 6$. Then $\mathcal{A} = \{\emptyset, \{\omega_1\}, ..., \{\omega_6\}, \{\omega_1, ..., \omega_6\}, \{\omega_2, \omega_3, \omega_4, \omega_5, \omega_6\}, ..., \Omega\}$ and $P(\{\omega_i\}) = 1/6$ for all $i$. The triple $(\Omega, \mathcal{A}, P)$ thus constitutes our probability space. Let the value of the discrete random variable $X$ correspond to the face showing or $X(\omega) \in \{x_1, ..., x_6\}$ or $X(\omega_i) = i, i = 1, ..., 6$. Table 2.2 houses the values of $X$'s probability mass function $p(x)$ along with the values of its distribution function $F(x)$. These functions are illustrated in Figure 2.4. ■

If the random variable $X$ is **continuous** (it can assume any value over some range), then there exists a piecewise continuous non-negative function $p(x)$, called the **probability density function** of $X$, such that

$$F(t) = P(X \le t) = \int_{-\infty}^{t} p(x)dx. \tag{2.2a}$$

(Note that since $p(x) \ge 0$ for all real $x \in (-\infty, +\infty)$, we require that $\int_{-\infty}^{+\infty} p(x)\, dx = 1$.) Here, (2.2a) represents the (cumulative) **distribution function** of $X$ (Figure 2.5). Key properties of $F(t)$ are the following:

a. $0 \le F(t) \le 1$.
b. $F(a) \le F(b)$ when $a < b$ ($F$ is nondecreasing in $t$).
c. $F(-\infty) = \lim\limits_{t \to -\infty} F(t) = 0$ and $F(+\infty) = \lim\limits_{t \to +\infty} F(t) = 1$.
d. $F$ is everywhere continuous from the right at each $t$.
e. For $a < b, P(a \le x \le b) = F(b) - F(a) = \int_{a}^{b} p(x)\, dx$.
f. If $F$ has a point of discontinuity at $t$, then $P(X = t)$ is the size of the jump exhibited by $F$ at $t$; but if $F$ is continuous at $t$, then $P(X = t) = 0$.

With $F$ a continuous function of $t$, its derivative exists at every point of continuity of $p(x)$ and, at each such point, $dF(t)/dt = p(t)$. (Thus, the integrand $p(x)$ in (2.2) must be the value of the derivative of $F$ at $x$.) So if we know $X$'s

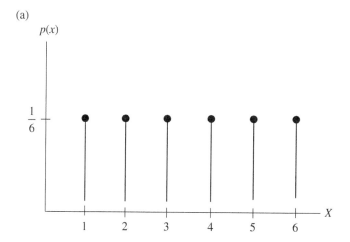

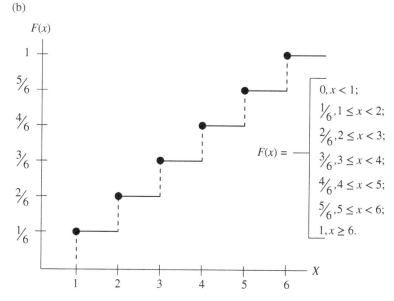

**Figure 2.4** (a) Probability mass function for $X$ and (b) distribution function for $X$.

**Figure 2.5** Distribution function of $X$.

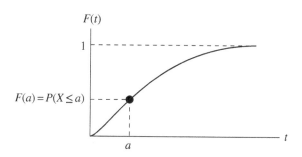

probability density function $p(x)$, then we can determine its distribution function using (2.2). Conversely, if the distribution function $F(t)$ is given, we can recover the probability density function $p(x)$ at each of its points of continuity by determining $dF(t)/dt = p(t)$.

**Example 2.6** Suppose $\Omega = [\alpha,\beta] = \{x|\alpha \le x \le \beta\}$. For random variable $X(x) = x$, let $A$ be the event "a randomly chosen value of $x \in (a,b) \subset [\alpha,\beta]$." Then $\mathcal{A}$ is the set generated by all intervals of the form $(a, b)$, all unions of the same, and the complements of the resulting sets. Let $p(x)$ be a **uniform probability density function** defined by

$$p(x) = \begin{cases} 0, x \le \alpha; \\ \dfrac{1}{\beta-\alpha}, \alpha < x < \beta; \\ 0, x \ge \beta \end{cases}$$

as shown in Figure (2.6a). Now, for $a,b \in [\alpha,\beta], a < b$,

$$P(a \le X \le b) = \int_a^b p(x)\,dx = \frac{b-a}{\beta-\alpha}.$$

In addition,

$$F(t) = \int_{-\infty}^t p(x)\,dx = \int_{-\infty}^\alpha (0)\,dx + \int_\alpha^t \left(\frac{1}{\beta-\alpha}\right) dx$$

$$= \frac{t-\alpha}{\beta-\alpha},$$

or

$$F(t) = \begin{cases} 0, t < \alpha; \\ \dfrac{t-\alpha}{\beta-\alpha}, \alpha \le t \le \beta; \\ 1, t > \beta \end{cases}$$

as shown in Figure (2.6b). Note that $F(\alpha) = 0$ and $F(\beta) = 1$. Moreover, for $\alpha < t < \beta$, $dF(t)/dt = 1/(\beta-\alpha)$, the probability density function. ∎

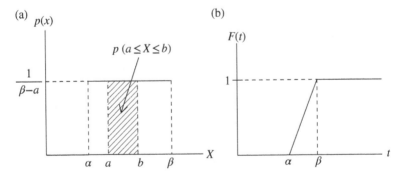

**Figure 2.6** (a) Probability density function for $X$ and (b) distribution function for $X$.

**Example 2.7**   Let the sample space be specified as $\Omega = \{x\,|\,-\infty < x < +\infty\}$ and let the $\sigma$-algebra $\mathcal{A}$ contain all intervals of the form $(a, b)$ for $a, b \in R$ along with all countable unions of the same and complements of the resultant sets. For random $X(x) = x$, let $A$ be an event in $\mathcal{A}$ and suppose $X$'s probability density function is of the **normal** form or $X \sim N(x;\mu,\sigma)$, with

$$p(x;\mu,\sigma) = \frac{1}{\sigma\sqrt{2\pi}} e^{-\frac{(x-\mu)^2}{2\sigma^2}}, \mu \in R, \sigma > 0. \tag{2.3}$$

(As $\mu$ and/or $\sigma$ is varied, we get an entirely new normal density function, that is, (2.3) represents a *two-parameter family of normal curves*.) Then

$$P(a \le X \le b) = \frac{1}{\sigma\sqrt{2\pi}} \int_a^b e^{-\frac{(x-\mu)^2}{2\sigma^2}} dx$$

as shown in Figure (2.7a). The distribution function corresponding to (2.3) appears as

(a)

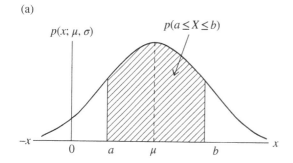

(b)

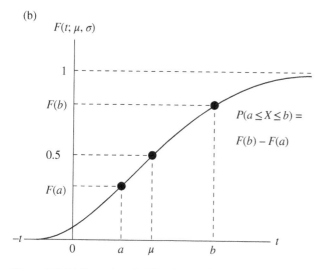

**Figure 2.7** (a) Normal probability density function and (b) normal distribution function.

(a)

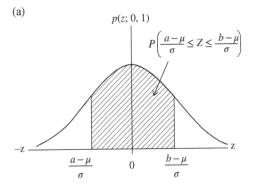

(b)

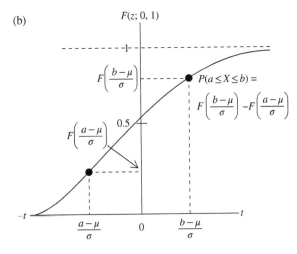

**Figure 2.8** (a) Standard normal probability density function and (b) standard normal distribution function.

$$F(t;\mu,\sigma) = \int_{-\infty}^{t} p(x;\mu,\sigma)\,dx = \frac{1}{\sigma\sqrt{2\pi}} \int_{-\infty}^{t} e^{-\frac{(x-\mu)^2}{2\sigma^2}}\,dx \qquad (2.4)$$

as shown in Figure (2.7b).

If we define a new random variable as $Z = (X - \mu)/\sigma$, then (2.3) is transformed to **standard normal** form or

$$p(z;0,1) = \frac{1}{\sqrt{2\pi}} e^{-\frac{1}{2}z^2} \qquad (2.3.1)$$

as shown in Figure (2.8a). Under this transformation,

$$P(a \le X \le b) = \int_{\frac{a-\mu}{\sigma}}^{\frac{b-\mu}{\sigma}} f(z;0,1)\,dz$$

$$= P\left(\frac{a-\mu}{\sigma} \le Z \le \frac{b-\mu}{\sigma}\right),$$

where again $Z \sim N(z;0,1)$. The distribution function corresponding to (2.3.1) is

$$F(t;\mu,\sigma) = F\left(\frac{t-\mu}{\sigma};0,1\right)$$

$$= \frac{1}{\sqrt{2\pi}} \int_{-\infty}^{\frac{t-\mu}{\sigma}} e^{-\frac{1}{2}z^2} dz \qquad (2.4.1)$$

as shown in Figure (2.8b). Then

$$P(a \le X \le b) = F\left(\frac{b-\mu}{\sigma};0,1\right) - F\left(\frac{a-\mu}{\sigma};0,1\right). \blacksquare$$

## 2.3 The Expectation of a Random Variable

### 2.3.1 Theoretical Underpinnings

Let $X$ be a real-valued **simple random variable** defined on the probability space $(\Omega, \mathcal{A}, P)$ and suppose that measurable $X$ has, say, a finite number of values $X(\omega) \in \{x_1, ...x_n\} \subset R, \omega \in \Omega$, with

$$X = \sum_{i=1}^{n} x_i \chi_{A_i},$$

where

$$\chi_{A_i}(\omega) = \begin{cases} 1, \omega \in A_i; \\ 0, \omega \notin A_i, \end{cases}$$

$\omega_i = X^{-1}(\{x_i\}), A_i = \{\omega \in \Omega | X(\omega) = x_i\}, i = 1,...,n$, and the events $\{A_1, ..., A_n\}$ form an $\mathcal{A}$-measurable partition of $\Omega$ (i.e., for events $A_i \in \mathcal{A}$, $i = 1,...,n, A_i \cap A_j = \emptyset$. $i \neq j$, and $\cup_{i=1}^{n} A_i = \Omega$).

The **integral of a simple random variable $X$ with respect to the measure $P$** is defined as

$$\int_{\Omega} X(\omega) \, dP(\omega) = \int_{\Omega} XdP = \sum_{i=1}^{n} x_i P(A_i) \qquad (2.5)$$

(since, for any event $A_i \in \mathcal{A}$, $E(\chi_{A_i}) = P(A_i)$) and termed the **expectation** of $X$ and denoted

$$E(X) = \int_{\Omega} XdP = \int_{R} xdP(x). \qquad (2.5.1)$$

*Thus, the expectation and integral operators are mutually interchangeable.* Here, $X$ is said to be "$P$-integrable" if the integral in (2.5.1) is finite. (See also Equation (C.1).)

If $X$ is a non-negative random variable on $(\Omega, \mathcal{A}, P)$, define

$$E(X) = \int_\Omega XdP = sup \int_\Omega YdP. Y \leq X, Y \text{ simple} \tag{2.6}$$

In fact, for non-negative $X$, there exists, via the monotone convergence theorem, a nondecreasing sequence of non-negative simple random variables $\{X_n, n \geq 1\}$ (i.e., $X_n(\omega) \leq X_{n+1}(\omega), \omega \in \Omega$) such that $X(\omega) = lim_{n \to \infty} X_n(\omega), \omega \in \Omega$. We can then define $E(X) = lim_{n \to \infty} E(X_n)$.

In addition, for $X$ an arbitrary random variable on $(\Omega, \mathcal{A}, P)$, set $X(\omega) = X^+(\omega) - X^-(\omega)$, where $X^+(\omega) = max\{X(\omega), 0\}$ (the positive part of $X$) and $X^-(\omega) = -min\{X(\omega), 0\}$ (the negative part of $X$). Since $X^+(\omega) \geq 0$ and $X^-(\omega) \geq 0$, define $E(X) = E(X^+) - E(X^-)$ and set

$$E(X) = \int_\Omega X(\omega)dP(\omega) = \int_\Omega X^+(\omega)dP(\omega) - \int_\Omega X^-(\omega)dP(\omega), \tag{2.7}$$

provided that *not both* $E(X^+) = \int_\Omega X^+ dP$ and $E(X^-) = \int_\Omega X^- dP$ are infinite. (If both $E(X^+), E(X^-)$ are infinite, $E(X)$ does not exist. If *only one* of the terms $E(X^+), E(X^-)$ is infinite, then $E(X)$ itself is infinite. Thus, $E(X)$ will be finite if both $E(X^+), E(X^-)$ are finite.) In sum, an arbitrary random variable $X$ on $(\Omega, \mathcal{A}, P)$ is integrable if its positive and negative parts $X^+$ and $X^-$, respectively, are both integrable or both $E(X^+), E(X^-)$ are finite. So if $X$ is integrable, then, as defined earlier $E(X) = E(X^+) - E(X^-) = \int_\Omega X^+ dP - \int_\Omega X^- dP = \int_\Omega Xdp$. For an arbitrary event $A \in \mathcal{A}$, it is customary to write

$$\int_A Xdp = \int_\Omega \chi_A Xdp. \tag{2.8}$$

### 2.3.2 Computational Considerations

Extremely useful versions of $E(X)$ are Equations (2.5) and (2.5.1). To operationalize these expressions, let us first consider the case where $X$ is a discrete random variable $(X(\omega) \in \{x_1, x_2, ...\}$ for $\omega \in \Omega)$ and $p(x) = P(X = x)$ is the probability mass function for $X$. The **expectation** (or mean of $X$) is then determined as

$$E(X) = \sum_j x_j p(x_j)(= \mu), \tag{2.9}$$

provided that the indicated sum is absolutely convergent, that is, $\sum_j |x_j| p(x_j)$ converges and is finite. Clearly, (2.9) represents a weighted mean, with the sum of the weights $\sum_j p(x_j) = 1$.

In general, if $g(x)$ is a single-valued real function of the discrete random variable $X$, then

$$E(g(x)) = \sum_j g(x_j)p(x_j).$$ (2.9.1)

$$\left( E(g(x)) \text{ exists if } \sum_j |g(x_j)| p(x_j) \text{ is finite.} \right)$$

Some important properties of the expectation (linear) operator are the following:

a. $E(c) = c$ ($c$ a constant).
b. $E(a \pm bX) = a \pm bE(X)$ ($a$, $b$ constants).
c. $E\left( \sum_{i=1}^n X_i \right) = \sum_{i=1}^n E(X_i)$.
d. $E[g(x) \pm h(x)] = E(g(x)) \pm E(h(x))$ ($g$, $h$ are functions of $X$).
e. $E[(a + bX)^n] = \sum_{i=0}^n \binom{n}{i} a^i b^{n-i} E(X^{n-i})$ ($a$, $b$ constants).

Suppose $X$ is a random variable defined on a probability space $(\Omega, \mathcal{A}, P)$ and $X^2$ is integrable or $E(X^2) < +\infty$. Then the **variance** of $X$ is defined as the expected value of the squared deviations from the expectation and denoted as

$$V(X) = E\left[(X - E(X))^2\right] = E(X^2) - E(X)^2 (= \sigma^2).$$ (2.10)

For $X$ a discrete random variable $(X(\omega) \in \{x_1, x_2, \dots\}, \omega \in \Omega)$, (2.10) appears as

$$V(X) = \sum_j (x_j - E(X))^2 p(x_j)$$ (2.10.1)

or

$$V(X) = \sum_j x_j^2 p(x_j) - E(X)^2.$$ (2.10.2)

In addition, if we take the positive square root of $V(X)$, we obtain the **standard deviation** of $X$ or $S(X) = \sqrt{V(X)}(=\sigma)$. Here, $S(X)$ serves as a measure of dispersion of the $x_j$'s around the mean of $X$.

A few key properties of $V(X)$ are the following:

a. $V(c) = 0$ ($c$ a constant).
b. $V(c + X) = V(X)$ ($c$ a constant).
c. $V(cX) = c^2 V(X)$ ($c$ a constant).
d. If $E(X^2)$ exists, then $E(X)$ exists and thus $V(X)$ exists. Hence, if $V(X)$ exists, it follows that $E(X)$ exists.

Given a discrete random variable $X$, we may express the distance from the mean in terms of standard deviations by forming the standardized variable

$Z = (X - \mu)/\sigma$. Then it is easily shown that $E(Z) = 0$ and $S(Z) = 1$. For any value $x_j$ of $X$, $z_j = (x_j - \mu)/\sigma$ indicates how many standard deviations $x_j$ is from the mean $\mu$.

If $X$ represents a continuous random variable with $X(x) = x$ and probability density function $p(x)$, then the **expectation** (or mean) of $X$ appears as

$$E(X) = \int_{-\infty}^{+\infty} x\, p(x)\, dx \tag{2.11}$$

(provided that the integral in (2.11) converges to a finite value). In general, if $g(X)$ is a single-valued real function of a continuous random variable $X$, then

$$E(g(x)) = \int_{-\infty}^{+\infty} g(x)p(x)\, dx \tag{2.11.1}$$

(provided $\int_{-\infty}^{+\infty} |g(x)| p(x)\, dx$ exists).

To obtain the **variance** of a continuous random variable $X$, we need to calculate

$$V(X) = \int_{-\infty}^{+\infty} (x - E(X))^2 p(x)\, dx \tag{2.12}$$

or

$$V(X) = E(X^2) - E(X)^2 = \int_{-\infty}^{+\infty} x^2 p(x) d(x) - E(X)^2. \tag{2.12.1}$$

## 2.4 Moments of a Random Variable

A **moment** of a random variable $X$ is defined as the expected value of some particular function of $X$. Specifically, moments of a random variable $X$ are specified in terms of having either zero or $E(X)$ as the reference point.

For a discrete random variable $X$, the **$r$th moment about zero** is

$$\mu'_r = E(X^r) = \sum_j x_j^r p(x_j) \tag{2.13}$$

(note that the first moment about zero is the mean of $X$ or $\mu'_1 = E(X) = \mu$) and the **$r$th central moment** of $X$ or the $r$th moment about the mean of $X$ is

$$\mu_r = E[(X - \mu)^r] = \sum_j (x_j - \mu)^r p(x_j). \tag{2.14}$$

If $X$ is a continuous random variable with probability density function $p(x)$, then, provided the following integrals exist, we may correspondingly define

$$\mu_r' = E(X^r) = \int_{-\infty}^{+\infty} x^r p(x) d(x) \tag{2.15}$$

and

$$\mu_r = E[(X-\mu)^r] = \int_{-\infty}^{+\infty} (x-\mu)^r p(x) d(x). \tag{2.16}$$

It is readily verified that

a. the zeroth central moment of $X$ is unity;
b. the first central moment of $X$ is zero; and
c. the second central moment of $X$ is the variance of $X$ or $\mu_2 = E[(X-\mu)^2] = V(X)$.

If we standardize the random variable $X$ to obtain $Z = (X-\mu)/\sigma$, then, since $E(Z) = 0$, the $r$th central moment of $Z$ can be expressed in terms of the $r$th central moment of $X$ as

$$\mu_r(Z) = E(Z^r) = E\left[\left(\frac{X-\mu}{\sigma}\right)^r\right]$$

$$= \frac{1}{\sigma^r} E[(X-\mu)^r] = \frac{\mu_r(X)}{\sigma^r}. \tag{2.17}$$

### Example 2.8

For the discrete probability distribution appearing in Table 2.3, it is readily determined that

$$\mu = E(X) = \sum_j x_j p(x_j) = 2.50;$$

$$\mu_2' = E(X^2) = \sum_j x_j^2 p(x_j) = 7.50;$$

and thus

$$\mu_2 = V(X) = \mu_2' - \mu^2 = 1.25.$$

**Table 2.3** $X$ a discrete random variable.

| $X$ | 1 | 2 | 3 | 5 |
|------|------|------|------|------|
| $p(x_j)$ | 0.2 | 0.3 | 0.4 | 0.1 |

Let the probability density function for a continuous random variable $X$ be given as

$$p(x) = \begin{cases} 2x, 0 < x < 1; \\ 0, \text{ elsewhere.} \end{cases}$$

Then

$$\mu = E(X) = \int_{-\infty}^{+\infty} x\,p(x)\,d(x)$$

$$= 2\int_{2}^{1} x^2\,dx = \frac{2}{3}x^3 \Big]_{0}^{1} = 0.666;$$

$$\mu_2' = E(X^2) = \int_{-\infty}^{+\infty} x^2 p(x)\,d(x)$$

$$= 2\int_{0}^{1} x^3\,dx = \frac{1}{2}x^4 \Big]_{0}^{1} = 0.5$$

so that $\mu_2 = V(X) = \mu_2' - \mu^2 = 0.055.$ ∎

## 2.5 Multiple Random Variables

### 2.5.1 The Discrete Case

Given a probability space $(\Omega, \mathcal{A}, P)$, let $X$ and $Y$ be distinct random variables defined on this space with $(X, Y): \Omega \to R^2$. For each outcome $\omega \in \Omega$, the bivariate random variable $(X, Y)$ assigns the value $(X(\omega), Y(\omega)) \in R^2$. The combination $(X, Y)$ defines the possible outcomes of some random experiment. In this regard, if the range of $(X, Y)$ is a discrete ordered set of points $\{(X_i, Y_j), i = 1, ..., n; j = 1, ..., m\}$ in $R^2$, then $(X, Y)$ is termed a **discrete bivariate random variable**. For convenience, we assume that both $X$ and $Y$ take on only a finite number of values (although assuming a countably infinite set of values is admissible).

Given a random variable $X: X_1, X_2, ..., X_n$ (with $X_1 < X_2 < \cdots < X_n$) and a second random variable $Y: Y_1, Y_2, ..., Y_m$ (with $Y_1 < Y_2 < \cdots < Y_m$), let the **joint probability** that $X = X_i$ and $Y = Y_j$ be denoted as $P(X = X_i, Y = Y_j) = f(X_i, Y_j), i = 1, ..., n; j = 1, ..., m$. Here, $f(X_i, Y_j)$ depicts the **probability mass** at the point $(X_i, Y_j)$. (To elaborate, let $A_r, A_s \in \mathcal{A}$ with $A_r = \{\omega \in \Omega | X(\omega) = X_r.\}, A_s = \{\omega \in \Omega | Y(\omega) = Y_s\}$. Then $A_r \cap A_s \in \mathcal{A}$ with $P(A_r \cap A_s) = P(X = X_r, Y = Y_s) = f(X_r, Y_s).$) In general, a function $f(X, Y)$ that assigns a probability $f(X_i, Y_j)$ within the ranges $\{X_i, i = 1, ..., n\}$ and $\{Y_j, j = 1, ..., m\}$ of the discrete random variables $X$ and $Y$, respectively, is called a **bivariate probability mass function** if

a. $f(X_i, Y_j) \geq 0$ for all $i, j$; and

b. $\sum_i \sum_j f(X_i, Y_j) = 1.$ (2.18)

The set of the $nm$ events $(X_i, Y_j)$, together with their associated probability $f(X_i, Y_j), i = 1, \ldots, n; j = 1, \ldots, m$, constitutes a **discrete bivariate probability distribution** of the random variables $X$ and $Y$.

For the bivariate probability mass function $f(X, Y)$, the joint probability that $X \leq X_r, Y \leq Y_s$ is provided by the **bivariate cumulative distribution function**

$$F(X_r, Y_s) = P(X \leq X_r, Y \leq Y_s) = \sum_{i \leq r} \sum_{j \leq s} f(X_i, Y_j).$$ (2.19)

(Now, let $A_r, A_s \in \mathcal{A}$ with $A_r = \{\omega \in \Omega | X(\omega) = X_r\}$ and $A_s = \{\omega \in \Omega | Y(\omega) = Y_s\}$. Then $A_r \cap A_s \in \mathcal{A}$ with $P(A_r \cap A_s) = P(X \leq X_r, Y \leq Y_s) = F(X_r, Y_s)$.)

While the bivariate probability distribution tells us something about the joint behavior of the random variables $X$ and $Y$, information about these two random variables taken individually can be obtained from their marginal probabilities. In this regard, for $X$ and $Y$ discrete random variables, the summation of the bivariate probability mass function $f(X, Y)$ over all $Y$ within $\{Y_j, j = 1, \ldots, m\}$ yields the univariate probability mass function $g(X)$ called the **marginal probability mass function of $X$**

$$g(X) = \sum_j f(X_i, Y_j).$$ (2.20)

Similarly, taking the summation of $f(X, Y)$ over all $X$ within $\{X_i, i = 1, \ldots, n\}$ produces the univariate **marginal probability mass function of $Y$**

$$h(Y) = \sum_i f(X_i, Y_j).$$ (2.21)

Then, from (2.20),

$$P(X = X_r) = g(X_r) = \sum_j f(X_r, Y_j);$$ (2.22)

and, from (2.21),

$$P(Y = Y_s) = h(Y_s) = \sum_i f(X_i, Y_s).$$ (2.23)

Once the marginal probability mass functions (2.22) and (2.23) are determined, we can also find

$$P(X_a \leq X \leq X_b) = \sum_{a \leq i \leq b} g(X_i); \tag{2.24}$$

$$P(Y_c \leq Y \leq Y_d) = \sum_{c \leq j \leq d} h(Y_j). \tag{2.25}$$

Using (2.20), we can define the set of all $X_i$'s together with their marginal probabilities $g(X_i), i = 1,\ldots,n$, as the **marginal probability distribution of** $X$—it depicts the probability distribution of a single discrete random variable $X$ when the levels of the random variable $Y$ are ignored. Obviously, we require that: (a) $g(X_i) \geq 0$ for all $i$ and (b) $\sum_i g(X_i) = 1$. (The marginal probability distribution of $Y$ is defined in a similar fashion.)

Next, our objective is to define conditional probability distributions of the discrete random variables $X$ and $Y$. To set the stage for this discussion, we first need to consider the notion of a conditional probability itself. To this end, for $(\Omega, \mathcal{A}, P)$ a probability space, let events $A, B \in \mathcal{A}$ with $P(B) > 0$. Suppose a simple event $\omega \in \Omega$ is randomly selected and it is determined that $\omega \in B$. What is the probability that $\omega$ is also in $A$? To answer this question, we need the concept of a **conditional probability**—the probability of occurrence of one event given that another event has definitely occurred. Then the probability of $A$ given $B$ (denoted $P(A|B)$ and read "the probability of $A$ given $B$") can be determined as follows: Since $\omega \in B$, event $B$ becomes the new or **effective sample space**, and thus our reduced or **effective probability space** is $\left(\tilde{\Omega}, \tilde{\mathcal{A}}, \tilde{P}\right)$, where $\tilde{\Omega} = B, \tilde{\mathcal{A}} = \{B \cap C | C \in \mathcal{A}\}$, and $\tilde{P} = P/P(B)$, with $\tilde{P}\left(\tilde{\Omega}\right) = 1$. Then the probability that $\omega \in A$ given $\omega \in B$ is $\tilde{P}(A \cap B) = P(A \cap B)/P(B) = P(A|B)$. To summarize,

$$P(A|B) = \frac{P(A \cap B)}{P(B)}, P(B) > 0; \tag{2.26}$$

and if $P(A) > 0$,

$$P(B|A) = \frac{P(A \cap B)}{P(A)}, P(A) > 0. \tag{2.27}$$

From (2.26) and (2.27), we can write

$$P(A \cap B) = P(A|B) \cdot P(B) = P(B|A) \cdot P(A). \tag{2.28}$$
[Multiplication law for probabilities]

Events $A, B \in \mathcal{A}$ are termed **independent** if, on any given trial of a random experiment, the occurrence of one of them in no way affects the probability of occurrence of the other. That is, **$A$ is independent of $B$** if an only if

$$P(A|B) = P(A), P(B) > 0; \tag{2.29}$$

and **$B$ is independent of $A$** if and only if

$$P(B|A) = P(B), P(A) > 0. \tag{2.30}$$

Note that under the (mutual) independence of events $A$ and $B$, (2.28) becomes, via (2.29) and (2.30),

$$P(A \cap B) = P(A) \cdot P(B).^3 \qquad (2.31)$$
[**Test for independent events**]

Now, back to our initial objective of defining a conditional probability distribution. Suppose $X$ and $Y$ are discrete random variables with bivariate probability mass function $f(X,Y)$ and marginal probability mass functions $g(X)$ and $h(Y)$, respectively. Given these relationships, we may now ask "What if, on a particular trial of a random experiment, we specify the level of $Y$ but the values of $X$ are allowed to be determined by chance?" That is, what is the probability that $X = X_i$ *given* that $Y = Y_j$ for fixed $j$? In a similar fashion, we may be interested in determining the probability that $Y = Y_j$ given that $X = X_i$ for fixed $i$. In this regard, if $(X,Y)$ is any point at which $h(Y) > 0$, then the **conditional probability mass function of $X$ given $Y$** is

$$g(X|Y) = \frac{f(X,Y)}{h(Y)}. \qquad (2.32)$$

Here, $g(X|Y)$ is a function of $X$ alone and, for each $X_i$ given $Y = Y_s$, the probability that $X = X_i$ given $Y = Y_s$ is

$$P(X = X_i | Y = Y_s) = g(X_i | Y_s) = \frac{f(X_i, Y_s)}{h(Y_s)}, h(Y_s) > 0, i = 1, \dots, n. \qquad (2.33)$$

Similarly, at any point $(X, Y)$ at which $g(X) > 0$, the **conditional probability mass function of $Y$ given $X$** (a function of $Y$ alone) is

$$h(Y|X) = \frac{f(X,Y)}{g(X)}. \qquad (2.34)$$

So for any $Y_j$ given $X = X_r$, the probability that $Y = Y_j$ given $X = X_r$ is

$$P(Y = Y_j | X = X_r) = h(Y_j | X_r) = \frac{f(X_r, Y_j)}{g(X_r)}, g(X_r) > 0, j = 1, \dots, m. \qquad (2.35)$$

Note that from (2.32) and (2.34), we can solve for $f(X, Y)$ as

$$f(X,Y) = g(X|Y) \cdot h(Y) = h(Y|X) \cdot g(X). \qquad (2.36)$$
[**Multiplication law for probability mass functions**]

From (2.33), we can define the set of all $X_i$'s together with their conditional probabilities $g(X_i | Y_j), i = 1, \dots, n$, as the **conditional probability distribution of $X$ given $Y = Y_j$**. This distribution depicts the probability distribution of a

---

3 What is the distinction between events that are *mutually exclusive* and events that are *independent*? Events $A$ and $B$ are mutually exclusive if they cannot occur together, that is, $A \cap B = \emptyset$ so that $P(A \cap B) = 0$. If events $A$ and $B$ are independent, they can occur together, that is, $A \cap B \neq \emptyset$ so that $P(A \cap B) = P(A) \cdot P(B)$.

single discrete random variable $X$ when the level of $Y$ is fixed at $Y_j$. For this distribution, we must have (a) $g(X_i|Y_j) \geq 0$ for all $i$ and (b) $\sum_i g(X_i|Y_j) = 1$. (The conditional probability distribution of $Y$ given $X = X_i$ is defined in an analogous fashion.)

Let $X$ and $Y$ be discrete random variables with bivariate probability mass function $f(X,Y)$ and marginal probability mass functions $g(X)$ and $h(Y)$, respectively. Then the **random variable $X$ is independent of the random variable $Y$** if

$$g(X|Y) = g(X) \tag{2.37}$$

for all $X$, $Y$ values for which both of these functions are defined. Similarly, the **random variable $Y$ is independent of the random variable $X$** if

$$h(Y|X) = h(Y) \tag{2.38}$$

for all $X$ and $Y$ at which these functions exist. Given (2.37) (or (2.38)), (2.36) renders

$$f(X,Y) = g(X) \cdot h(Y), \tag{2.39}$$

that is, **$X$ and $Y$ are independent random variables** if and only if their joint probability mass function can be written as the product of their individual marginal probability mass functions $g(X)$ and $h(Y)$, respectively. Hence, under independence,

$$P(X = X_i, Y = Y_j) = P(X = X_i) \cdot P(Y = Y_j) \tag{2.39.1}$$

for all points $(X, Y)$.

**Example 2.9**  Given the bivariate probability mass function appearing in Table 2.4, we can readily determine that

$$F(X_2, Y_2) = P(X \leq X_2, Y \leq Y_2) = \sum_{i \leq 2} \sum_{j \leq 2} f(X_i, Y_j)$$

$$= \sum_{i \leq 2} [f(X_i, Y_1) + f(X_i, Y_2)]$$

$$= f(X_1, Y_1) + f(X_1, Y_2) + f(X_2, Y_1) + f(X_2, Y_2)$$

$$= 0.07 + 0.14 + 0.05 + 0.25 = 0.51;$$

Table 2.4 Bivariate probability mass function $f(X, Y)$.

| $f(X_i, Y_j), i = 1,2,3; j = 1,2$ | | | |
|---|---|---|---|
| Y<br>X | $Y_1$ | $Y_2$ | $g(X_i)$ |
| $X_1$ | 0.07 | 0.14 | 0.21 |
| $X_2$ | 0.05 | 0.25 | 0.30 |
| $X_3$ | 0.21 | 0.28 | 0.49 |
| $h(Y_j)$ | 0.33 | 0.67 | 1.00 |

$$P(X = X_2) = g(X_2) = \sum_j f(X_2, Y_j) = f(X_2, Y_1) + f(X_2, Y_2)$$

$$= 0.05 + 0.25 = 0.30;$$

$$P(Y = Y_1) = h(Y_1) = \sum_j f(X_i, Y_1) = f(X_1, Y_1) + f(X_2, Y_1) + f(X_3, Y_1)$$

$$= 0.07 + 0.05 + 0.21 = 0.33;$$

$$P(X_1 \leq X \leq X_2) = \sum_{1 \leq i \leq 2} g(X_i) = g(X_1) + g(X_2) = 0.51;$$

$$P(X = X_3 | Y = Y_1) = g(X_3 | Y_1) = \frac{f(X_3, Y_1)}{h(Y_1)} = \frac{0.21}{0.33} = 0.6363; \text{ and}$$

$$P(Y = Y_2 | X = X_1) = h(Y_2 | X_1) = \frac{f(Y_2, X_1)}{g(X_1)} = \frac{0.14}{0.21} = 0.6667.$$

Are $X$ and $Y$ independent random variables? One way to answer this question is to check to see if $f(X_2, Y_2) = g(X_2) \cdot h(Y_2)$. From Table 2.4, $f(X_2, Y_2) = 0.25$ while $g(X_2) = 0.30$ and $h(Y_2) = 0.67$. Then $g(X_2) \cdot h(Y_2) = 0.20 \neq f(X_2, Y_2)$. Hence, $X$ and $Y$ are not independent random variables. ■

### 2.5.2 The Continuous Case

Suppose the sample space $\Omega \subseteq R^2$ consists of a class of events representable by all open and closed rectangles. Then an event in $\Omega$ can be written as, say, $A = \{(X, Y) | a \leq X \leq b, c \leq Y \leq d\}$, where $X$ and $Y$ are continuous random variables. In addition, let $P$ be a (joint) probability measure that associates with each $A \in \Omega$ a number $P(A \in R)$. In general, a function $f(X, Y)$ that defines the probability measure,

$$P(A) = P(a \leq X \leq b, c \leq Y \leq d) = \int_a^b \int_c^d f(x, y) \, dy \, dx \qquad (2.40)$$

is called a **bivariate probability density function** if

a. $f(x, y) \geq 0$ for all real $x$ and $y$ such that $-\infty < x < +\infty$, $-\infty < y < +\infty$, and $f(x, y) > 0$ for $(x, y) \in A$; and

b. $\int_{-\infty}^{+\infty} \int_{-\infty}^{+\infty} f(x, y) \, dy \, dx = 1$ \qquad (2.41)

So if $\Omega$ is made up of all open and closed rectangles in $R^2$ and $P(A)$ is determined via (2.40), then the random variables $X$ and $Y$ follow a **continuous bivariate probability distribution**. (Note that the probability of an event containing a single point is always zero, for example, $P(X = a, Y = b) = 0$.)

Equation (2.40) defines the probability that the continuous random variables $X$ and $Y$ with bivariate probability density $f(x, y)$ assumes a value within the set $A = \{(X, Y) | a \leq X \leq b, c \leq Y \leq d\}$. A related function (derived from $f$) that yields the (joint) probability that $X$ assumes a value less than or equal to some number $t$ and $Y$ takes on a value less than or equal to a number $s$ is the **bivariate cumulative distribution function** for the continuous random variables $X$ and $Y$ or

$$F(t,s) = P(X \leq t, Y \leq s) = \int_{-\infty}^{t} \int_{-\infty}^{s} f(x,y)\, dy\, dx \qquad (2.42)$$

$F(t, s)$ is a continuous function of $t$ and $s$ and, at every point of continuity of $f(x,y), \partial^2 F(t,s)/\partial s\, \partial t = f(s,t)$. Hence, knowing the bivariate probability density function $f(x, y)$ enables us to determine the associated bivariate cumulative distribution function $F(t, s)$ using (2.42). Conversely, knowing $F(t, s)$ enables us to determine the probability density $f(s, t)$ at each of its points of continuity by finding $\partial^2 F/\partial s\, \partial t$.

Given the continuous random variables $X$ and $Y$ with bivariate probability density function $f(x, y)$, the **marginal probability density function of $X$** is

$$g(x) = \int_{-\infty}^{+\infty} f(x,y)\, dy \qquad (2.43)$$

and is a function of $x$ alone with (a) $g(x) \geq 0$ and (b) $\int_{-\infty}^{+\infty} g(x)\, dx = 1$. In addition, the **marginal probability density function of $Y$** is

$$h(y) = \int_{-\infty}^{+\infty} f(x,y)\, dx \qquad (2.44)$$

and is a function of $y$ alone with (a) $h(y) \geq 0$ and (b) $\int_{-\infty}^{+\infty} h(y)\, dy = 1$.

Given the joint cumulative distribution function $F(t, s)$, **the marginal cumulative distribution functions of $X$ and $Y$** are

$$P(X \leq t) = F(t) = \int_{-\infty}^{t} \int_{-\infty}^{+\infty} f(x,y)\, dy\, dx = \int_{-\infty}^{t} g(x)\, dx \qquad (2.45)$$

and

$$P(Y \leq s) = F(s) = \int_{-\infty}^{s} \int_{-\infty}^{+\infty} f(x,y)\, dx\, dy = \int_{-\infty}^{s} h(y)\, dy, \qquad (2.46)$$

respectively, where $g(x)$ and $h(y)$ are determined from (2.43) and (2.44), respectively. If $f(x, y)$ is the bivariate probability density function for the random variables $X$ and $Y$ and the marginal densities $g(x)$ and $h(y)$ are known, then we may define, for $Y$ fixed at $y$, the **conditional probability density function for $X$ given $Y$** as

$$g(x|y) = \frac{f(x,y)}{h(y)}, h(y) > 0; \tag{2.47}$$

and for $X$ fixed at $x$, the **conditional probability density function for $Y$ given $X$** is

$$h(y|x) = \frac{f(x,y)}{g(x)}, g(x) > 0. \tag{2.48}$$

Here, $g(x|y)$ is a function of $x$ alone with properties (a) $g(x|y) \geq 0$ and (b) $\int_{-\infty}^{+\infty} g(x|y)\,dx = 1; h(y|x)$ is a function of $y$ alone, and (a) $h(y|x) \geq 0$ and (b) $\int_{-\infty}^{+\infty} h(y|x)\,dy = 1$. How may we interpret (2.47)? If $Y = y$ but the value of $X$ is unknown, then the function $g(x|y)$ gives the probability density of $X$ along a fixed line $Y = y$ in the $X,Y$-plane. (A similar interpretation can be offered for (2.48).)

Once (2.47) and (2.48) are determined, we may calculate probabilities such as

$$P(a \leq X \leq b|y) = \int_a^b g(x|y)\,dx \tag{2.49}$$

and

$$P(c \leq Y \leq d|x) = \int_c^d h(y|x)\,dy. \tag{2.50}$$

Suppose $X$ and $Y$ are continuous random variables with bivariate probability density function $f(x, y)$ and marginal probability densities $g(x)$ and $h(y)$, respectively. Then the **random variable $X$ is independent of the random variable $Y$** if

$$g(x|y) = g(x) \tag{2.51}$$

for all values of $X$ and $Y$ for which both of these functions exist. In similar fashion, we may state that the **random variable $Y$ is independent of the random variable $X$** if

$$h(y|x) = h(y), \tag{2.52}$$

again for all $X$ and $Y$ values for which both of these functions are defined. From (2.47) and (2.51) we can write

$$f(x,y) = g(x) \cdot h(y), \tag{2.53}$$

that is, **$X$ and $Y$ are independent random variables** if and only if their joint probability density function $f(x, y)$ can be written as the product of their individual marginal probability densities $g(x)$ and $h(y)$, respectively.

**Example 2.10**   Given the probability density function

$$f(x,y) = \begin{cases} x+y, 0 < x < 1 \text{ and } 0 < y < 1; \\ 0, \text{ elsewhere,} \end{cases}$$

we can readily find

$$P(0.333 < X < 0.5, 0.5 < y < 1) = \int_{0.333}^{0.5} \int_{0.5}^{1} (x+y) \, dy \, dx$$

$$= \int_{0.333}^{0.5} \left\{ \left( xy + \frac{1}{2}y^2 \right) \Big]_{0.5}^{1} \right\} dx = \int_{0.333}^{0.5} (0.5x + 0.375) \, dx$$

$$= (0.25x^2 + 0.375x)]_{0.333}^{0.5} = 0.097;$$

$$g(x) = \int_{-\infty}^{+\infty} (x+y) \, dy = \int_{0}^{1} (x+y) \, dy = \left( xy + \frac{1}{2}y^2 \right) \Big]_{0}^{1} = x + \frac{1}{2};$$

$$h(y) = \int_{-\infty}^{+\infty} (x+y) \, dx = \int_{0}^{1} (x+y) \, dx = \left( \frac{1}{2}x^2 + xy \right) \Big]_{0}^{1} = \frac{1}{2} + y;$$

$$g(x|y) = \frac{x+y}{(1/2)+y}; h(y|x) = \frac{x+y}{x+(1/2)}.$$

When $X = x = 2$, $h(y|2) = 0.80 + 0.40y$ so that

$$P(0.333 \le Y \le 0.5 | x = 2) = \int_{0.333}^{0.5} (0.80 + 0.40y) \, dy$$

$$= (0.80y + 0.20y^2)]_{0.333}^{0.5} = 0.16.$$

Are $X$ and $Y$ independent random variables? If they are, then $f(x,y) = g(x) \cdot h(y)$. Clearly, this is not the case. ∎

**Example 2.11**   Given the probability density function

$$f(x,y) = \begin{cases} 4xy, 0 < x < 1 \text{ and } 0 < y < 1; \\ 0, \text{ elsewhere,} \end{cases}$$

determine the associated cumulative distribution function.
   We thus need to find

$$F(t,s) = 4 \int_{-\infty}^{t} \int_{-\infty}^{s} xy \, dy \, dx = 4 \int_{0}^{t} \int_{0}^{s} xy \, dy \, dx$$

$$= 4 \int_{0}^{t} \left\{ \frac{1}{2}xy^2 \right]_{0}^{s} \right\} dx = 2s^2 \int_{0}^{t} x \, dx = s^2 t^2$$

or

$$F(t,s) = \begin{cases} 0, t < 0 \text{ and } s < 0; \\ s^2 t^2, 0 \le t \le 1 \text{ and } 0 \le s \le 1; \\ 0, t > 1 \text{ and } s > 1. \end{cases}$$

Then, as expected, $\partial^2 F / \partial s \partial t = 4st$. ∎

### 2.5.3 Expectations and Moments

In what follows, it is assumed that all specified expectations exist. For $X$ and $Y$ discrete random variables with bivariate probability mass function $f(X,Y)$, the **$r$th moment of the random variable $X$ about zero** is

$$E(X^r) = \sum_i \sum_j X_i^r f(X_i, Y_j) = \sum_i X_i^r g(X_i); \tag{2.54}$$

and the **$s$th moment of the random variable $Y$ about zero** is

$$E(Y^s) = \sum_i \sum_j Y_j^s f(X_i, Y_j) = \sum_j Y_j^s h(Y_j). \tag{2.55}$$

In addition, the **$r$th and $s$th product or joint moment of $X$ and $Y$ about the origin** $(0,0)$ is

$$E(X^r Y^s) = \sum_i \sum_j X_i^r Y_j^s f(X_i, Y_j) \tag{2.56}$$

and, for $r$ and $s$ non-negative integers, the **$r$th and $s$th product or joint moment of $X$ and $Y$ about the mean** is

$$E[(X - \mu_X)^r (Y - \mu_Y)^s] = \sum_i \sum_j (X_i - \mu_X)^r (Y_j - \mu_Y)^s f(X_i, Y_j), \tag{2.57}$$

where $\mu_X = E(X)$ and $\mu_Y = E(Y)$.

If $X$ and $Y$ are continuous random variables with joint probability density function $f(x, y)$, then **the $r$th moment of $X$ about zero** is

$$E(X^r) = \int_{-\infty}^{+\infty} \int_{-\infty}^{+\infty} x^r f(x,y) \, dy \, dx = \int_{-\infty}^{+\infty} x^r g(x) dx; \tag{2.58}$$

and the **$s$th moment of $Y$ about zero** is

$$E(Y^s) = \int_{-\infty}^{+\infty} \int_{-\infty}^{+\infty} y^s f(x,y) \, dy \, dx = \int_{-\infty}^{+\infty} y^s h(y) dy. \tag{2.59}$$

In addition, the **$r$th and $s$th product or joint moment of $X$ and $Y$ about the origin** $(0,0)$ is

$$E(X^r Y^s) = \int_{-\infty}^{+\infty} \int_{-\infty}^{+\infty} x^r y^s f(x,y) dy \, dx \tag{2.60}$$

and, for $r$ and $s$ non-negative integers, the **$r$th and $s$th product or joint moment of $X$ and $Y$ about the mean** is

$$E[(X-\mu_X)^r(Y-\mu_Y)^s] = \int_{-\infty}^{+\infty}\int_{-\infty}^{+\infty}(x-\mu_X)^r(y-\mu_Y)^s f(x,y)dy\,dx. \qquad (2.61)$$

Some additional considerations regarding expectations are the following:

1. If $\varsigma$ is a linear function of the random variables $X$ and $Y$, $\varsigma = aX \pm bY$, where $a$ and $b$ are constants, then, for either the discrete or continuous case,

$$E(\varsigma) = E(aX \pm bY) = a\,E(X) \pm b\,E(Y)$$

2. If in (2.57) and (2.61) we set $r = s = 1$, then the expression

$$E[(X-\mu_X)(Y-\mu_Y)] = E(XY) - \mu_X\mu_Y = COV(X,Y) = \sigma_{XY} \qquad (2.62)$$

is called the **covariance** of the random variables $X$ and $Y$—it depicts the joint variability of the random variables $X$ and $Y$. Note that $COV(aX, bY) = ab\sigma_{XY}$. While the sign of $\sigma_{XY}$ indicates the *direction* of the relationship between the discrete or continuous random variables $X$ and $Y$ ($X$ and $Y$ are positively related when $\sigma_{XY} > 0$; they are negatively related when $\sigma_{XY} < 0$), the **coefficient of correlation** between the random variables $X$ and $Y$,

$$\rho_{XY} = \frac{\sigma_{XY}}{\sigma_X\sigma_Y}, -1 \le \rho_{XY} \le 1, \qquad (2.63)$$

measures the *strength* and the *direction* of the *linear* relation between $X$ and $Y$, where $\sigma_X$ and $\sigma_Y$ are, respectively, the standard deviations of $X$ and $Y$. When $\rho_{XY} = 1$ (resp. $\rho_{XY} = -1$), we have perfect positive (resp. negative) association between $X$ and $Y$.

3. If $X$ **and** $Y$ **are independent** discrete or continuous random variables, then $\sigma_{XY} = 0$ and, from (2.62), $E(XY) = \mu_x\mu_Y$. If $X$ **and** $Y$ **are uncorrelated** or $\rho_{XY} = 0$, it does not necessarily follow that $X$ and $Y$ are independent random variables.

4. If $\varsigma = aX + bY$, then the **variance of** $\varsigma$ is

$$V(\varsigma) = V(aX + bY) = a^2\sigma_X^2 + b^2\sigma_Y^2 + 2ab\sigma_{XY}, \qquad (2.64)$$

where $\sigma_X^2 = V(X)$ and $\sigma_Y^2 = V(Y)$ are the variances of the random variables $X$ and $Y$, respectively. If $\varsigma = aX - bY$,

$$V(\varsigma) = V(aX - bY) = a^2\sigma_X^2 + b^2\sigma_Y^2 - 2ab\sigma_{XY}. \qquad (2.65)$$

In addition, if $X$ and $Y$ are independent random variables, then

$$V(aX \pm bY) = a^2\sigma_X^2 + b^2\sigma_X^2 \qquad (2.66)$$

5. For random variables $X$ and $Y$,

$$E(XY) = \mu_X\mu_Y + \sigma_{XY}. \qquad (2.67)$$

If $X$ and $Y$ are independent, then $E(XY) = \mu_X\mu_Y$ (as indicated in item (3)).

6. For random variables $X$ and $Y$,

$$V(XY) = E\{[XY - E(XY)]^2\} = E\{[XY - (\mu_X\mu_Y + \sigma_{XY})]^2\}$$
$$= \mu_Y^2\sigma_X^2 + \mu_X^2\sigma_Y^2 + 2\mu_X\mu_Y\sigma_{XY} - \sigma_{XY}^2$$
$$+ E[(X - \mu_X)^2(Y - \mu_Y)^2] + 2\mu_Y E[(X - \mu_X)^2(Y - \mu_Y)] \qquad (2.68)$$
$$+ 2\mu_X E[(X - \mu_X)(Y - \mu_Y)^2].$$

If $X$ and $Y$ are independent random variables,

$$V(XY) = \mu_Y^2\sigma_X^2 + \mu_X^2\sigma_Y^2 + \sigma_X^2\sigma_Y^2. \qquad (2.68.1)$$

7. We previous defined a Hilbert space $H$ (Section 1.11.3) as a complete inner product space, where the inner product $(f,g) = \int f(x)g(x)dx \in R$ for $f,g \in H$ and the norm $\|f\| = (f,f)^{1/2}$. Given this setting, we now turn to the specification of the **Hilbert space of random variables** $H_R$. To this end, let $(\Omega, \mathcal{A}, P)$ be a probability space with event $A \in \mathcal{A}$. Then the indicator function

$$\chi_A(\omega) = \begin{cases} 1, \omega \in A; \\ 0, \omega \notin A \end{cases}$$

is a random variable with $E(\chi_A) = P(A)$. In addition, if $X$ is a simple random variable or

$$X(\omega) = \sum_{i=1}^{n} x_i\chi_{A_i}, A_i \in \mathcal{A}, i = 1,\dots,n,$$

then $E(X) = \sum_{i=1}^{n} x_i P(A_i)$.

Next, let $S$ be the set of simple random variables defined on $(\Omega, \mathcal{A}, P)$. With $S$ an inner product space and for simple random variables, $X, Y \in S$, the norm of $X$ is

$$\|X\| = (X,X)^{\frac{1}{2}} = (E|X|^2)^{\frac{1}{2}};$$

and the inner product of $X$ and $Y$ on $S$ is defined as

$$(X,Y) = E(XY) = E\left(\sum_{i=1}^{n}\sum_{j=1}^{n} x_i\chi_{A_i}y_j\chi_{B_j}\right)$$
$$= \sum_{i=1}^{n}\sum_{j=1}^{n} x_iy_j P(A_i \cap B_j)$$

for $\{A_i, i = 1,\dots,n\}, \{B_j, j = 1,\dots,n\} \in \mathcal{A}$. Since $S$ might not be complete (in the sense that all Cauchy sequences of random variables defined on $S$ must converge in $S$), how do we make the leap from $S$ to $H_R$? The completion of $S$ may be effected by augmenting $S$ by a set of random variables in order to form $H_R$ by rendering $S$ dense in $H_R$.

Let $X$ and $Y$ be discrete random variables with $g(X|Y)$ the conditional probability mass function of $X$ given $Y$ and $h(Y|X)$ the conditional probability mass function of $Y$ given $X$. Then the **conditional expectation of $X$ given $Y = Y_s$**[4] is defined as

$$E(X|Y_s) = \sum_i X_i g(X_i|Y_s); \tag{2.69}$$

and the **conditional variance of $X$ given $Y = Y_s$** is

$$V(X|Y_s) = E\{[X - E(X|Y_s)]^2|Y_s\} = E(X^2|Y_s) - E(X|Y_s)^2, \tag{2.70}$$

where

$$E(X^2|Y_s) = \sum_i X_i^2 g(X_i|Y_s).$$

Similarly, the **conditional expectation of $Y$ given $X = X_r$** is

$$E(Y|X_r) = \sum_j Y_j h(Y_j|X_r); \tag{2.71}$$

and the **conditional variance of $Y$ given $X = X_r$** is

$$V(Y|X_r) = E\{[Y - E(Y|X_r)]^2|X_r\} = E(Y^2|X_r) - E(Y|X_r)^2, \tag{2.72}$$

with

$$E(Y^2|X_r) = \sum_j Y_j^2 h(Y_j|X_r).$$

If $X$ and $Y$ are continuous random variables with $g(x|y)$ and $h(y|x)$ representing conditional probability density functions for $X$ given $Y$ and $Y$ given $X$, respectively, then the **conditional expectation of $X$ given $Y = y$** is

$$E(X|y) = \int_{-\infty}^{+\infty} x g(x|y)\, dx; \tag{2.73}$$

and the **conditional variance of $X$ given $Y = y$** is

$$V(X|y) = E\{[X - E(x|y)]^2|y\} = E(X^2|y) - E(X|y)^2, \tag{2.74}$$

where

$$E(X^2|y) = \int_{-\infty}^{+\infty} x^2 g(x|y)\, dx. \tag{2.75}$$

---

4 Appendix 2.A provides a discussion of the theoretical foundation for conditional expectations.

In similar fashion, the **conditional expectation of $Y$ given $X = x$** is defined as

$$E(Y|x) = \int_{-\infty}^{+\infty} y\, h(y|x)\, dy; \qquad (2.76)$$

and the **conditional variance of $Y$ given $X = x$** is

$$V(Y|x) = E\left\{[Y - E(Y|x)]^2 |x\right\} = E\left(Y^2|x\right) - E(Y|x)^2, \qquad (2.77)$$

where

$$E\left(Y^2|x\right) = \int_{-\infty}^{+\infty} y^2\, h(y|x)\, dy. \qquad (2.78)$$

If $X$ and $Y$ are **independent discrete or continuous random variables**, then the preceding sets of conditional means and variances equal their unconditional counterparts, for example, under the independence of $X$ and $Y$, (2.73) reduces to $E(X|y) = E(X)$ and (2.74) becomes $V(X|y) = V(X)$.

**Example 2.12**   Let's generalize the univariate normal distribution presented in Example 2.7 involving the random variable $X$ to the bivariate case where we have both $X$ and $Y$ as random variables. Specifically, let the random variable $(X, Y)$ have the joint probability density function

$$f(x,y) = \frac{1}{2\pi\sigma_x\sigma_y(1-\rho^2)^{\frac{1}{2}}} e^{-\frac{1}{2}Q}, \ -\infty < x, y < +\infty, \qquad (2.79)$$

where

$$Q = \frac{1}{1-\rho^2}\left[\left(\frac{x-\mu_x}{\sigma_x}\right)^2 - 2\rho\left(\frac{x-\mu_x}{\sigma_x}\right)\left(\frac{y-\mu_y}{\sigma_y}\right) + \left(\frac{y-\mu_y}{\sigma_y}\right)^2\right], \qquad (2.80)$$

and where $\mu_x, \mu_y, \sigma_x, \sigma_y$, and $\rho$ are all parameters, with $-\infty < \mu_x, \mu_y < +\infty, \sigma_x > 0$, $\sigma_y > 0$, and $|\rho| < 1$ ((2.79) is undefined if $\rho = \pm1$). Here, $\mu_x = E(X)$, $\mu_y = E(Y)$, $\sigma_x^2 = V(X), \sigma_y^2 = V(Y)$, and $\rho$ is the coefficient of correlation between $X$ and $Y$. If Equations (2.79) and (2.80) hold, then $X$ and $Y$ are said to follow a **bivariate normal distribution** with joint probability density function (2.79), or $X$ and $Y$ are $N(x, y; \mu_x, \mu_y, \sigma_x, \sigma_y, \rho)$.

Looking to the properties of (2.79),

1. $f(x,y) > 0$.
2. The probability that a point $(X, Y)$ will lie within a region $A$ of the $x, y$-plane is

$$P[(x,y) \in A] = \int_A\int f(x,y)\, dydx.$$

3. As required,

$$\int_{-\infty}^{+\infty}\int_{-\infty}^{+\infty} f(x,y)\, dydx = 1; \text{ and}$$

4. if in (2.79) we set $U = (x - \mu_x)/\sigma_x$, and $V = (y - \sigma_y)/\sigma_y$, then $f(x, y)$ can be expressed in terms of the single parameter $\rho$.

In addition, for $(X, Y)$ bivariate normal, the **marginal probability density function of $X$** (denoted $N(x; \mu_x, \sigma_x)$) can be obtained from (2.79) by integrating out $y$ or

$$g(x) = \int_{-\infty}^{+\infty} f(x,y)\, dy = \frac{1}{\sigma_x\sqrt{2\pi}} e^{-\frac{1}{2}\left[\frac{(x-\mu_x)}{\sigma_x}\right]^2}, \quad -\infty < x < +\infty; \tag{2.81}$$

and the **marginal probability density function of $Y$** (denoted $N(y; \mu_y, \sigma_y)$) is obtained from (2.79) by integrating out $x$ or

$$h(y) = \int_{-\infty}^{+\infty} f(x,y)\, dx = \frac{1}{\sigma_y\sqrt{2\pi}} e^{-\frac{1}{2}\left[\frac{(y-\mu_y)}{\sigma_y}\right]^2}, \quad -\infty < y < +\infty. \tag{2.82}$$

(Note that if the marginal distributions of the random variables $X$ and $Y$ are each univariate normal, this does not imply that $(X, Y)$ is bivariate normal.)

In this section 2.5.3 we noted that the random variables $X$ and $Y$ are independent, then they must also be uncorrelated or $\rho_{XY} = 0$. However, the converse of this statement does not generally apply. Interestingly, it does apply if $X$ and $Y$ follow a bivariate normal distribution. That is, if $X$ and $Y$ are bivariate normal with probability density function (2.79), then $X$ and $Y$ are **independent random variables** if and only if $\rho_{XY} = \rho = 0$. When $\rho = 0$ ($X$ and $Y$ are independent), (2.79) factors as $f(x,y) = g(x) \cdot h(y)$.

Next, if the random variable $(X, Y)$ is bivariate normal, then the conditional distribution of $X$ given $Y = y$ is univariate $N\left(x; \mu_x + \rho\left(\sigma_x/\sigma_y\right)\left(y - \mu_y\right), \sigma_x\sqrt{1-\rho^2}\right)$. Here, the **conditional mean of $X$ given $Y = y$** is $E(X|Y = y) = \mu_x + \rho\left(\sigma_x/\sigma_y\right)\left(y - \mu_y\right)$, and the **conditional variance of $X$ given $Y = y$** is $V(X|Y = y) = \sigma_x^2(1 - \rho^2)$. The **conditional probability density function of $X$ given $Y = y$** is obtained from the joint and marginal densities in (2.79) and (2.81), respectively, for $h(y) \neq 0$, as

$$g(x|y) = \frac{f(x,y)}{h(y)}$$

$$= \frac{1}{\sigma_x\sqrt{2\pi}\sqrt{1-\rho^2}} e^{-\frac{\left[x - \mu_x - \rho\left(\sigma_x/\sigma_y\right)\left(y - \mu_y\right)\right]^2}{2\sigma_x^2(1-\rho^2)}}, \quad -\infty < x < +\infty. \tag{2.83}$$

Similarly, the **conditional distribution of $Y$ given $X = x$** is univariate $N\left(y; \mu_y + \rho\left(\sigma_y/\sigma_x\right)(x - \mu_x), \sigma_y\sqrt{1-\rho^2}\right)$. Thus, the **conditional mean of $Y$ given**

$X = x$ is $E(Y|X = x) = \mu_x + \rho(\sigma_y/\sigma_x)(x - \mu_x)$, and the **conditional variance of** $Y$ **given** $X = x$ is $V(Y|X = x) = \sigma_y^2(1 - \rho^2)$. Hence, the **conditional probability density function of** $Y$ **given** $X = x$ is obtained from the joint and marginal densities in (2.79) and (2.82), respectively, given $g(x) \neq 0$, as

$$h(y|x) = \frac{f(x,y)}{g(x)}$$

$$= \frac{1}{\sigma_y\sqrt{2\pi}\sqrt{1-\rho^2}} e^{-\frac{\left[y - \mu_y - \rho(\sigma_y/\sigma_x)(x-\mu_x)\right]^2}{2\sigma_x^2(1-\rho^2)}}, \quad -\infty < y < +\infty. \blacksquare$$

(2.84)

### 2.5.4   The Multivariate Discrete and Continuous Cases

In some of the preceding sections, we exclusively considered bivariate random variables $(X, Y)$. Now, let us expand our discussion to the case of dealing with the multivariate random vector $X = (X_1, X_2, \ldots, X_n)$, where each $X_i, i = 1, \ldots, n$, is a random variable. To this end, let $(\Omega, \mathcal{A}, P)$ be a probability space and suppose that $X : \Omega \to R^n$ is a random vector. If $X$ is discrete, then the **joint probability mass function of** $X$ is defined as

$$f(x) = f(x_1, \ldots, x_n)$$

$$= P(X_1 = x_1, \ldots, X_n = x_n) : R^n \to R$$

(2.85)

for each $x = (x_1, \ldots, x_n) \in R^n$. Then for any event $A \subset R^n$,

$$P(X \in A) = \sum_{x \in A} f(x).$$

(2.86)

If $X$ is a continuous random vector, then the **joint probability density function** is a non-negative, integrable function $f(x) = f(x_1, \ldots, x_n) : R^n \to R$ that satisfies

$$P(X \in A) = \int_A \cdots \int f(x)dx = \int_A \cdots \int f(x_1, \ldots, x_n)dx_n \ldots dx_1,$$

(2.87)

where the limits of integration are specified so that integration occurs over all vectors $X \in A$.

Given the multivariate probability mass function $f(X)$, the joint probability that $X_i \le Y_i, i = 1, \ldots, n$, where $Y = (Y_1, \ldots, Y_n) \in R^n$, is provided by the **discrete multivariate cumulative distribution function** $F : R^n \to [0,1]$, with

$$F(y_1, \ldots, y_n) = P(X \le Y)$$

$$= P(X_1 \le y_1, \ldots, X_n \le y_n).$$

(2.88)

In addition, for the multivariate probability density function $f(x)$, the **continuous multivariate cumulative distribution function** $F : R^n \to [0,1]$ is written as

$$
\begin{aligned}
F(y) &= P(X_1 \le y_1, \ldots, X_n \le y_n) \\
&= \int_{-\infty}^{y_1} \ldots \int_{-\infty}^{y_n} f(x_1, \ldots, x_n)\, dx_n \ldots dx_1.
\end{aligned}
\tag{2.89}
$$

Let $\emptyset(x) = \emptyset(x_1, \ldots, x_n)$ be a real-valued function defined on $\Omega$. With $\emptyset(X)$ a random variable, the **expectation of** $\emptyset(X)$ in the discrete case is

$$
E[\emptyset(X)] = \sum_{x \in \Omega} \emptyset(x)f(x);
\tag{2.90}
$$

and for $X$ continuous,

$$
E[\emptyset(X)] = \int_{-\infty}^{+\infty} \ldots \int_{-\infty}^{+\infty} \emptyset(x)f(x)dx.
\tag{2.91}
$$

As was presented earlier (Sections 2.5.1 and 2.5.2 covered, respectively, the discrete and continuous cases), we can define the **discrete multivariate marginal probability mass function** of a subset of coordinates of $X$ by summing the joint probability mass function over all possible values of the remaining coordinates, for example, the marginal distribution of $(X_1, \ldots, X_k) \in R^k, k < n$, can be determined as

$$
\begin{aligned}
g(x_1, \ldots, x_k) &= \sum f(x). \\
(x_{k+1}, \ldots, x_n) &\in R^{n-k}
\end{aligned}
\tag{2.92}
$$

The **continuous multivariate marginal probability mass function**, again taken for a subset of coordinates of $X$, is obtained by integrating the joint probability density function over all possible values of the other coordinates, that is, the marginal density of $(X_1, \ldots, X_k) \in R^k, k < n$, is specified as

$$
g(x_1, \ldots, x_k) = \int_{-\infty}^{+\infty} \ldots \int_{-\infty}^{+\infty} f(x)dx_n \ldots dx_{k+1}.
\tag{2.93}
$$

The **multivariate conditional probability mass function** or **probability density function** of the subset $(X_1, \ldots, X_k) \in R^k, k < n$, of coordinates of $X$, *given* the values of the remaining coordinates, is obtained by dividing the joint probability mass function by the marginal probability mass function (in the discrete case) or by dividing the joint probability density function by the marginal probability density function (for the continuous case). In this regard, the conditional probability mass function or probability density function of $(X_{k+1}, \ldots, X_n)$ given $X_1 = x_1, \ldots, X_k = x_k$ is a function of $(x_{k+1}, \ldots, x_n)$ defined by

$$
f(x_{k+1}, \ldots, x_n | x_1, \ldots, x_k) = \frac{f(x)}{g(x_1, \ldots, x_k)}, g(x_1, \ldots, x_k) > 0.
\tag{2.94}
$$

**Example 2.13**   Let the random variables $X_1, X_2, ..., X_n$ constitute the components of the $(n \times 1)$ random vector $X$ and let the means of these random variables, denoted $\mu_1, \mu_2, ..., \mu_n$, respectively, represent the components of the $(n \times 1)$ mean vector $\mu$. In addition, let $\Sigma$ serve as the $n$th order variance–covariance matrix of $X_1, ..., X_n$, where

$$\sum = E\left[(X-\mu)(X-\mu)'\right] = E\begin{bmatrix} (X_1-\mu_1)^2 & \cdots & (X_n-\mu_n)(X_1-\mu_1) \\ (X_1-\mu_1)(X_2-\mu_2) & \cdots & (X_n-\mu_n)(X_2-\mu_2) \\ \vdots & & \vdots \\ (X_1-\mu_1)(X_n-\mu_n) & \cdots & (X_n-\mu_n)^2 \end{bmatrix}$$

$$= \begin{bmatrix} \sigma_1^2 & \sigma_{12} & \cdots & \sigma_{1n} \\ \sigma_{21} & \sigma_2^2 & \cdots & \sigma_{2n} \\ \vdots & \vdots & \vdots & \\ \sigma_{n1} & \sigma_{n2} & \cdots & \sigma_n^2 \end{bmatrix} = \begin{bmatrix} \sigma_1^2 & \rho_{12}\sigma_1\sigma_2 & \cdots & \rho_{1n}\sigma_1\sigma_n \\ \rho_{21}\sigma_2\sigma_1 & \sigma_2^2 & \cdots & \rho_{21}\sigma_2\sigma_n \\ \vdots & \vdots & & \vdots \\ \rho_{n1}\sigma_n\sigma_1 & \rho_{n2}\sigma_n\sigma_2 & \cdots & \sigma_n^2 \end{bmatrix}$$

$$(2.95)$$

is a symmetric positive definite matrix and $\rho_{kl} = \sigma_{kl}/\sigma_k\sigma_l$, with $k, l = 1, ..., n$.

Then the joint probability density function for the **multivariate normal distribution** appears as

$$f(X; \mu, \Sigma) = N(X; \mu, \Sigma)$$

$$= (2\pi)^{-\frac{n}{2}}|\Sigma|^{-\frac{1}{2}}e^{-\frac{1}{2}(X-\mu)'\Sigma^{-1}(X-\mu)}, \quad -\infty < \mu_i < +\infty, \; i = 1, ..., n,$$

$$(2.96)$$

where a prime denotes vector (matrix) transposition.

Given this expression, the following can be demonstrated:

1. The **marginal probability density function of each** $X_i$ is $N(X_i; \mu_i, \sigma_i)$, $i = 1, ..., n$.
2. The **conditional probability density function of** $X_k$, **given fixed values of the remaining variables,** is

$$N\left(E(X_k | X_1, ..., X_{k-1}, X_{k+1}, ..., X_n), \sigma_{X_k | X_1, ..., X_{k-1}, X_{k+1}, ..., X_n}^2\right).$$

3. If $R$ represents an $n$th order correlation matrix with components $\rho_{kl}, k, l = 1, ..., n$, then (2.96) can be written as

$$f(U) = (2\pi)^{-\frac{n}{2}}(\sigma_1\sigma_2...\sigma_n)^{-1}|R|^{-\frac{1}{2}}e^{-\frac{1}{2}U'R^{-1}U}, \qquad (2.97)$$

where $U_i = (X_i - \mu_i)/\sigma_i, i = 1, \ldots n$. Furthermore, if $\rho_{kl} = 0, k \neq l$ (the jointly distributed normal random variables are uncorrelated), then $R = I_n$ (the identity matrix—a matrix having ones along the main diagonal and zeros elsewhere) and

$$f(U) = (2\pi)^{-\frac{n}{2}} (\sigma_1 \sigma_2 \ldots \sigma_n)^{-1} e^{-\frac{1}{2} u'u}. \tag{2.97.1}$$

If it is also true that $\mu_i = 0$ and $\sigma_i = \sigma$ for all $i$, then $U_i = X_i/\sigma, i = 1, \ldots, n$, and thus (2.97.1) appears as

$$f(U) = (2\pi)^{-\frac{n}{2}} \sigma^{-n} e^{-\frac{1}{2} u'u}. \tag{2.97.2}$$

In addition, if $\sigma = 1$, then U follows a **multivariate standard normal distribution** with probability density function

$$f(U) = (2\pi)^{-\frac{n}{2}} e^{-\frac{1}{2} u'u}. \tag{2.97.3}$$

4. Given (2.96), if the random variables $X_1, \ldots, X_n$ are mutually independent, then

$$f(X; \mu, \Sigma) = \Pi_{i=1}^{n} f(X_i; \mu_i, \sigma_i) = \Pi_{i=1}^{n} (2\pi)^{-\frac{1}{2}} \sigma_i^{-\frac{1}{2}} e_i^{-\frac{1}{2} \left[ \frac{(X_i - \mu_i)}{\sigma_i} \right]^2}, \tag{2.98}$$

where $f(X_i; \mu_i, \sigma_i)$ is univariate normal. Note that (2.98) also results if $\Sigma$ is a diagonal matrix ($\Sigma$ is diagonal if the normally distributed random variables $X_i, i = 1, \ldots, n$, are uncorrelated or have zero covariance). Hence, with respect to Equation (2.98), the variables $X_i, i = 1, \ldots, n$, are independent if and only if $\Sigma$ is a diagonal matrix. Thus, zero covariance is equivalent to independence under multivariate normality. ∎

## 2.6 Convergence of Sequences of Random Variables

Let $\{X_n\}_{n=1}^{\infty}$ (or simply $\{X_n\}$ for short) denote a sequence of random variables defined on the probability space $(\Omega, \mathcal{A}, P)$. An important question that emerges in the area of probability theory is "Does there exist a **limiting random variable** $X$ to which the sequence approaches as $n \to \infty$?" Our objective in this section is to characterize the manner in which $\{X_n\} \to X$ as $n \to \infty$ since, as we shall now see, there are many varieties of convergence criteria for sequences of random variables.

Being already familiar with the concept of "pointwise convergence" of real numbers (Section 1.5), the reader may be tempted to define the convergence of $\{X_n\}$ to $X$ as $n \to \infty$ as $lim_{n \to \infty} X_n(\omega) = X(\omega), \omega \in \Omega$. The mode of convergence just described is termed the **sure convergence** of $\{X_n\}$ to $X$. However, in terms of the requirements of probability theory, this form of convergence of random variables is "too strong." In this regard, we shall now look to a set of "weaker" concepts pertaining to the convergence of a sequence of random variables.

### 2.6.1 Almost Sure Convergence

We may weaken the notion of "sure convergence" by requiring only that $\{X_n\}$ **converges almost surely** (a.s.) or **converges with probability 1** to $X$ as $n \to \infty$. That is, let $\{X_n\}$ and $X$ be random variables defined on the same probability space $(\Omega, \mathcal{A}, P)$ and let (measurable) set $A = \{\omega \in \Omega | X_n(\omega) \to X$ $(\omega)$ as $n \to \infty\}$. Then $\{X_n\}$ converges a.s. to $X$ as $n \to \infty$ (written $X_n \overset{a.s.}{\to} X$) if $P(A) = 1$. In short, $X_n \overset{a.s.}{\to} X$ if $P(X_n \to X) = 1$. A sufficient condition for a.s. convergence is:

> **Sufficient Condition for a.s. Convergence** If $\sum_{n=1}^{\infty} P(|X_n - X| \geq \varepsilon) < +\infty$ for all $\varepsilon > 0$, then $X_n \overset{a.s.}{\to} X$.

Alternative expressions for a.s. convergence are (a) $\{X_n\}$ converges to $X$ a.s. if $P(\{\omega \in \Omega | \lim_{n \to \infty} |X_n(\omega) - X(\omega)| = 0\}) = 1$; and (b) $\{X_n\}$ converges to $X$ a.s. if $P(\{\omega \in \Omega | \lim_{n \to \infty} X_n(\omega) \neq X(\omega)\}) = 0$. In addition, we may observe briefly the following:

1. The random variables $\{X_n\}$ and $X$ are all defined on the same probability space and are generally taken to be highly dependent.
2. Almost sure convergence or convergence with probability 1 is the strongest form of convergence in the context of probability theory. In this regard, it is often said that a.s. convergence implies that $\{X_n\}$ "converges strongly" to $X$ as $n \to \infty$.
3. The definition of a.s. convergence implies that, for any $\varepsilon > 0$, there exists a value $N_\varepsilon$ such that $|X_n - X| < \varepsilon$ for every $n \geq N_\varepsilon$. Hence, the probability distribution of $X_n - X$ becomes increasingly concentrated about zero as $n \to \infty$, that is, the values of $|X_n - X|$ approach zero as $n \to \infty$ (Figure 2.9).

### 2.6.2 Convergence in $L^p, p > 0$

An alternative way to weaken the "sure convergence" concept is to require that $\{X_n\}$ **converges in $L^p$** or **converges in $p$th mean** to $X$ as $n \to \infty$. That is, for $p$ a positive constant, $\{X_n\}$ and $X$ are random variables defined on the same probability space, and the $p$th absolute moments $E(|X_n|^p)$ and $E(|X|^p)$ of $X_n$ and $X$, respectively, exist, then $\{X_n\}$ converges in $L^p$ to $X$ (expressed as $X_n \overset{L^p}{\to} X$) if $E(|X_n - X|^p) \to 0$ as $n \to \infty$. In short, $X_n \overset{L^p}{\to} X$ if $\lim_{n \to \infty} E(|X_n - X|^p) = 0$ for finite $E(|X_n|^p)$ and $E(|X|^p)$.

$X_n - X$

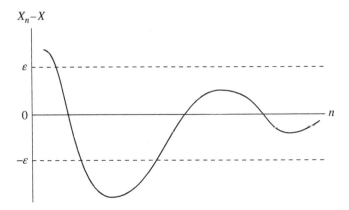

**Figure 2.9** $P\left(\lim_{n\to\infty} X_n = X\right) = 1$ or $X \overset{\text{a.s.}}{\to} X$.

For $p=1$, we have **convergence in mean**; and for $p = 2$, we have **convergence in mean square** or **convergence in quadratic mean**.[5] In short, the sequence $\{X_n\}$ converges to $X$ in mean square (denoted $X_n \overset{\text{m.s.}}{\to} X$) if $\lim_{n\to\infty} E(X_n - X)^2 = 0$. In fact, a necessary and sufficient condition for the mean-square limit to exist is the

**Cauchy Criterion for m.s. Convergence**: A sequence of random variables $\{X_n\}\left(E(X_n^2) < +\infty\right)$ converges to a random variable $X$ $(E(X^2) < +\infty)$ in mean square if and only if $\lim_{n,m\to\infty} E(X_n - X_m)^2 = 0$.

We note in passing the following:

1. For $c \in R, X_n \overset{\text{m.s.}}{\to} c$ if and only if $E(X_n) \to c$ and $V(X_n) \to 0$.
2. Mean-square convergence does not imply a.s. convergence; a.s. convergence does not imply mean-square convergence.
3. For $\{X_n\} \subset H_R$ (the Hilbert space of random variables), mean-square convergence is equivalent to $\|X_n - X\| \to 0$ as $n \to \infty$. In fact, the existence of a random variable $X \in H_R$ is guaranteed if $\{X_n\}$ is a Cauchy sequence in $H_R$.
4. For $p \geq q, X_n \overset{L^p}{\to} X$ implies $X_n \overset{L^q}{\to} X$. (So for $p = 2$ and $q = 1$, we see that convergence in mean-square implies convergence in mean.)

---

5 More formally, a sequence $\{X_n\}$ of random variables, with $E(|X_n|) < +\infty$, **converges in $L^1$** to a random variable $X$, with $E(|X|) < +\infty$, if $\lim_{n\to\infty} E(|X_n - X|) = 0$. (If this limit is zero, then $\{X_n\}$ is said to "converge strongly" to $X$.) A sequence $\{X_n\}$ of random variables, with $E(X_n^2) < +\infty$, **converges in $L^2$** to a random variable $X$, with $E(X^2) < +\infty$, if $\lim_{n\to\infty} E(X_n - X)^2 = 0$. (Mean-square convergence implies that $\{X_n\}$ "converges strongly" to $X$ as $n \to \infty$).

5. If $lim_{n\to\infty} E(X_n - X)^2 = 0$ (convergence in $L^2$), then $lim_{n\to\infty} E(X_n^2) = E(X^2)$ and $lim_{n\to\infty} E(X_n) = E(X)$ (see also the discussion offered in Section 2.6.5).

### 2.6.3 Convergence in Probability

Suppose $\{X_n\}$ and $X$ are random variables defined on the same probability space $(\Omega, \mathcal{A}, P)$. $\{X_n\}$ **converges in probability** or **converges in measure** to $X$ if, for every $\varepsilon > 0$, $P(\{\omega \in \Omega \,||X_n(\omega) - X(\omega)| \geq \varepsilon\}) \to 0$ as $n \to \infty$ (written $X_n \overset{P}{\to} X$). In short, $X_n \overset{P}{\to} X$ if $lim_{n\to\infty} P(|X_n - X| \geq \varepsilon) = 0$ for all $\varepsilon > 0$. Equivalently, $X_n \overset{P}{\to} X$ if $lim_{n\to\infty} P(|X_n - X| < \varepsilon) = 1$ for each $\varepsilon > 0$. (As this latter expression indicates, $|X_n - X|$ should be arbitrarily close to zero with probability arbitrarily close to 1. So if $X_n \overset{P}{\to} X$, then, with high probability, $X_n \approx X$ as $n \to \infty$.)

It is important to note the following:

1. Convergence in probability means that there exists a subsequence on which a.s. convergence occurs, that is, if $X_n \overset{P}{\to} X$, then there exists a subsequence $\{X_{n_j}\}_{j=1}^{\infty} \subset \{X_n\}_{n=1}^{\infty}$ such that $X_{n_j} \overset{a.s.}{\to} X$.
2. For $c \in R, X_n \overset{P}{\to} c$ if, for all $\varepsilon > 0$, $lim_{n\to\infty} P(|X_n - c| \geq \varepsilon) = 0$.
3. Consider now the following relationships between our first three types of convergence of a sequence of random variables. (Note that the symbol "$\Rightarrow$" means "implies."):

$$X_n \overset{a.s.}{\to} X \Rightarrow X_n \overset{P}{\to} X \text{ (generally } < \neq );$$

$$X_n \overset{m.s.}{\to} X \Rightarrow X_n \overset{P}{\to} X \text{ (generally } < \neq );$$

$$X_n \overset{a.s.}{\to} X \nLeftrightarrow X_n \overset{m.s.}{\to} X \text{ (convergence a.s. does not imply convergence in mean square and vice versa).}$$

While a.s. convergence, mean-square convergence, and convergence in probability are often referred to as forms of "strong convergence," convergence in probability is the "weakest" type of convergence among these three convergence concepts.

4. Let the random variables $\{X_n\}, X, \{Y_n\}$, and $Y$ all be defined on the same probability space. If $X_n \overset{P}{\to} X$ and $Y_n \overset{P}{\to} Y$, then $(aX_n + bY_n) \overset{P}{\to} (aX + bY)$ for all constants $a$ and $b$.
5. Convergence in probability requires that the joint distributions of $X_n$ and $X$ be known.
6. Convergence in probability means that the distribution of $X_n - X$ becomes increasingly concentrated about zero as $n \to \infty$; most assuredly, all values of $X_n - X$ eventually fall within the interval $(-\varepsilon, \varepsilon)$ as $n \to \infty$ (Figure 2.10).

### 2.6.4 Convergence in Distribution

Let $F_n(t) = P(X_n \leq t)$ and $F(t) = P(X \leq t)$ denote the cumulative distribution functions of the random variables $\{X_n\}$ and $X$, respectively. The sequence of

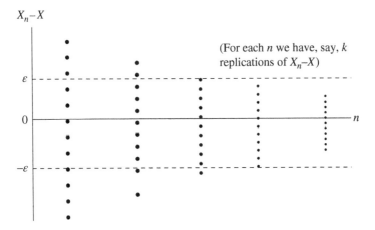

**Figure 2.10** $X_n \xrightarrow{P} X$.

random variables $\{X_n\}$ is said to **converge in distribution** or **converge in law** to the random variable $X$ (denoted $X_n \xrightarrow{d} X$) if $F_n(t) \to F(t)$ as $n \to \infty$ for all $t$ at which the **limiting distribution function** $F$ is continuous. (Remember that $F$ is continuous at $X = t$ if $P\ (X = t) = 0$.) In short, $X_n \xrightarrow{d} X$ if $lim_{n \to \infty} F_n(t) = F(t)$ at every continuity point $t$ of $F$. That is, $X_n \xrightarrow{d} X$ if, for large $n$, the distribution function of $X_n$ is close to that of $X$.

It is important to note the following:

1. The random variables $\{X_n\}$ and $X$ need not be defined on the same probability space.
2. While the preceding definition of convergence in distribution is stated in terms of the random variables $\{X_n\}$ converging to $X$, it is actually the cumulative distribution functions that converge.
3. $X_n \xrightarrow{P} X \Rightarrow X_n \xrightarrow{d} X (\nLeftarrow)$. Hence, convergence in distribution is the "weakest" of all forms of convergence and thus is referred to as "weak convergence." (Note that the notion of "strength of convergence" should *not* be associated with the *importance* of the type of convergence. Each has its place in probability theory.)
4. If $X$ is a constant random variable $(P(X = c) = 1, c \in R)$, then $X_n \xrightarrow{d} c \Rightarrow X_n \xrightarrow{P} c$.

A summary of the relationship between the various modes of convergence is provided by Figure 2.11.

### 2.6.5 Convergence of Expectations

We now consider the following tangential question associated with the convergence of a sequence of random variables. Specifically, given that $X_n \to X$ (in some meaningful fashion) as $n \to \infty$, can we also conclude that

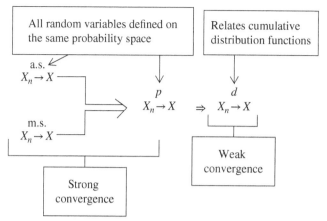

(Convergence a.s., convergence in mean square, and convergence in probability all imply convergence in distribution. In general, no other implications hold.)

**Figure 2.11** Relationship between the different convergence concepts.

$E(X_n) \to E(X)$ as $n \to \infty$? The answer depends upon the mode of convergence. That is,

1. If $X_n \overset{\text{m.s.}}{\to} X$ and the expectations $E(X_n)$ and $E(X)$ exist, then $E(X_n) \to E(X)$ as $n \to \infty$.
2. If $X_n \overset{d}{\to} X$, it does not generally follow that $E(X_n) \to E(X)$. However, $X_n \overset{d}{\to} X$ if and only if $E[f(X_n)] \to E[f(X)]$ as $n \to \infty$ for all bounded continuous real-valued functions $f$. In particular, if $X_n \overset{d}{\to} X$ and $|X_n| < M < +\infty$, then $E(X)$ exists and $E(X_n) \to E(X)$ as $n \to \infty$.
3. If either $X_n \overset{\text{a.s.}}{\to} X$ or $X_n \overset{p}{\to} X$, it does not follow that $E(X_n) \to E(X)$ as $n \to \infty$ (even if we assume that $E(X_n)$ and $E(X)$ exist). However, if we introduce some auxiliary conditions into our discussion, then we can be assured that $E(X_n) \to E(X)$ as $n \to \infty$. To this end, let us restate, respectively, the Monotone Convergence Theorem and the Dominated Convergence Theorem (Section 1.12.5) in terms of random variables (since random variables are indeed real-valued functions). These restated theorems are Theorems 2.6.1 and 2.6.2, respectively.

**Theorem 2.6.1** Monotone Convergence Theorem for Random Variables
For all $\omega \in \Omega$, let $\{X_n\}_{n=1}^{\infty}$ be a nondecreasing sequence of nonnegative random variables $(0 \leq X_1 \leq X_2 \leq \cdots \leq X_n \leq \cdots)$. If $lim_{n \to \infty} X_n(\omega) = X(\omega) \in [0, +\infty]$ exists, then $E(X_n) \to E(X)$ as $n \to \infty$.

**Theorem 2.6.2**  Dominated Convergence Theorem for Random Variables

If either $X_n \overset{a.s.}{\to} X$ or $X_n \overset{P}{\to} X$, and all the random variables $X_n$ are dominated in absolute value by a positive random variable $Y$ having a finite expectation ($|X_n| \le Y$ for all $n$ and $E(Y) < +\infty$), then $E(X_n) \to E(X)$ as $n \to \infty$.

### 2.6.6  Convergence of Sequences of Events

Our objective in this section is to introduce a way of determining if a sequence of events $\{A_n\}_{n=1}^{\infty}$ in a probability space $(\Omega, \mathcal{A}, P)$ occurs infinitely often. To this end, for events $A_1, A_2, ...$, the event

$$A = \cap_{n=1}^{\infty} \cup_{m=n}^{\infty} A_m = \{\omega \in \Omega | \omega \in A_n \text{ infinitely often}\}$$

is termed "$A_n$ infinitely often" and written "$A_n$ i.o." In this regard, $A$ is the upper limit of $\{A_n\}$ or $A = \lim_{n \to \infty} \sup A_n$ and thus is a member of $\mathcal{A}$. Hence, $A$ is the collection of all those outcomes that appear in infinitely many events, or $A$ is the event that occurs if and only if an infinite number of events $A_n$ occur.

It should be apparent that $A = \lim_{n \to \infty} \sup A_n$ is a **tail event** of the sequence $\{A_n\}$. Such events are determined solely by the $A_n$ for arbitrarily large values of $n$. Moreover, the **tail $\sigma$-algebra** associated with the sequence $\{A_n\}$ will be denoted as $\sigma(A_n, n = 1, ..., +\infty)$, with $\mathcal{A} = \cap_{n=1}^{\infty} \sigma(A_n, n = 1, ..., +\infty)$. As we shall see later, the probability of a tail event is either 0 or 1, depending on whether the series of the constituent probabilities $\{P(A_n)\}_{n=1}^{\infty}$ converges or diverges.

Given these considerations, we now look to the Borel–Cantelli (BC) lemmas. In this regard, we start with the first BC lemma (BC1), Lemma 2.6.1.

**Lemma 2.6.1**  BC1

If $\{A_n\}_{n=1}^{\infty}$ is a sequence of events in a probability space $(\Omega, \mathcal{A}, P)$ and $\sum_{n=1}^{\infty} P(A_n) < +\infty$, then $P(A) = P(A_n \text{ i.o.}) = 0$.

That is, if $\sum_{n=1}^{\infty} P(A_n)$ converges, then the probability that $\omega \in \Omega$ belongs to infinitely many $A_n$'s is zero.

To state the second Borel–Cantelli lemma (BC2), we need to consider the notion of an independent sequence of events. Specifically, $\{A_n\}$ is an **independent sequence of events** if each of its finite subsequences is independent. This definition takes us to Lemma 2.6.2.

**Lemma 2.6.2**  BC2

If $\{A_n\}_{n=1}^{\infty}$ is an independent sequence of events in a probability space $(\Omega, \mathcal{A}, P)$ and $\sum_{n=1}^{\infty} P(A_n) = +\infty$, then $P(A) = P(A_n \text{ i.o.}) = 1$.

To summarize,

BC1: If the sum of the probabilities of $\{A_n\}$ is finite (i.e., $\sum_{n=1}^{\infty} P(A_n) < +\infty,$),
then the probability that infinitely many events occur is 0 or, stated alternatively, the set $\{\omega \in \Omega | \omega$ is repeated infinitely many times$\}$ occurs with probability zero.

BC2: If the sum of the probabilities of $A_n$ is infinite (i.e., $\sum_{n=1}^{\infty} P(A_n) = +\infty,$),
then the probability that infinitely many events occur is unity.

### 2.6.7 Applications of Convergence of Random Variables

To see how these various modes of convergence of sequences of random variables are utilized in practice, let us now turn to an assortment of theorems that relate to the sums or averages of the random variables $\{X_n\}_{n=1}^{\infty}$. We state first the following:

> **Strong Law of Large Numbers.** Let $\{X_n\}$ be a sequence of independent and identically distributed (i.i.d.) random variables defined on the same probability space $(\Omega, \mathcal{A}, P)$. Let $E(X_i) = \mu$ and $E|X_i| < +\infty$ for all $i \geq 1$. Define the sum $S_n = \sum_{i=1}^{n} X_i$ or, in terms of sample averages, $M_n = S_n/n$. Then, in an "almost sure sense,"
>
> a. $M_n \xrightarrow{\text{a.s.}} \mu$ or $P(lim_{n \to \infty} M_n = \mu) = 1;$
>    or, in a "mean-square sense,"
> b. $M_n \xrightarrow{\text{m.s.}} \mu$ or $lim_{n \to \infty} E(|M_n - \mu|^2) = 0.$

As part (a) reveals, the averages converge to a common mean $\mu$ with probability 1 as $n$ increases without bound. So for $n$ large enough, all of these averages will eventually be very close to $\mu$.

For the strong law of large numbers to hold, we require that $\{X_n\}$ converges a.s. or converges with probability 1 as $n \to \infty$. Let us now weaken this requirement and assume convergence in probability rather than a.s. convergence or convergence with probability 1. To this end, we have the following:

> **Weak Law of Large Numbers.** Let $\{X_n\}$ be a sequence of i.i.d. random variables defined on the same probability space $(\Omega, \mathcal{A}, P)$ with $E(X_i) = \mu$ for all $i \geq 1$. Then for every $\varepsilon > 0$, $lim_{n \to \infty} P(|M_n - \mu| \geq \varepsilon) = 0$.

(If we require that the random variables $X_n$ be only independent instead of i.i.d., then we need the additional restriction that $V(X_i) = \sigma^2 < +\infty$.) Thus, the averages $M_n$ converge in probability to $\mu$ as $n \to \infty$ or $M_n \xrightarrow{P} \mu$. So as $n$ becomes large, the random variables $X_n$ become more and more concentrated about a common mean value $\mu$; $M_n$ is with high probability very close to $\mu$.

We next examine a Theorem 2.7.1 that informs us that a large sum of i.i.d. random variables, appropriately normalized, has an approximate standard normal distribution. That is,

**Theorem 2.7.1**    Central Limit Theorem

Let $\{X_n\}$ be a sequence of i.i.d. random variables with $E(X_i) = \mu$, $0 < V(X_i) = \sigma^2 < +\infty$, and $M_n = S_n/n$. If $G_n(t)$ denotes the cumulative distribution function of $Z_n = \sqrt{n}(M_n - \mu)/\sigma$ and $G_n(t)$ represents the cumulative distribution function of the standard normal variate $Z$, then for any $t$, $-\infty < t < +\infty$,

$$\lim_{n \to \infty} G_n(t) = G(t) = \frac{1}{\sqrt{2\pi}} \int_{-\infty}^{t} e^{-y^2/2} dy,$$

that is, $\sqrt{n}(M_n - \mu)/\sigma$ has a limiting $N(y;0,1)$ distribution.

As this theorem indicates, $Z_n = (S_n - n\mu)/\sigma\sqrt{n} = \sqrt{n}(M_n - \mu)/\sigma$ converges in distribution to $Z \sim N(z;0,1)$ or $Z_n \xrightarrow{d} Z \sim N(z;0,1)$—the normalized average of $n$ i.i.d. random variables with finite variance tends in distribution to a normalized, normally distributed random variable.

An alternative expression for the preceding limit is for $a$, $b \in R(-\infty < a < b < +\infty)$,

$$\lim_{n \to \infty} P\left(a < \frac{S_n - n\mu}{\sigma/\sqrt{n}} < b\right) = \frac{1}{\sqrt{2\pi}} \int_{a}^{b} e^{-y^2/2} dy$$

$$= G(b) - G(a).$$

## 2.7    A Couple of Important Inequalities

Within the area of probability theory, two very useful inequalities emerge. First,

> **Markov's Inequality.** Let $X$ be a non-negative random variable (i.e., $P(X \geq 0) = 1$). Then for any positive $c \in R$,

$$P(X \geq c) \leq \frac{1}{c} E(X.)$$

Thus, the probability that $X$ takes on a value at least as large as $c$ is no more than the mean of $X$ divided by $c$. So if $E(X)$ is small, then it is unlikely that $X$ will be very large.

Next,

> **Chebyshev's Inequality.** Let $X$ be a random variable with mean $\mu$ and variance $\sigma^2$. Then for any positive $c \in R$,

$$P(|X - \mu| \geq c) \leq \frac{\sigma^2}{c^2}.$$

Thus, a random variable $X$ tends to be concentrated about its mean if the probability that $X$ is far away from its mean is small, that is, if the variance of a random variable $X$ is small, then it is unlikely that $X$ will be very far from its mean.

## Appendix 2.A    The Conditional Expectation *E(X|Y)*

Suppose $X$ is a random variable on the probability space $(\Omega, \mathcal{A}, P)$. We determined earlier that, for events $A, B \in \mathcal{A}$, the conditional probability of $A$ given $B$ was defined as $P(A|B) = P(A \cap B)/P(B), P(B) > 0$, where the reduced or effective probability space was denoted as $(\tilde{\Omega}, \tilde{\mathcal{A}}, \tilde{P})$, with probability measure $\tilde{P} = P/P(B)$. In light of this result, we can set

$$E(X|B) = \frac{1}{P(B)} \int_B X dP$$

for integrable $X$. Given this expression, let us now look to the calculation of $E(X|Y)$, where $X, Y$ are random variables defined on the same probability space $(\Omega, \mathcal{A}, P)$. That is, if a simple event $\omega \in \Omega$ is chosen and $Y(\omega)$ is determined, then, given $Y(\omega)$, we are interested in finding the value of $X(\omega)$.

Suppose $Y$ is discrete. Then we can define the expected value of $X$ with respect to the probability measure $P(\cdot|Y = y)$ as

$$E(X|Y = y) = \int_{\Omega} X(\omega) \, dP(\omega|Y = y)$$

$$= \frac{1}{P(\{Y = y\})} \int_{\{Y = y\}} X(\omega) \, dP(\omega).$$

If $X$ is $P$-integrable, then the preceding expression is defined for all $y$ such that $P(\{Y = y\}) > 0$. Clearly, $E(X|Y = y)$ is a particularization of the random variable $E(X|Y)$. Hence, our best estimate of $X$ will be the average value of $X$ over a partition of $\Omega$.

To this end, given that $Y$ is discrete, it induces the $\mathcal{A}$-partition of $\Omega, B_j = \{\omega | Y(\omega) = Y_j\}, j \geq 1$, where $Y(\omega) = \sum_{j \geq 1} y_j \chi_{B_j}(\omega)$ is a simple random variable defined on $(\Omega, \mathcal{A}, P)$ with $Y = y_j$ on $B_j$, and the $B_j$s are disjoint events satisfying $\cup_j B_j = \Omega$. Let $\sigma(B_j, j \geq 1)$ depict the $\sigma$-algebra generated by the partition $\{B_j\}, j \geq 1$, and define $E(X|Y)(\omega) = E(X|B_j)$ if $\omega \in B_j$ and $P(B_j) > 0$. If $P(\{Y = y_j\}) > 0$ for all $j \geq 1$, set

$$E(X|Y)(\omega) = \sum_{j \geq 1} E(X|Y = y_j) \chi_{\{Y = y_j\}}(\omega).$$

To summarize, given $\sigma(B_j, j \geq 1) = \sigma(Y)$, suppose

a. $E(X|\sigma(Y))$ is $\sigma(Y)$-measurable; and
b. for event $B \in \sigma(Y)$,

$$\int_B E(X|\sigma(Y))(\omega)\, dP(\omega) = \int_B X(\omega)\, dP(\omega).$$

Then the $\mathcal{A}(Y)$-measurable random variable $E(X|Y)$ can be viewed as $E(X|\sigma(Y)) = E(X|\mathcal{A}(Y))$—the conditional expectation of $X$ with respect to a $\sigma$-algebra.

In general, for $(\Omega, \mathcal{A}, P)$ a probability space, $X$ an integrable random variable on $\Omega$, and $\mathcal{U} \subset \mathcal{A}$ a sub-$\sigma$-algebra, the **conditional expectation** $E(X|\mathcal{U})$ is *any* random variable on $\Omega$ such that

a. $E(X|\mathcal{U})$ is $\mathcal{U}$-measurable; and
b. $\int_B E(X|\mathcal{U})\, dP = \int_B X\, dP$ for all events $B \in \mathcal{U}$.[6]

Moreover, if $Y$ is *any* $\mathcal{U}$-measurable random variable satisfying

$$\int_B Y\, dP = \int_B X\, dP \text{ for all } B \in \mathcal{U},$$

then $Y = E(X|\mathcal{U})$ a.s. in $(\Omega, \mathcal{A}, P)$. Any such $Y$ is termed a **version** of $E(X|\mathcal{U})$. If $\mathcal{U}$ is generated by a random variable $Y$ on $(\Omega, \mathcal{A}, P)$, simply insert $E(X|Y)$ in place of $E(X|\mathcal{U})$.

As the preceding discussion indicates, conditional expectations are not unique. But since any two conditional expectations of $X$ with respect to $\mathcal{U}$ are a.s. equal as random variables on $(\Omega, \mathcal{A}, P)$, we can, for all intents and purposes, view conditional expectations as unique. In fact, from this point on, all general results involving conditional expectations should be interpreted as holding a.s. Stated more formally, if $X$ is an integrable random variable on $(\Omega, \mathcal{A}, P)$, then for each sub-$\sigma$-algebra $\mathcal{U} \subset \mathcal{A}$, the conditional expectation $E(X|\mathcal{U})$ exists and is unique up to $\mathcal{U}$-measurable sets of probability zero. The justification for this assertion is Theorem 2.A.1.

**Theorem 2.A.1**  Radon–Nikodým Theorem
Let $v$ and $\mu$ be finite measures on the measurable space $(\Omega, \mathcal{A})$ with $v$ absolutely continuous with respect to $\mu$ (typically written $v < <\mu$). Then there is a non-negative $\mathcal{A}$-measurable function $f$ (unique up to $\mu$-null sets) on $\Omega$ such that

---

6 This equality is sometimes referred to as the **partial-averaging property** of a conditional expectation, that is, $E(X|\mathcal{U})$ and X render the same value when "averaged" over only that part of $\Omega$ containing the sets in the conditioning sub-$\sigma$-algebra $\mathcal{U}$.

$$v(A) = \int_A f d\mu \text{ for all } A \in \mathcal{A}.$$

If $h$ is another such function, then $f = h$ almost everywhere (a.e.).

The function $f$ usually appears as $dv/d\mu$ and is termed the **Radon–Nikodỳm derivative**.

To see how this theorem supports the existence of conditional expectation, suppose $X \geq 0$ with $\mu = P$ and let

$$v(A) = \int_A X dP \text{ for all } A \in \mathcal{U} \subset \mathcal{A}.$$

Since $v \ll \mu, dv/d\mu \in \mathcal{U}$ and for $A \in \mathcal{U}$,

$$v(A) = \int_A X dP = \int_A \frac{dv}{d\mu} dP.$$

For $A = \Omega, dv/d\mu \geq 0$ is integrable and consequently can be deemed a version of $E(X|\mathcal{U})$.

How should we interpret $E(X|\mathcal{U})$? Remember that our task is to obtain an estimate of the random variable $X$ from the information available in the sub-$\sigma$-algebra $\mathcal{U}$. Thus, the given condition (a) requires that $E(X|\mathcal{U})$ be constructed from knowledge about the sets in $\mathcal{U}$; and condition (b) requires that our estimate be consistent with $X$ when integration is taken over the sets in $\mathcal{U}$. Hence, we may regard $E(X|\mathcal{U})$ as the expectation of $X$ taken with respect to all possible information about $X$ that can be discerned from the events in $\mathcal{U}$. For instance, if $\mathcal{U}$ is generated by a countable partition $\{B_j\}, j \geq 1$, with $P(B_j) > 0$ for all $j$ and $X$ is a random variable on $(\Omega, \mathcal{A}, P)$, then for all $j$ and all $\omega \in B_j$, $E(X|\mathcal{U})(\omega) =$ average value of $X$ taken over all $B_j$. Now, if the number of $B_j$s is "large," then the countable partition of $\mathcal{U}$ enables us to obtain considerable information about $X$ from the $B_j$s. In this instance, the partial-averaging property taken over many sets in $\mathcal{U}$ tells us that $E(X|\mathcal{U})$ is a relatively "good" estimator for $X$. But if the partition of $\mathcal{U}$ has but a scant number of sets, then the aforementioned property relegates $E(X|\mathcal{U})$ to providing only a somewhat unrefined estimate of $X$.

Looking to the basic properties of $E(X|\mathcal{U})$, we assume that $X$ and $Y$ are integrable random variables on the probability space $(\Omega, \mathcal{A}, P)$ and that $\mathcal{U}$ and $\mathcal{G}$ are sub-$\sigma$-algebras of $\mathcal{A}$ with $\mathcal{G} \subset \mathcal{U} \subset \mathcal{A}$. Then the following results hold a.s.:

1. (**Linearity**) if $a$, $b$ are real numbers, then $E(aX + bY|\mathcal{U}) = aE(X|\mathcal{U}) + bE(Y|\mathcal{U})$.
2. If $X \geq 0$, then $E(X|\mathcal{U}) \geq 0$.
3. If $X = c = $ constant, then $E(X|\mathcal{U}) = c$.
4. If $\mathcal{U} = \{\emptyset, \Omega\}$ is the trivial $\sigma$-algebra, then $E(X|\mathcal{U}) = E(X)$.
5. If $X \leq Y$, then $E(X|\mathcal{U}) \leq E(Y|\mathcal{U})$.

6. If $X$ is $\mathcal{U}$-measurable and $XY$ is integrable, then $E(XY|\mathcal{U}) = XE(Y|\mathcal{U})$.
7. If $X$ is independent of $\mathcal{U}$, then $E(X|\mathcal{U}) = E(X)$.
8. (**Tower law**) $E(X|\mathcal{G}) = E(E(X|\mathcal{U})|\mathcal{G}) = E(E(X|\mathcal{G})|\mathcal{U})$.
9. If $X$ is $\mathcal{U}$-measurable, then $E(X|\mathcal{U}) = X$ and thus $E(E(X|\mathcal{U})) = E(X)$, that is, $E(X|\mathcal{U})$ is an unbiased estimator of $X$.
10. $E(E(X|\mathcal{U})Y|\mathcal{U}) = E(X|\mathcal{U}) \cdot E(Y|\mathcal{U})$.
11. If the random variables $X_n, n \geq 1$, are non-negative and the sequence $\{X_n, n \geq 1\}$ increases a.s. to $X$, then the sequence $\{E(X_n|\mathcal{U}), n \geq 1\}$ increases a.s. to $E(X|\mathcal{U})$.
12. (**Conditional Chebyshev's inequality**) for real $a$, $P(|X| \geq a|\mathcal{U}) \leq a^{-2} E(X^2|\mathcal{U})$.
13. (**Conditional Cauchy–Schwarz inequality**) $E(XY|\mathcal{U})^2 \leq E(X^2|\mathcal{U}) \cdot E(Y^2|\mathcal{U})$.
14. If $\mathcal{G} \subset \mathcal{U}$ and $E(X|\mathcal{U}) \in \mathcal{G}$, then $E(X|\mathcal{G}) = E(X|\mathcal{U})$.
15. If $X$ and $Y$ are random variables with $E(Y|\mathcal{U}) = X$ and $E(X^2) = E(Y^2) < +\infty$, then $X = Y$ a.s.
16. If $\mathcal{G} \subset \mathcal{U}$, $E(E(X|\mathcal{U})|\mathcal{G}) = E(X|\mathcal{G})$.
17. $|E(X|\mathcal{U})| \leq E(|X||\mathcal{U})$ and $|E(X|\mathcal{U})|^2 \leq E(|X|^2|\mathcal{U})$.
18. If $X_n \to X$ in $L^2$, then $E(X_n|\mathcal{U}) \to E(X|\mathcal{U})$ in $L^2$.

# 3

# Mathematical Foundations 3

Stochastic Processes, Martingales, and Brownian Motion

## 3.1 Stochastic Processes

Generally speaking, a stochastic process is a collection of random variables representing the set of random outcomes resulting from the evolution of some system over time. More formally, a **stochastic (random) process** is a parameterized collection of random variables $\{X_t\}_{t \in T}$ defined on a given probability space $(\Omega, \mathcal{A}, P)$, where $X_t$ assumes values (called **states**) within a **state space** $S$ for each $t$ in a **parameter space** (index set) $T$. (Note that an alternative notation often used is $\{X(t), t \in T\}$.) Hence, a stochastic process is a real-valued random function $X_t : T \to S$, where, in general, $S$ and $T$ are nonempty spaces on which are defined $\sigma$-algebras of measurable subsets of $S$ and $T$. More specifically, if $T$ is designated as **time space**, so that $X_t$ is the state of the process at time $t \in T$, then, if we work in **discrete** time, take $T = N$ (the non-negative integers); or if we work in **continuous** time, set $T = [0, +\infty)$. Moreover, the state space $S$ is either (1) countable ($N$), in which case $X(t), t \in T$, has a discrete distribution; or (2) an interval in $R$ (it is usually $R^+ = [0, +\infty)$), in which case $X(t), t \in T$, has a continuous distribution.

For example, a discrete-time, discrete-state process may involve flipping a fair coin successively to generate the set of outcomes $X_t, t = 1, 2, \ldots$. A continuous-time, discrete-state process can involve the number of vehicles $X_t$ passing through a tollgate over a given time period $[0, n]$. In addition, the values $X_t$ of the Dow Jones industrial average (DOW), recorded over a time interval $[0, n]$ taken from the beginning of trading to the end of trading on a particular day, describe a continuous-time, continuous-state process. Clearly, a continuous process is defined at all instants of time over $[0, n]$.

Let us be a bit more explicit in our notation. After all, a random process is actually defined on a probability space $(\Omega, \mathcal{A}, P)$. In this regard, we can express

*Stochastic Differential Equations: An Introduction with Applications in Population Dynamics Modeling*, First Edition. Michael J. Panik.
© 2017 John Wiley & Sons, Inc. Published 2017 by John Wiley & Sons, Inc.

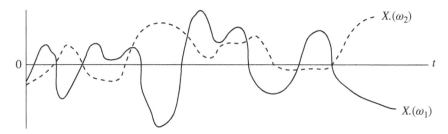

**Figure 3.1** Sample paths of the stochastic process $\{X_t\}_{t \in T}$.

the stochastic process as $X(t, \omega)$, a real-valued function of the pair $(t, \omega)$ from $T \times \Omega$ to $S$, or, alternatively, as $X_t(\omega)$. So for each fixed $t \in T$, we have a random variable $\omega \in \Omega \rightarrow X_t(\omega) \in R$ ( or $X(t, \cdot) \in R$); and for each fixed $\omega \in \Omega$, we have a real-valued function $t \in T \rightarrow X_t(\omega) \in R$ called the **sample path** of the process. The sample path is often denoted as $X.(t)$ or $X(\cdot, \omega)$. If we think of $t$ as time and $\omega$ as, say, a treatment level of an ongoing experiment (or process of observation), then $X_t(\omega)$ can be viewed as the outcome at time $t$ of the experimental level $\omega$. Thus, for each fixed $\omega = \omega_1 \in \Omega$, sample path $X.(\omega_1)$ tracks the random outcomes of an experiment as time evolves. In addition, if we rerun the experiment by picking a different $\omega = \omega_2 \in \Omega$, we get a different sample path $X.(\omega_2)$ (Figure 3.1).

Let $\{X_t\}_{t \in T}$ and $\{Y_t\}_{t \in T}$ be stochastic processes defined on the same probability space $(\Omega, \mathcal{A}, P)$. These processes are said to be **stochastically equivalent** or **indistinguishable** if for all $t \in T$, $X_t = Y_t$ almost surely (a.s.), that is, $P(\{\omega | X_t(\omega) = Y_t(\omega)\}) = 1$. Hence, $\{X_t\}_{t \in T}$ is termed a **version** or **modification** of $\{Y_t\}_{t \in T}$ and vice versa. In particular, the finite-dimensional distributions of these two processes coincide.

### 3.1.1 Finite-Dimensional Distributions of a Stochastic Process

For a stochastic process $\{X_t\}_{t \in T}$ defined on a probability space $(\Omega, \mathcal{A}, P)$, the joint distribution of the random vector $(X(t_1), ..., X(t_n))$ for finitely many time points $t_1 < \cdots < t_n$ is called the **finite-dimensional distribution** of the process. Specifically, the **finite-dimensional distributions** of $\{X_t\}_{t \in T}$ are given by

$$
\begin{aligned}
F_{t_1}(x_1) &= P(X(t_1) \leq x_1) \\
F_{t_1, t_2}(x_1, x_2) &= P(X(t_1) \leq x_1, X(t_2) \leq x_2); ...; \\
F_{t_1, ..., t_n}(x_1, ..., x_n) &= P(X(t_1) \leq x_1, ..., X(t_n) \leq x_n),
\end{aligned}
\tag{3.1}
$$

where $t_i \in T$, $x_i \in R$, and $n \geq 1$. Hence, a stochastic process defined on a probability space has associated with it a set of finite-dimensional distributions.

Why are such distributions important? To fully comprehend the behavior of a stochastic process $\{X_t\}_{t \in T}$, we need to obtain information about the family of finite-dimensional distributions that it defines. For instance, if we know the distribution of $(X(t_1), ..., x(t_n))$ for every choice of, say, $n \in N_+$ and $(t_1,...,t_n) \in T^n$, then, ostensibly, we can determine $P((X(t_1),...,X(t_n)) \in A)$ for every measurable set $A \subseteq S^n$.

Looked at from the converse perspective, we might be interested in constructing random processes with specified finite-dimensional distributions. However, certain random processes might not be readily constructed from a given sequence $\{X_i, i = 1,...,n\}$ of random variables. In this situation, we need to focus on whether or not such processes can even exist. Luckily, the question of the existence of stochastic processes exhibiting certain prespecified distributional characteristics has been settled by Kolmogorov (1933) (see Baudoin (2010), Billingsley (2012), Øksendal (2013), and Daniell (1918)). Specifically, Kolmogorov demonstrates that if we specify the finite-dimensional distributions in a *consistent* fashion, then there exists a stochastic process $\{X_t\}_{t \in T}$ defined on a compatible probability space that possesses the specified finite-dimensional distributions.

What are the consistency requirements for the finite-dimensional distributions? The said system of distributions is taken to be **consistent** if the following two conditions hold:

a. If $\{\pi(1),...,\pi(n)\}$ is a permutation of the numbers $1,...,n$, then,

for arbitrary time points $t_1, ..., t_n \in T$ and $n \geq 1$,

$$F_{t_{\pi(1)},...,t_{\pi(n)}}\left(x_{\pi(1)},...,x_{\pi(n)}\right) = F_{t_1,...,t_n}(x_1,...,x_n).$$

b. For $m < n$ and arbitrary $t_{m+1},...,t_n \in T$,

$$F_{t_1,...,t_m,t_{m+1},...,t_n}(x_1,...,x_m,\infty,...,\infty) = F_{t_1,...,t_m}(x_1,...,x_m).$$

$$(3.2)$$

(Note that (3.2a) holds because $(x_1, ..., x_n)$ and $(x_{\pi(1)}, ..., x_{\pi(n)})$ represent the same $n$-tuple of sample values.) Intuitively, the finite-dimensional distributions of a stochastic process $\{X_t\}_{t \in T}$ are consistent in the sense that for any two sets of time points, with one being a subset of the other, the joint distribution taken over the smaller set of time points is the restriction of the one defined over the larger set of time points.

So as alluded to earlier, the justification for the existence of a stochastic process with given finite-dimensional distributions is provided by Theorem 3.1.

**Theorem 3.1**   Kolmogorov Existence Theorem

If $\mathcal{P}$ is a family of finite-dimensional distributions satisfying the consistency conditions (3.2a and b), then there exists a probability space $(\Omega, \mathcal{A}, P)$ and a stochastic process $\{X_t\}_{t \in T}$ defined on that space such that $\mathcal{P}$ is the collection of finite-dimensional distributions of $\{X_t\}_{t \in T}$.[1]

### 3.1.2 Selected Characteristics of Stochastic Processes

Let $\{X_t\}_{t \in T}$ be a stochastic process defined on a probability space $(\Omega, \mathcal{A}, P)$. This process is as follows:

a. **Continuous** if for almost all $\omega \in \Omega$, the function $t \in T \rightarrow X_t(\omega)$ is continuous. Thus the process is continuous if all its paths are continuous. Moreover, it is **right-continuous** (resp. **left-continuous**) if $X_t(\omega)$ is right-continuous (resp. left-continuous) for almost all $\omega \in \Omega$ and $t \in T$. Thus, all paths of a right-continuous (resp. left-continuous) process are right-continuous (resp. left-continuous).

b. **Integrable** if, for every $t \geq 0, X_t$ is an integrable random variable.

c. **Measurable** if the mapping $(t, \omega) \rightarrow X_t(\omega)$ is measurable with respect to the $\sigma$-algebra $\mathcal{B}(T) \times \mathcal{A}$, where $\mathcal{B}(T)$ is the family of all Borel subsets of $T$.

d. **Square-integrable** if $E|X_t|^2 < +\infty$ for every $t \geq 0$.

e. **Increasing** if $X_s \leq X_t$ for any $s < t$. Thus, a process is increasing if all its paths are increasing.

f. **(Strictly) stationary** if its finite-dimensional distributions are invariant under time displacements, that is, for $t_i, t_{i+h} \in [t_0, n]$ and $h \in R$, $F_{t_1+h,\ldots,t_n+h}(x_1,\ldots,x_n) = F_{t_1,\ldots,t_n}(x_1,\ldots,x_n)$.

   That is, the random variables $X(t_1), \ldots, X(t_n)$ have the same joint probability distribution as the random variables $X(t_1 + h), \ldots, X(t_n + h)$ for any positive integer $n$ and all $h \in T$ so that the probability structure of $\{X_t\}_{t \in T}$ is invariant under parameter translation.

g. **Gaussian** if its finite-dimensional distributions are normal distributions.

h. **Continuous in probability** if for any $s \in T$ and $\varepsilon > 0, P(|X_t - X_s| > \varepsilon) \rightarrow 0$ if $t \in T, t \rightarrow s$.

i. A **quadratic variation process** if $X_t(\omega) : \Omega \rightarrow R$ is continuous and the **second variation** of the process, $(X, X)_t^2$, is defined as

$$(X, X)_t^2(\omega) = \lim_{\Delta t_k \rightarrow 0} \sum_{t_k \leq t} |X_{t_{k+1}}(\omega) - X_{t_k}(\omega)|^2,$$

where $0 = t_1 < t_2 < \cdots < t_n = t$ and $\Delta t_k = t_{k+1} - t_k$.

An additional feature is worth mentioning. Specifically, a stochastic process $\{X_t\}_{t \in T}$ is termed a **Markov process** if it exhibits the so-called **Markov property**—the change at time $t$ is determined only by the value of the process at

---

1 An alternative (equivalent) statement of this theorem, specified in terms of probability measures defined on the Borel $\sigma$-algebra on $R$, appears in Appendix 3.A.

time $t$ and not by the values at any times before $t$. Under this property, to make predictions of the future behavior (outcome) of a system, it suffices to consider only the present state of the system and not its past performance or history. Thus, the current state of a system is critical, but how it arrived at that state is irrelevant. Thus, a Markov process "has no memory" (more on this notion in Section 3.5.2).

### 3.1.3 Filtrations of $\mathcal{A}$

Let $(\Omega, \mathcal{A}, P)$ be a probability space and, for $T = N$ or $T = [0, +\infty)$, let $\mathcal{A}_t = \sigma(X_s, s \in T, s \leq t)$ depict a $\sigma$-algebra of events defined in terms of the outcomes of a stochastic process up to time $t$, that is, $\mathcal{A}_t$ represents the history of the process up to time $t$. Hence, the $\sigma$-algebra $\mathcal{A}_t$ classifies all the information about the process available at time $t \in T$ and, for a given event $A \in \mathcal{A}_t$, observing the process up to time $t$ enables us to determine if that event has or has not occurred. We shall define $\mathcal{A}_\infty = \sigma(\cup_{t \in T} \mathcal{A}_t)$—it is the $\sigma$-algebra generated by $\cup_{t \in T} \mathcal{A}_t$ that encodes our information over all time points $t$. Clearly, $\mathcal{A}_t \subseteq \mathcal{A}_\infty, t \in T$.

A **filtration** on $(\Omega, \mathcal{A}, P)$ is an increasing family $\mathfrak{I} = \{\mathcal{A}_t\}_{t \in T}$ of sub-$\sigma$-algebras of $\mathcal{A}$—it enables us to assess how the process information is arranged (grouped) into a family of $\sigma$-algebras. So for $s, t \in T$, if $\mathcal{A}_s$ and $\mathcal{A}_t$ are $\sigma$-algebras included in $\mathcal{A}$ and if $0 \leq s \leq t < +\infty$, then $\mathcal{A}_s \subseteq \mathcal{A}_t \subseteq \mathcal{A}$. (Observe that a filtration on $\mathcal{A}$ is bounded above by $\mathcal{A}$.) To elaborate, if $\mathcal{A}_s$ and $\mathcal{A}_t$ are two $\sigma$-algebras on $(\Omega, \mathcal{A}, P)$, then $\mathcal{A}_s \subset \mathcal{A}_t$ if and only if $\mathcal{A}_t$ contains more information about the process than does $\mathcal{A}_s$; the $\mathcal{A}_s$ information set is a subset of the $\mathcal{A}_t$ information set. (Remember that for a discrete-time filtration, $T = N$; and for a continuous-time filtration, $T$ is an interval in $R$.)

A probability space $(\Omega, \mathcal{A}, P)$ endowed with a filtration $\mathfrak{I}$ is termed a **filtered probability space** and denoted $(\Omega, \mathcal{A}, \mathfrak{I}, P)$. Also, a stochastic process $\{X_t\}_{t \in T}$ on $(\Omega, \mathcal{A}, P)$ is said to be **adapted** to the filtration $\mathfrak{I}$ if, for each $t \in T, X_t$ is $\mathcal{A}_t$-measurable (the information at time $t$ carried by $\mathcal{A}_t$ determines the value of the random variable $X_t$). Equivalently, $\{X_t\}_{t \in T}$ is adapted to $\mathfrak{I}$ if, for each $t \in T, \sigma(X_s, s \in T, s \leq t) \subseteq \mathcal{A}_t$. Sometimes, an adapted process is called a **nonanticipating process**—a process that "cannot see into the future." Thus, $\{X_t\}_{t \in T}$ is $\mathcal{A}_t$-**adapted** if and only if, for each $t \in T, X_t$ is known at time $t$ when the information carried by $\mathcal{A}_t$ is known. Given that the filtration $\mathfrak{I}$ represents the evolution of knowledge about some random system through time, it follows that if $\{X_t\}_{t \in T}$ is an adapted process, then $X_t$ depends only on the systems behavior prior to time $t$, with no prior information lost.

Suppose that the expected value of a random process $\{X_t\}_{t \in T}$ is conditioned on the history of the process up to time $s < t, E(X_t | \mathcal{A}_s)$. If $E(X_t | \mathcal{A}_s)$ is, in turn, conditioned on the process history up to time $r < s$, then the outcome is the same as if we had conditioned on the earlier time $r$ to begin with. That is, the **tower**

**property** holds or $E(E(X_t|\mathcal{A}_s)|\mathcal{A}_r) = E(X_t|\mathcal{A}_r), r < s < t$. This is because the $\mathcal{A}_r \subset \mathcal{A}_s$ or there is less information carried by $\mathcal{A}_r$ relative to that carried by $\mathcal{A}_s$.

A stochastic process $\{X_t\}_{t \in T}$ is always adapted to its **natural filtration** $\mathcal{A}_t^X = \sigma(X_s, s \leq t)$—the filtration where $\mathcal{A}_t$ is generated by all values of $X_s$ up to time $s = t$. Since $\mathcal{A}_t$ is the smallest $\sigma$-algebra for which all variables $X_s, s \leq t$, are measurable, it follows that $F^X = \{\mathcal{A}_t^X\}_{t \in T}$ is the smallest filtration to which $\{X_t\}_{t \in T}$ is adapted.

What is the importance of filtrations? The information carried by a stochastic process that evolves over time is modeled by filtrations. If the system is observed up to time $t$, then one is able to determine if an event $A$ has occurred, that is, if $A \in \sigma(X_s, s \leq t)$, where $\sigma(X_s, s \leq t)$ is the smallest $\sigma$-algebra for which all the random variables $X(t_1), ..., X(t_n), 0 \leq t_1 \leq \cdots \leq t_n \leq t$, are measurable. In fact, if $\{X_t\}_{t \in T}$ is a stochastic process defined on $(\Omega, \mathcal{A}, P)$, then $\mathcal{A}_t = \sigma(\mathcal{A}_s, s \leq t)$ is itself a filtration.

Suppose $(\Omega, \mathcal{A}, \mathcal{I}, P)$ is a filtered probability space. The filtration $\mathcal{I} = \{\mathcal{A}_t\}_{t \in T}$ is **complete** if the probability space $(\Omega, \mathcal{A}, P)$ is complete[2] and if $\mathcal{A}_0$ (the collection of all subsets of $\mathcal{A}$ that have a zero probability) contains all the $P$-null sets (i.e., if $A \in \mathcal{A}$ and $P(A) = 0$, then $A \in \mathcal{A}_0$).[3] Moreover, the filtration $\mathcal{I}$ is said to be **right-continuous** if $\mathcal{A}_t = \mathcal{A}_{t+}$, where $\mathcal{A}_{t+} = \bigcap_2 \{\mathcal{A}_s, s \in T, s > t\}$ for all $t \in T$. Obviously, $\mathcal{A}_{t+}$ is also a filtration and $\mathcal{A}_t \subseteq \mathcal{A}_{t+}, t \in T$.

When the probability space $(\Omega, \mathcal{A}, P)$ is complete in that $\mathcal{A}$ contains all $P$-null sets (so that the filtration $\mathcal{I}$ is complete), then the filtration $\mathcal{I}$ is said to satisfy the **usual conditions** if it is right-continuous and $\mathcal{A}_0$ contains all $P$-null sets. In what follows, we shall always work on a given complete probability space $(\Omega, \mathcal{A}, P)$ with a filtration $\mathcal{I} = \{\mathcal{A}_t\}_{t \in T}$ satisfying the *usual conditions*.[4]

To reflect the dynamic aspects of a filtration, the concept of progressive measurability is needed. Specifically, a continuous stochastic process $\{X_t\}_{t \geq 0}$ that is adapted to a filtration $\mathcal{I} = \{\mathcal{A}_t\}_{t \geq 0}$ is said to be **progressively measurable** with respect to $\mathcal{I}$ if for every $t \in T = [0, +\infty)$, and for all events $A \in \mathcal{B}(R)$, $\{(s, \omega) \in T \times \Omega, X_s(\omega) \in A\} \in \mathcal{B}(T) \times \mathcal{A}_t$. While a process may be adapted but not progressively measurable, an adapted and continuous stochastic process is inherently progressively measurable. Thus, a continuous stochastic process that is adapted to a filtration $\mathcal{I}$ is also progressively measurable with respect to $\mathcal{I}$.

---

2 In general, a class $\mathcal{A}$ of subsets of $\Omega$ is **complete with respect to a measure** $\mu$ if $E \subset F, F \in \mathcal{A}$, and $\mu(F) = 0$ implies $E \in \mathcal{A}$.

3 A set $A \in \Omega$ is termed a ***P*-null set** if there exists a set $B \in \mathcal{A}$ with $A \subset B$ and $P(B) = 0$.

4 This assumption is not all that restrictive since any filtration can be made complete and right-continuous. Given a filtered probability space $(\Omega, \mathcal{A}, \mathcal{I}, P)$, we first complete the probability space $(\Omega, \mathcal{A}, P)$ and then add to every $_{t+}, t \in T$, all the $P$-null sets. This new filtration, called the **augmentation of** $\mathcal{I}$, also satisfies the *usual conditions*. Typically, $\mathcal{I}$ is right-continuous from the start.

## 3.2 Martingales

### 3.2.1 Discrete-Time Martingales

Suppose $\mathcal{F} = \{\mathcal{A}_t\}_{t \in T}$ is a filtration on a probability space $(\Omega, \mathcal{A}, P)$. Generally speaking, a **martingale** on a filtered probability space $(\Omega, \mathcal{A}, \mathcal{F}, P)$ is a sequence $\{X_t\}_{t \in T}$ of integrable random variables on $(\Omega, \mathcal{A}, P)$ that are adapted to the filtration $\mathcal{F}$. More specifically, for $\{X_t\}_{t=1}^{\infty}$ a sequence of random variables on a probability space $(\Omega, \mathcal{A}, P)$ and with $\mathcal{F} = \{\mathcal{A}_t\}_{t=1}^{\infty}$ a sequence of $\sigma$-algebras in $\mathcal{A}$, the sequence $\{X_t\}_{t \in T}$ is a **discrete-time martingale** if

a. $\mathcal{A}_t \subset \mathcal{A}_{t+1}$ (the $\mathcal{A}_t$ form a filtration $\mathcal{F} = \{\mathcal{A}_t\}_{t=1}^{\infty}$);
b. $X_t$ is $\mathcal{A}_t$-measurable (the $X_t$'s are adapted to the filtration);
c. $E(|X_t|) < +\infty$ (the $X_t$'s are integrable); and
d. $E(X_{t+1}|\mathcal{A}_t) = X_t$ a.s. for all $t \geq 1$ (the **martingale property**).[5]

Clearly, the sequence of random variables $\{X_t\}_{t=1}^{\infty}$ is defined as a martingale relative to the sequence $\mathcal{F} = \{\mathcal{A}_t\}_{t=1}^{\infty}$ of $\sigma$-algebras, that is, $X$ is a martingale with respect to the $\mathcal{A}_t$'s.

As the martingale property indicates, the average value of $X_{t+1}$ taken over $\mathcal{A}_t$ is just $X_t$. Thus, a "fair" or "unbiased" estimate of $X_{t+1}$ relative to $\mathcal{A}_t$ is $X_t$; $X_t$ is the best prediction the process can make given the information contained in $\mathcal{A}_t$ up to time $t$. (For example, if $X_t$ represents the fortune of a gambler after the $t$th play of a game of chance and if $\mathcal{A}_t$ represents the information about the game at that time, then this property tells the gambler that the expected fortune after the next play is the same as the current fortune. Hence, a martingale $X_t$ describes the fortune at time $t$ of a gambler who is betting on a fair game.)

Remember that the sequence $\{X_t\}_{t=1}^{\infty}$ of random variables is defined to be a martingale with respect to the filtration $\mathcal{F} = \{\mathcal{A}_t\}_{t=1}^{\infty}$. When the filtration is not mentioned or specified, we may take as the $\sigma$-algebra the sets $\mathcal{G}_t = \sigma(X_1,...,X_t)$—$\mathcal{G}_t$ is the $\sigma$-algebra generated by $X_t$. In fact, $\mathcal{G}$ is the smallest $\sigma$-algebra on $\Omega$ for which $X_t$ is measurable. Since $\mathcal{G}_t \subset \mathcal{G}_{t+1}$ (the $\mathcal{G}_t$'s form a filtration), $X_t$ is $\mathcal{G}_t$-measurable (the $X_t$'s are adapted to the filtration), and $\mathcal{A}(X_t) \subset \mathcal{A}_t \subset \mathcal{A}$, it follows that $\mathcal{G}_t \subset \mathcal{A}_t$ so that $\{\mathcal{G}_t\}_{t=1}^{\infty}$ is a filtration on $(\Omega, \mathcal{A}, P)$. Furthermore, if the martingale property $E(X_{t+1}|\mathcal{A}_t) = X_t$ holds, then $E(X_{t+1}|\mathcal{G}_t) = E(E(X_{t+1}|\mathcal{A}_t)|\mathcal{G}_t) = E(X_{t+1}|\mathcal{G}_t) = X_t$ a.s. So for $\mathcal{G}_t = \sigma(X_1,...,X_t)$, we can write $E(X_{t+1}|\mathcal{A}_t)E(X_{t+1}|X_1,...,X_t) = X_t$. So with $\{X_t\}_{t=1}^{\infty}$ adapted to $\{\mathcal{G}_t\}_{t=1}^{\infty}$, it follows that $\{X_t\}_{t=1}^{\infty}$ is a martingale on the filtered probability space $(\Omega, \mathcal{A}, \{\mathcal{G}_t\}_{t=1}^{\infty}, P)$.

---

5 Sometimes, this property appears as $E(X_t|\mathcal{A}_s) = X_s$ a.s. for all $1 \leq s < t < +\infty$. That is, if $\mathcal{A}_s$ is our latest information set, then $X_s$ is known and thus constitutes the best forecast of future values of $X_t$.

The following few related points merit our attention:

1. Under the martingale property $X_t$ is a version of $E(X_{t+1}|A_t)$ and, with $X_t$ being $A_t$-measurable, the martingale property can be written

$$\int_A X_{t+1}\, dP = \int_A X_t\, dP, A \in A_t.$$

2. Suppose $\{X_t\}_{t=1}^{\infty}$ is a sequence of independent integrable random variables on the probable space $(\Omega, A, P)$ with $E(X_t) = 0$ for all $t$. If $A_t = \sigma(X_1,...,X_t)$, then $A_t \subset A$ for all $t$; thus, $\{A_t\}_{t=1}^{\infty}$ is a filtration on $(\Omega, A, P)$. Let $Y_t - \sum_{i=1}^{t} X_i$ for all $t$. Since $|Y_t| \le \sum_{i=1}^{t} |X_i|$ and each $X_i$ is integrable and $A_t$-measurable, it follows that $Y_t$ is integrable and $A_t$-measurable. For $i \le t, X_{t+1}$ is independent of both $X_i$ and $A_t$ so that $E(X_{t+1}|A_t) = E(X_{t+1}) = 0$. Since $Y_{t+1} = Y_t + X_{t+1}, E(Y_{t+1}|A_t) = E(Y_t + X_{t+1}|A_t) = E(Y_t|A_t) + E(X_{t+1}|A_t) = Y_t + 0 = Y_t$ so that $\{Y_t\}_{t=1}^{\infty}$ is a martingale adapted to the filtration $\{A_t\}_{t=1}^{\infty}$. Thus, the martingale $\{Y_t\}_{t+1}^{\infty}$ can be viewed as a representation of a sequence of independent integrable random variables on a probability space $(\Omega, A, P)$.

3. Given that $A_t \subset A_{t+1} \subset \cdots \subset A_{t+j}, j \ge 1$, for event $A \in A_t$, $\int_A X_t\, dP = \int_A X_{t+1}\, dP = \cdots = \int_A X_{t+j}\, dP$. With $X_t$ being $A_t$-measurable, $X_t$ is a version of $E(X_{t+j}|A_t)$ so that $E(X_{t+j}|A_t) = X_t$ a.s. For $A = \Omega$, $E(X_t) = \int_\Omega X_t\, dP = \int_\Omega E(X_{t+j}|A_t)\, dP = \int_\Omega X_{t+j}\, dP = E(X_{t+j}), j \ge 1$. Hence, if $\{X_t\}_{t=1}^{\infty}$ is a martingale on a filtered probability space, then $E(X_t) = E(X_{t+1}) = E(X_{t+2}) = \cdots$.

4. Suppose $X$ is an integrable random variable on a probability space $(\Omega, A, P)$ and the $A_t$'s are nested $\sigma$-algebras in $A$ (i.e., $A_t \subset A_{t+1} \subset \cdots$). If $X_t = E(X_t|A_t)$, then clearly parts (a)–(c) of the definition of a discrete-time martingale are satisfied. In addition, since $E(X_{t+1}|A_t) = E(E(X|A_{t+1})|A_t) = E(X|A_t) = X_t$, it follows that $\{X_t\}_{t=1}^{\infty}$ is a discrete-time martingale adapted to the filtration $\{A_t\}_{t=1}^{\infty}$.

5. The process $\{\tilde{X}_t, 0 \le t < +\infty\}$ defined by setting $\tilde{X}_0 = X_0$ with $\tilde{X}_t = X_0 + A_1(X_1 - X_0) + A_2(X_2 - X_1) + \cdots + A_n(X_t - X_{t-1}), t \ge 1$, is termed the **martingale transform** of $\{X_t\}_{t\in T}$ **by** $\{A_t\}_{t\in T}$. Clearly, this transform represents a general device for constructing new martingales from old ones, that is, the transform of a martingale is also a martingale. More formally, we have Theorem 3.2.

**Theorem 3.2** Martingale Transform Theorem
Let $\{X_t\}_{t\in T}$ be a martingale adapted to the filtration $\mathcal{I} = \{A_t\}_{t\in T}$ with $\{A_t, 1 \le t < +\infty\}$ a sequence of bounded random variables that are

nonanticipating with respect to $\mathcal{I}$. Then the sequence of martingale transforms $\{\tilde{X}_t\}_{t\in T}$ is itself a martingale adapted to $\mathcal{I}$.

If in the martingale property $E(X_{t+1}|\mathcal{A}_t) = X_t$ "=" is replaced by "$\leq$" or "$\geq$," then $\{X_t\}_{t=1}^{\infty}$ is called **a supermartingale** or **submartingale**. In this regard, a stochastic process $\{X_t\}_{t=1}^{\infty}$ that is adapted to the filtration $\mathcal{I} = \{\mathcal{A}_t\}_{t=1}^{\infty}$ and such that $\{-X_t\}_{t=1}^{\infty}$ is a submartingale is termed a supermartingale. A stochastic process $\{X_t\}_{t=1}^{\infty}$ adapted to $\mathcal{I}$ that is simultaneously a submartingale and a supermartingale must obviously be a martingale.

A concept closely allied with a martingale is that of a stopping time. Specifically, a random variable $\tau : \Omega \to T = N \cup \{+\infty\}$ is termed a **stopping time** for the filtration $\mathcal{I} = \{\mathcal{A}_t\}_{t\in T}$ if $\{w|\tau(w) \leq t\} \in \mathcal{A}_t$ for all $0 \leq t < +\infty$. That is, think of a stopping time as a random variable that posits a criterion for determining whether or not $\tau \leq t$ has occurred on the basis of information about $\mathcal{A}_t$. For instance, we may view $\tau$ as a rule for stopping play when engaging in a gambling game; it is independent of the outcome of a play of a game that has yet to be undertaken.

Oftentimes, we are interested in the behavior of a random process $\{Y_t\}_{t\in T}$ exactly at the stopping time $\tau$. If $\tau < +\infty$ a.s., then the **stopped process** $Y_\tau$ is defined by setting $Y_\tau = \sum_{k=0}^{\infty} \chi_{(\tau = k)} Y_k$. We note briefly that if $\{X_t\}_{t\in T}$ is a martingale adapted to $\mathcal{I}$, then the stopped process $\{X_{min(t,\tau)}\}_{t\in T}$ is also a martingale adapted to $\mathcal{I}$.

### 3.2.1.1 Discrete-Time Martingale Convergence

We noted previously in Section 3.1 that if $\{X_t\}_{t=1}^{\infty}$ is a martingale on the probability space $(\Omega, \mathcal{A}, P)$ and adapted to the filtration $\mathcal{I} = \{\mathcal{A}_t\}_{t=1}^{\infty}$, then $\{X_t(\omega)\}_{t=1}^{\infty}$ is a sample path for each $\omega \in \Omega$. A question that naturally arises is "Does the sequence of sample paths become less erratic over time?" That is, "Does $lim_{t\to\infty} X_t(\omega) = Z$ exist, where $Z$ is some long-run stable value?"

To answer these questions, let us first develop some conceptual material. To this end, suppose $L^1(\Omega, \mathcal{A}, P)$ denotes the space of integrable random variables (see Section 1.11.3 and Equations (1.20) and (1.20.1)). Next, a collection of integrable random variables $\{X_k\}_{k\in\mathcal{K}}$ is $L^1$-**bounded** if $sup_{k\in\mathcal{K}} E(|X_k|) < +\infty$. Finally, if $\{X_t\}_{t=1}^{\infty}$ is a sequence of integrable random variables and there exists an integrable random variable $Z$ such that $lim_{t\to\infty}(|X_t - Z|) = 0$, then the sequence $\{X_t\}_{t=1}^{\infty}$ converges to $Z$ in $L^1(\Omega, \mathcal{A}, P)$. Based on these considerations, we can now state Theorem 3.3.

**Theorem 3.3** If $\{X_t\}_{t=1}^{\infty}$ is an $L^1$-bounded martingale on a probability space $(\Omega, \mathcal{A}, P)$, then there exists an integrable random variable $Z$ such that $lim_{t\to\infty} X_t = Z$ a.s.

To strengthen the conditions for convergence of a martingale sequence $\{X_t\}_{t=1}^{\infty}$, let us consider the concept of uniform integrability. (Uniform integrability is used to prove convergence in $L^1$.) Specifically, a sequence of integrable random variables $\{X_t\}_{t=1}^{\infty}$ on a probability space $(\Omega, \mathcal{A}, P)$ is **uniformly integrable** if either

a. $\lim\limits_{m\to\infty} \left( \sup\limits_t E(|X_t|); |X_t| > m \right) = \lim\limits_{m\to\infty} \left( \sup\limits_t \int_{\{|X_t|>m\}} |X_t|\, dP \right) = 0;^6$ or

b. $\{X_t\}_{t=1}^{\infty}$ is bounded on $L^1(\Omega, \mathcal{A}, P)$ and for all $\varepsilon > 0$ there exists a $\delta > 0$

such that, for an event $A \in \mathcal{A}, P(A) < \delta$ implies $\int_A |X_t|\, dP < \varepsilon$ for all $t$.

$$(3.3)$$

(As an example, think of a collection of random variables $X_t$ dominated by an integrable random variable $Y$, that is, $|X_t| < Y$, where $E(Y) < +\infty$.) So if uniform integrability is effected, then Theorem 3.3 can be strengthened a bit, as Theorem 3.4 reveals.

**Theorem 3.4** Let $\{X_t\}_{t=1}^{\infty}$ denote a martingale on the probability space $(\Omega, \mathcal{A}, P)$ that is adapted to the filtration $\{\mathcal{A}_t\}_{t=1}^{\infty}$. If $\{X_t\}_{t=1}^{\infty}$ is uniformly integrable, then (1) there exists an integrable random variable $Z$ on $(\Omega, \mathcal{A}, P)$ such that $X_t \to Z$ a.s. in $L^1(\Omega, \mathcal{A}, P)$ as $t \to \infty$; and (2) $X_m = E(Z|\mathcal{A}_m)$ a.s. in $(\Omega, \mathcal{A}, P)$.[7]

If $X$ is an integrable random variable, then a characteristic of the conditional expectation of $X$ is provided by Theorem 3.6.

**Theorem 3.6** If $(\Omega, \mathcal{A}_0, P)$ is a probability space and $X \in L^1$, then $\{E(X|\mathcal{A})|\mathcal{A}$ a $\sigma$-algebra $\mathcal{A}_o\}$ is uniformly integrable. More broadly, if $X \in L^1$, and the $\mathcal{A}_t$ are arbitrary $\sigma$-algebras within $\mathcal{A}_o$, then the random variables $E(X|\mathcal{A}_t)$ are uniformly integrable.

Additional equivalent convergence results for a martingale sequence $\{X_t\}_{t=1}^{\infty}$ are the following:

i. If $X_t \to X$ in $L^1$, then $X_t = \{X|\mathcal{A}_t\}$.
ii. $X_t$ is uniformly integrable.

---

6 Note that the sequence of sets $\{|X_t| > m\}_{m=1}^{\infty}$ decreases to $\emptyset$. Also, an $m$ chosen large enough renders (a) to $\sup_t E(|X_t|) \le m + 1 < +\infty$.

7 The concept of uniform integrability is also useful when discussing the convergence of sequences of random variables that are not necessarily martingales, for example,

**Theorem 3.5** Let $\{X_t\}_{t\in T}$ be a sequence of integrable random variables with $Z$ an integrable random variable. Then the sequence $\{X_t\}_{t\in T}$ converges to $Z$ in $L^1$ (i.e., $\lim_{t\to\infty} E(|X_t - Z|) = 0$ if and only if $\{X_t\}_{t\in T}$ is uniformly integrable).

iii. $X_t$ converges a.s. and in $L^1$.
iv. $X_t$ converges in $L^1$.
v. There is an integrable random variable $Z$ such that $X_t = E(Z|A_t)$.
vi. Suppose $A_t \uparrow A_\infty$ (i.e., $A_t$ is an increasing sequence of $\sigma$-algebras and the union $\cup_{t=1}^\infty A_t$ generates the $\sigma$-algebra $A_\infty$). Then as $t \to \infty$, $E(X|A_t) \to E(X|A_\infty)$ a.s. and in $L^1$.

In fact, to amplify (2): if $\mathcal{J} = \{A_t\}_{t=1}^\infty$ is a filtration on the probability space $(\Omega, A, P)$ and $Z$ is an integrable and $A_t$-measurable random variable, then the process $\{E(Z|A_t)\}_{t=1}^\infty$ is a martingale with respect to $\mathcal{J}$.

We next offer a particularization of the dominated convergence theorem (Section 1.12.5). Specifically, we state Theorem 3.7.

**Theorem 3.7** Dominated Convergence Theorem for Conditional Expectations
Suppose $\{X_t\}_{t=1}^\infty$ is a martingale on a probability space $(\Omega, A, P)$. Let $X_t \to X$ a.s. and $|X_t| \leq Z$ for all $t$, with $E(Z) < +\infty$. If $A_t \uparrow A_\infty$, then $E(X_t|A_t) \to E(X|A_\infty)$ a.s.

In closing this portion of our discussion of martingales, we present a theorem that will be particularly useful in Chapter 4. But first a definition. The martingale $\{X_t\}_{t \geq 0}$ is said to be $L^2$-**bounded** if $E(|X_t|^2) \leq B < +\infty, B \in R$, for all $t \geq 1$. Under this condition, $\{X_t\}_{t \geq 0}$ converges a.s. or with probability 1 as an $L^2$ sequence. More formally, we state Theorem 3.8.

**Theorem 3.8** $L^2$-bounded Martingale Convergence Theorem
Suppose $\{X_t\}_{t \geq 0}$ is a martingale that is $L^2$-bounded for all $t \geq 0$. Then there exists a random variable $X$, with $E(|X_t|^2) < B, B \in R$, such that $P(\lim_{t \to \infty} X_t = X) = 1$ and $\lim_{t \to \infty} \|X_t - X\|_2 = 0$.

Next follows some convergence results for submartingales and supermartingales. We state first Theorem 3.9.

**Theorem 3.9** Let $\{X_t\}_{t=1}^\infty$ be a submartingale on the probability space $(\Omega, A, P)$ (Doob, 1953). If $k = \sup_t E(|X_t|) < +\infty$, then $X_t \to X$ a.s., where $X$ is a random variable satisfying $E(|X|) < k$.

For $\{X_t\}_{t=1}^\infty$ a submartingale, the following results are equivalent:

i. $X_t$ is uniformly integrable.
ii. $X_t$ converges a.s. and in $L^1$.
iii. $X_t$ converges in $L^1$.

A special case of Theorem 3.9 is provided by Theorem 3.10.

**Theorem 3.10** If $X_t > 0$ is a supermartingale on the probability space $(\Omega, \mathcal{A}, P)$, then, as $t \to \infty$, $X_t \to X$ a.s. and $E(X) \le E(X_o)$.

### 3.2.2 Continuous-Time Martingales

We now turn our attention to continuous-time martingales. Specifically, for $\{\mathcal{A}_t\}_{t \ge 0}$ a filtration on the probability space $(\Omega, \mathcal{A}, P)$ and $\{X_t\}_{t \ge 0}$ a collection of integrable random variables that are $\mathcal{A}_t$-measurable (i.e., $\{X_t\}_{t \ge 0}$ is adapted to the filtration), $\{X_t\}_{t \ge 0}$ is a **continuous-time martingale** if $E(X_t | \mathcal{A}_s) = X_s$ for all $0 \le s \le t$.

It is important to note that, under an appropriate conversion, a discrete-time martingale can be used to generate some of the preceding results for a continuous-time martingale. For example (Dineen, 2013), suppose $Z$ is an integrable random variable on the probability space $(\Omega, \mathcal{A}, P)$ and $\{\mathcal{A}_t\}_{t \ge 0}$ is a filtration on $(\Omega, \mathcal{A}, P)$. Let $E(Z | \mathcal{A}_t) = X_t$ for all $t$. For $0 \le s \le t$, $E(X_t | \mathcal{A}_s) = E(E(Z | \mathcal{A}_t) | \mathcal{A}_s) = E(Z | \mathcal{A}_s) = X_s$ so that $\{X_t\}_{t \ge 0}$ is a martingale on $(\Omega, \mathcal{A}, P)$. Conversely, let $\{X_t\}_{t \ge 0}$ be a martingale adapted to the filtration $\{\mathcal{A}_t\}_{t \ge 0}$ and suppose there exists a strictly increasing sequence of real numbers $\{t_n\}_{n=1}^{\infty}$, with $t_n \to \infty$ as $n \to \infty$, such that $\{X_{t_n}\}_{n=1}^{\infty}$ is uniformly integrable. (Given that $\{X_t\}_{t \ge 0}$ is a continuous-time martingale, $\{X_{t_n}\}_{n=1}^{\infty}$ is a discrete-time martingale for $t_n \to \infty$ as $n \to \infty$.) By Theorem 3.4, there exists an integrable random variable $Z$ such that $E(|X_{t_n} \to Z|) \to 0$ as $n \to \infty$ and $E(Z | \mathcal{A}_{t_n}) = X_{t_n}$. If $t_n > t > 0$ $E(Z | \mathcal{A}_t) = E(E(Z | \mathcal{A}_{t_n}) | \mathcal{A}_t) = E(X_{t_n} | \mathcal{A}_t) = X_t$. Then it can be shown that $lim_{t \to \infty} E(|X_t - Z|) = 0$ and that (3.3a) holds.

Continuous-time martingales also admit stopping times. That is, given a filtration $\mathcal{I} = \{\mathcal{A}_t\}_{t \ge 0}$, $\tau : \Omega \to R \cup \{+\infty\}$ is a **stopping time** adapted to $\mathcal{I}$ provided that the set $\{w \in \Omega | \tau(w) \le t\} \in \mathcal{A}_t$ for all $t \ge 0$. In addition, if $\{X_t\}_{t \in [0, +\infty)}$ is any collection of random variables, the **stopped variable** $X_\tau$ on $\{w \in \Omega | \tau(w) < +\infty\}$ is defined by setting $X_\tau(w) = X_t(w)$ for $\tau(w) = t$.

We next have Theorem 3.11, an extension of Doob's stopping time theorem in discrete time to continuous time. That is,

**Theorem 3.11** Doob's Continuous-Time Stopping Theorem (Doob, 1953). Let $\{X_t\}_{t \ge 0}$ be a continuous martingale adapted to the filtration $\mathcal{I} = \{\mathcal{A}_t\}_{t \ge 0}$ that satisfies the *usual conditions* (see Section 3.1.3). If $\tau$ is a stopping time for $\mathcal{I}$, then the process $\{Y_t\}_{t \ge 0} = \{X_{min(t, \tau)}\}_{t \ge 0}$ is also a continuous martingale adapted to $\mathcal{I}$.

### 3.2.2.1 Continuous-Time Martingale Convergence

We briefly look to the issue of the convergence of continuous-time martingales. To facilitate this discussion we first examine a couple of requisite definitions pertaining to the sample paths of a martingale. In particular, a path $X_t(\omega)$ is said to be **right-continuous** if for almost all $\omega \in \Omega$ the function $X_t(\omega)$ is right-continuous for $t \geq 0$; it is said to be **left limit** if for almost all $\omega \in \Omega$ the left limit $\lim_{s \uparrow t} X_s(\omega)$ exists and is finite for all $t \geq 0$. Given these considerations, we now offer the following important convergence theorem. Specifically, we have Theorem 3.12.

**Theorem 3.12** Doob's Convergence Theorem (Doob, 1953)
Let $\mathcal{I} = \{\mathcal{A}_t\}_{t \geq 0}$ be a filtration on a probability space $(\Omega, \mathcal{A}, P)$ and let $\{X_t\}_{t \geq 0}$ be a martingale with respect to the filtration $\mathcal{I}$ whose paths are right-continuous and left limit. Then the following are equivalent:

a. $\{X_t\}_{t \geq 0}$ converges in $L^1$ when $t \to \infty$.
b. As $t \to \infty$, $\{X_t\}_{t \geq 0}$ converges a.s. to an integrable and $\mathcal{A}_t$-measurable random variable $Z$ that satisfies $X_t = E(Z|\mathcal{A}_t)$, $t \geq 0$.
c. $\{X_t\}_{t \geq 0}$ is uniformly integrable.

### 3.2.3 Martingale Inequalities

For every martingale $\{X_t\}_{t \in T}$, the process $|X_t|^p, p \geq 1$, is a submartingale whenever $X_t \in L^p$. (The process $-|X_t|^p, p \geq 1$, is a supermartingale if $X_t \in L^p$.) For a real-valued martingale, $X_t, X_t^+ = max(X_t, 0)$ and $X_t^- = max(-X_t, 0)$ are submartingales. In this regard, the following martingale estimates are useful for forming approximations:

1. **Discrete-Time Martingale Inequalities**

   a. If $\{X_t\}_{t=1}^{\infty}$ is a submartingale, then $P\left( \max_{1 \leq k \leq t} X_k \geq \lambda \right) \leq \frac{1}{\lambda} E(X_t^+)$

   holds for any $\lambda > 0, t \geq 1$.
   b. If $\{X_t\}_{t=1}^{\infty}$ is a martingale with $E(|X_t|^p) < +\infty$ and $1 < p < +\infty$, then

   $$E\left( \max_{1 \leq k \leq t} |X_k|^p \right) \leq \left( \frac{p}{p-1} \right)^p E(|X_t|^p), t \geq 1.$$

   (3.4)

2. **Continuous-Time Martingale Inequalities**
   Suppose $\{X_t\}_{t \geq 0}$ is a stochastic process with continuous sample paths a.s.

a. If $\{X_t\}_{t \geq 0}$ is a submartingale, then

$$P\left(\max_{0 \leq s \leq t} X_s \geq \lambda\right) \leq \frac{1}{\lambda} E\left(X_t^+\right)$$

holds for all $\lambda > 0, t \geq 0$. $\qquad(3.5)$

b. If $\{X_t\}_{t \geq 0}$ is a martingale with $E(|X_t|^p) < +\infty$ and $1 < p < +\infty$, then

$$E\left(\max_{0 \leq s \leq t} |X_s|^p\right) \leq \left(\frac{p}{p-1}\right)^p E(|X_t|^p), t \geq 0.$$

(Note that (3.4a) is a generalization of Chebyshev's inequality.)

## 3.3 Path Regularity of Stochastic Processes

While the Kolmogorov theorem (see Section 3.1.1 and Appendix 3.A) addresses the issue of the existence of stochastic processes, it does not reveal anything about the paths of such processes. The Kolmogorov continuity theorem given later states that, under some rather mild conditions, we can employ processes whose paths are fairly regular. To start, we define a function $f : R_+ \to R$ as **Hölder with exponent $\gamma$** (or $\gamma$-**Hölder**) if there exist constants $c, \gamma > 0$ such that $|f(t) - f(s)| \leq c|t - s|^\gamma$ for $s, t \in R_+$. A function that satisfies this inequality is termed a **Hölder function**. Such functions are continuous. (For $f$ a continuous function, $|f(t) - f(s)| \to 0$ as $s \to t$. Hölder continuity enables us to determine the *rate* of this convergence, that is, $|f(t) - f(s)| \to 0$ as $s \to t$ at least as fast as $|t - s|^\gamma \to 0$.)

Given these considerations, we may now state Theorem 3.13.

**Theorem 3.13** Kolmogorov Continuity Theorem
For real $\alpha, \varepsilon, c > 0$, if the stochastic process $\{X_t\}_{t \in [0, 1]}$ defined on a probability space $(\Omega, \mathcal{A}, P)$ satisfies $E(|X_t - X_s|^\alpha) \leq c|t - s|^{1 + \varepsilon}$ for $s, t \in [0, 1]$, then there exists a modification (or version) of $\{X_t\}_{t \in [0, 1]}$ that is a continuous process and whose paths are $\gamma$-Hölder for every $\gamma \in [0, \varepsilon/\alpha]$.

As just indicated, it is possible to obtain versions of stochastic processes whose paths are highly regular. The Kolmogorov continuity theorem provides a sufficient condition that allows us to work with continuous versions of stochastic processes. But for martingales, the possibility of the existence of regular versions depends upon the regularity properties of the filtration with respect to which the martingale property holds. In this regard, let $\mathcal{I} = \{\mathcal{A}_t\}_{t \geq 0}$

be a filtration on a probability space $(\Omega, \mathcal{A}, P)$ and let the following assumptions hold:

1. If event $A \in \mathcal{A}$ satisfies $P(A) = 0$, then every subset of $A$ is in $\mathcal{A}_0$.
2. The filtration is right-continuous (i.e., for every $t \geq 0$, $\mathcal{A}_t \cap_{\varepsilon > 0} \mathcal{A}_{t+\varepsilon}$).

Then the filtered probability space $(\Omega, \mathcal{A}, \mathcal{I}, P)$ is said to satisfy the *usual conditions* (also see Section 3.1.3).

Armed with these regularity considerations, we can now state Theorem 3.14.

**Theorem 3.14** Doob's Regularization Theorem (Doob, 1953)

Let $\left(\Omega, \mathcal{A}, \{\mathcal{A}_t\}_{t \geq 0}, P\right)$ be a filtered probability space that satisfies the *usual conditions* and let $\{X_t\}_{t \geq 0}$ be a supermartingale with respect to the filtration $\mathcal{I} = \{\mathcal{A}_t\}_{t \geq 0}$. Assume the function $t \mapsto E(X_t)$ is right-continuous. Then there exists a modified process $\{\tilde{X}_t\}_{t \geq 0}$ of $\{X_t\}_{t \geq 0}$ with the following properties:

1. $\{\tilde{X}_t\}_{t \geq 0}$ is adapted to the filtration $\mathcal{I}$.
2. The paths of $\{\tilde{X}_t\}_{t \geq 0}$ are locally bounded, right-continuous, and left limited.
3. $\{\tilde{X}_t\}_{t \geq 0}$ is a supermartingale with respect to the filtration $\mathcal{I}$.

## 3.4  Symmetric Random Walk

Let $\{X_t\}_{t=1}^{\infty}$ denote a sequence of independent random variables on the probability space $(\Omega, \mathcal{A}, P)$, where $\Omega = \{-1, 1\}$, $\mathcal{A} = 2^{\Omega}$, and $X_t$ assume only the value $\pm 1$ with probability $1/2$, that is, $P(\{1\}) = (1/2) = P(\{-1\})$. It is easily verified that $E(X_t) = 0, V(X_t) = 1$. This sequence is termed a **symmetric random walk** since the indicated probabilities are each $1/2$.

Why the nomenclature "random walk?" We may consider the outcomes of this random experiment as a sequence of steps, each determined at random and independent of previous steps, taken in either the forward or backward direction, for example, if $X_t = 1$, a step forward is taken; and if $X_t = -1$, a step backward is taken.

To view the outcomes of this experiment as constituting a random process, let $\Omega = 2^n$, $\mathcal{A}$ be the $\sigma$-algebra generated by the sequence $\{X_t\}_{t=1}^{\infty}$ of independent random variables, and $P$ be a probability measure[8] (specified above). Then $\{X_t\}_{t=1}^{\infty}$ is defined on the probability space $(\Omega, \mathcal{A}, P)$. If we set

---

8 Actually, $P$ is a *product* (*probability*) measure. That is, let $(\Omega_i, \mathcal{A}_i, P_i)$, $i = 1, 2$, be two probability spaces and let $\Omega = \Omega_1 \times \Omega_2$ and consider $\mathcal{A}_1 \otimes \mathcal{A}_2 = \sigma(\mathcal{A}_1 \times \mathcal{A}_2)$ as the **product $\sigma$-algebra** on $\Omega$. If we define $P$ on the class $\varepsilon = \{A_1 \times A_2 | A_i \in \mathcal{A}_i, i = 1, 2\}$ as $P(A_1 \times A_2) = P_1(A_1) \cdot P_2(A_2)$, then $P$ can be extended to a unique probability measure on $\sigma(\varepsilon) = \sigma(\mathcal{A}_1 \times \mathcal{A}_2)$. Then the **product probability measure** $P = P_1 \otimes P_2$ is the probability measure on $\sigma(\mathcal{A}_1 \times \mathcal{A}_2)$.

$Y_t = \sum_{i=1}^{t} X_i, t \geq 1$, then the sequence $\{Y_t\}_{t=1}^{\infty}$, a **partial sum process** associated with $\{X_t\}_{t=1}^{\infty}$, is a martingale.

To verify the martingale assertion, let $Y_t = \sum_{i=1}^{t} X_i$, $\mathcal{A}_t = \sigma(X_1...X_t)$, $t \geq 1$, with $Y_t \in \mathcal{A}_t$, $E(|Y_t|) < +\infty$, and $X_{t+1}$ be independent of $\mathcal{A}_t$.

(Note that $\mathcal{A}_t$ is the smallest filtration to which $Y_t$ is adapted.) Since conditional expectation is linear (see Appendix 2.A) and $Y_{t+1} = Y_t + X_{t+1}$, it follows that $E(Y_{t+1}|\mathcal{A}_t) = E(Y_t|\mathcal{A}_t) + E(X_{t+1}|\mathcal{A}_t) = Y_t + E(X_{t+1}) = Y_t$ so that $\{Y_t\}_{t=1}^{\infty}$ is a martingale with respect to the filtration $\{\mathcal{A}_t\}_{t=1}^{\infty}$.

## 3.5 Brownian Motion

### 3.5.1 Standard Brownian Motion

A one-dimensional standard Brownian motion **(SBM) process** (or **Wiener process**) is a stochastic process $\{W_t\}_{t \geq 0}$ on a probability space $(\Omega, \mathcal{A}, P)$ with the following properties:

i. The SBM process begins at $t_0 = 0$ or $P(W_0 = 0) = 1$.

ii. For $0 = t_0 \leq t_1 \leq \cdots \leq t_k$, the increments (displacements)

   $W(t_1), W(t_2) - W(t_1), ..., W(t_k) - W(t_{k-1})$ are independent random variables.

iii. For $0 \leq s < t$, the increments $W_t - W_s \sim N(0, t-s)$, that is,

$$P((W_t - W_s) \in B) = \frac{1}{\sqrt{2\pi(t-s)}} \int_B e^{-x^2\frac{}{2(t-s)}} dx.$$

(3.6)

where $B$ is a Borel subset of $R$.[9] Equivalently, $W_t - W_s \sim \sqrt{t-s}N(0,1)$.

Property (3.6i) is a convention. To determine the position of a Brownian particle in one dimension, we start at $t = 0$, with the initial position specified as $W_0 = 0$. Property (3.6ii) indicates that the increments $W(t_1), W(t_2) - W(t_1), ..., W(t_{k-1}) - W(t_{k-2})$ occurring during the time intervals $[t_0, t_1], ..., [t_{k-2}, t_{k-1}]$, respectively, do not affect the increment $W(t_k) - W(t_{k-1})$ that obtains during the time interval $[t_{k-1}, t_k]$, that is, the SBM process is assumed to be *without memory*. (For instance, the path of a pollen particle traverses in order to get to its current position does not influence its future

---

9 The SBM process just defined is the namesake of the Scottish botanist Robert Brown (1828), who first described the phenomenon of pollen particles moving erratically while suspended in water. This led Brown to conclude that the particles exhibited a "seemingly random movement." It was later explained that the pollen particles were being impacted by the rapidly moving water molecules.

location.) Property (3.6iii) indicates that (a) $W_t - W_s, 0 \le s < t$, has a zero mean (if we think of $W_t$ as the height above a horizontal time-axis of pollen particles at time $t$, then a zero mean indicates that, at time $t + 1$, the particle's height is just as likely to increase as it is to decrease, with no upward or downward drift); and (b) the variance $t - s$ of an SBM process increases with the length of the time interval $[s, t]$ (the pollen particle moves away from its position at time $s$, and there is no tendency for the particle to return to that position, that is, the SBM process lacks any propensity for *position reversion*).

Do SBM processes satisfying these properties exist? The answer is a definite *yes*, that is, there exists a probability space $(\Omega, \mathcal{A}, P)$ and a stochastic process $\{W_t\}_{t \ge 0}$ defined on this space such that Properties (3.6i)–(3.6iii) hold. In fact, these properties lead to a consistent set of finite-dimensional distributions via the Kolmogorov existence theorem (Section 3.1.1 and Appendix 3.A). In this regard, for $0 < t_1 < \cdots < t_k$, let $\mu_{t_1,...,t_k}$ be the distribution function of $(S_1,...,S_k) \in R^k$, where $S_i = \sum_{j=1}^{i} X_j$ and $X_1,...,X_k$ are independent, normally distributed random variables with zero means and variances $t_1, t_2 - t_1, ..., t_k - t_{k-1}$, respectively. Then it can be shown (e.g., Billingsley, 2012) that for $0 < t_1 < \cdots < t_k$, the $\mu_{t_1,...,t_k}$ satisfy the consistency conditions of the Kolmogorov existence theorem; thus, there exists a stochastic process $\{W_t\}_{t \ge 0}$ corresponding to $\mu_{t_1,...,t_k}$. In sum, if $W_0 = 0$, then there exists on the probability space $(\Omega, \mathcal{A}, P)$ a process $\{W_t\}_{t \ge 0}$ whose finite-dimensional distributions possess Properties (3.6i)–(3.6iii).

The importance of an SBM process $\{W_t\}$ is that it serves to represent the cumulative effect of process noise. That is, if $W_s$ and $W_t, 0 \le s < t$, mark the position of the process at times $s$ and $t$, respectively, then the increment $W_t - W_s$ reflects *pure noise* over the interval $[s, t]$.

It is important to note that a slight modification of the preceding definition of an SBM process can also be offered—one that explicitly incorporates the notion of a filtration. Specifically, let $(\Omega, \mathcal{A}, P)$ be a probability space with filtration $\{\mathcal{A}_t\}_{t \ge 0}$. A one-dimensional SBM is an $\{\mathcal{A}_t\}$-adapted process $\{W_t\}_{t \ge 0}$ with the following properties:

    i. $P(W_0 = 0) = 1$.

    ii. For $0 \le s < t$, the increment $W_t - W_s$ is independent of $\mathcal{A}_s$.      (3.7)

    iii. For $0 \le s < t$, the increment $W_t - W_s$ is $N(0, t - s)$.

So given the filtration $\{\mathcal{A}_t\}_{t \ge 0}$ for $0 \le s < t$, there is at least as much information available at time $t$ as there is at time $s$. Thus, $\mathcal{A}_s \subset \mathcal{A}_t$ or *information accumulates*. Also, the information available at time $t$ is sufficient to evaluate $W_t$ at time $t$ is $\mathcal{A}_t$-measurable. Also, finally, for $0 \le t < u$, the future increment $W_u - W_t$ is independent of $\mathcal{A}_t$—any increment in the SBM process subsequent to time $t$ is independent of the information existing at that time.

While the filtration $\{\mathcal{A}_t\}_{t\geq 0}$ is a key element of this alternative specification of the properties (3.7) of an SBM process, we can readily incorporate a specialized type of filtration into our discussion of SBM even if (3.6) holds. To this end, let $\mathcal{A}_t^w = \sigma(W_s, 0 \leq s < t)$ represent the $\sigma$-algebra generated by $\{W_s | 0 \leq s < t\}$ so that $\{\mathcal{A}_t^w\}_{t\geq 0}$ is the **natural filtration** generated by $\{W_t\}_{t\geq 0}$. Here, $\{W_t\}_{t\geq 0}$ under (3.6) is an SBM process with respect to the natural filtration $\{\mathcal{A}_t^w\}_{t\geq 0}$. (If $\mathcal{A}_t^w \subset \mathcal{A}_t$, $t \geq 0$, and $W_t - W_s$ is independent of $\mathcal{A}_s$, $0 \leq s < t$, then $\{W_t\}_{t>0}$ is an SBM process with respect to the filtration $\{\mathcal{A}_t\}_{t\geq 0}$.)

Suppose $\{W_t\}_{t\geq 0}$ is an SBM process defined on a probability space $(\Omega, \mathcal{A}, P)$ and let $(\Omega, \tilde{\mathcal{A}}, P)$ denote the completion of $(\Omega, \mathcal{A}, P)$. Thus, $\{W_t\}_{t\geq 0}$ is an SBM process on the complete probability space $(\Omega, \tilde{\mathcal{A}}, P)$. For $\mathcal{N}$ the collection of $P$-null sets (see Section 3.1.3), let $\tilde{\mathcal{A}}_t = \sigma(\mathcal{A}_t^w \cup \mathcal{N})$, $t \geq 0$, depict the **augmentation with respect to $P$ of** $\{\mathcal{A}_t^w\}_{t\geq 0}$. Since $\{\tilde{\mathcal{A}}_t\}_{t\geq 0}$ is a filtration on $(\Omega, \tilde{\mathcal{A}}, P)$ satisfying the *usual conditions* (Section 3.1.3), it follows that $\{W_t\}_{t\geq 0}$ is an SBM process on $(\Omega, \tilde{\mathcal{A}}, P)$ with respect to $\{\tilde{\mathcal{A}}_t\}_{t\geq 0}$. So given an SBM process defined on a probability space $(\Omega, \mathcal{A}, P)$, one can construct a complete probability space with a filtration satisfying the usual conditions. In this regard, unless otherwise stated, we can assume that we have an SBM process defined on a complete probability space $(\Omega, \mathcal{A}, P)$ with a filtration $\{\mathcal{A}_t\}_{t\geq 0}$ satisfying the usual conditions.

Looking to some of the salient features of SBM processes we have the following:

1. The increments of the SBM process are **stationary**, that is, the distribution of $W_t - W_s$, $0 \leq s < t$, depends only upon the difference $t - s$. (By virtue of Properties (3.6i and iii), the distribution of $W_t - W_s$ is characterized by noting that $W_t \sim N(0, t)$.)

2. For $0 \leq s < t$, it is readily determined that $E(W_t) = 0, E(W_t^2) = V(W_t) = t$ (SBM is a **quadratic variation process** with quadratic variation equal to $t$), and, via property (3.6ii), $E(W_s W_t) = E(W_s(W_t - W_s)) + E(W_s^2) = E(W_s)$ $E(W_t - W_s) + E(W_s^2) = s$. In general, $E(W_s W_t) = min\{s, t\}$.

3. If we represent the value of $W_t$ at $\omega \in \Omega$ by $t \mapsto W(t, w)$, then the sample path functions of SBM processes can be denoted as $W(\cdot, w)$. For each $w$, $W(0,w) = 0$ and, a.s., $t \mapsto W(t, w)$ is continuous in $t$. More specifically, for every $\omega \in \Omega$, the sample path $t \mapsto W(t, w)$ is Hölder-continuous for each $0 < \gamma < (1/2)$; it is nowhere Hölder-continuous for any $\gamma > (1/2)$.

4. If the SBM process $\{W_t\}_{t\geq 0}$ is measurable, then each sample path function $W(\cdot, w)$ is $\mathcal{A}$-measurable.

5. SBM process sample paths $W(\cdot, w)$, $\omega \in \Omega$, are highly irregular and, a.s., are of unbounded variation.[10]

6. Sample paths of SBM processes are, a.s., nowhere differentiable (see Appendix 3.B).

7. **Symmetry property**: the process $\{-W_t\}_{t \geq 0}$ is an SBM.

8. **Scaling property**: for every real $c > 0$, the SBM process $\{W_{ct}\}_{t \geq 0}$ follows the same probability law as the SBM process $\{\sqrt{c}W_t\}_{t \geq 0}$.

9. **Time-inversion property**: a.s., $lim_{t \to 0}(W_t/t) = 0$, that is, the SBM process $\{tW_{1/t}\}_{t \geq 0}$ follows the same probability law as the process $\{W_t\}_{t \geq 0}$.

10. For any $h > 0$, the process $\{W_{t+h} - W_t\}_{t \geq 0}$ is an SBM.

11. If $\{W_t\}_{t \geq 0}$ is an SBM process, then $W_t \sim N(0, t)$ and, for real numbers $a$ and $b$, $a \leq b$, the probability that the sample path assumes values between $a$ and $b$ at time $t$ is

$$P(a \leq W_t \leq b) = \frac{1}{\sqrt{2\pi t}} \int_a^b e^{\frac{-x^2}{2t}} dx;$$

and for the function $f : R \to R$,

$$E(f(W_t)) = \frac{1}{\sqrt{2\pi t}} \int_{-\infty}^{+\infty} f(W_t) e^{\frac{-x^2}{2t}} dx.$$

(Note that if $f = \chi_{[a,b]}(x)$, where

$$\chi_{[a,b]}(x) = \begin{cases} 1, x \in [a,b]; \\ 0, x \notin [a,b], \end{cases}$$

then the preceding expression obtains.)

12. An SBM process is a continuous-time martingale. That is, if $\{W_t\}_{t \geq 0}$ is an SBM process and $A_t = \sigma(W_s, 0 \leq s < t)$, then, for $W_t = (W_t - W_s) + W_s, E(W_t|A_s) = E(W_t - W_s|A_s) + E(W_s|A_s) = E(W_t - W_s) + W_s = 0 + W_s$ (since $W_t - W_s$ is independent of $A_s$ and $W_s$ is $A_s$-measurable) and $E((W_t - W_s)^2|A_s) = t - s$ a.s. since the increment $W_t - W_s$ is independent of the past and normally distributed with mean 0. Conversely, let $\{W_t\}_{t \geq 0}$ be a stochastic process with $\{A_t\}_{t \geq 0}$ an increasing family of $\sigma$-algebras such that $W_t$ is $A_t$-measurable and the preceding two equalities hold a.s. for $0 \leq s < t$. Then $\{W_t\}_{t \geq 0}$ is an SBM process. In fact, since an

---

10 A right-continuous function $f : [0, t] \to R$ is a function of **bounded variation** if $sup \sum_{j=1}^k |f(t_j) - f(t_{j-1})| < +\infty$. If the supremum is infinite, then $f$ is said to be of **unbounded variation**.

SBM process $\{W_t\}_{t\geq 0}$ has independent increments, it is actually a square integrable martingale.

The preceding discussion on SBM processes has focused on what we called *standard BM (SBM)*. A more general BM process has us modify (3.6iii) to the following: for $0 \leq s < t$, $W_t - W_s$ is normally distributed with $E(W_t - W_s) = \mu(t-s)$ and $E((W_t - W_s)^2) = \sigma^2(t-s), \sigma^2 \neq 0$, where $\mu$ (called the **drift coefficient**) and $\sigma^2$ (the **variance coefficient**) are real constants. If $\mu = 0$ and $\sigma^2 = 1$, we have a **normalized BM process**. In fact, for any BM process $\{W_t\}_{t\geq 0}$ with drift $\mu$ and variance $\sigma^2$,

i. $\{W_t\}_{t\geq 0}$ is a BM process with drift $-\mu$ and variance $\sigma^2$;
ii. for constants $a, b > 0$, $\{aW_{bt}\}_{t\geq 0}$ is a BM process with drift $ab\mu$ and variance $a^2b\sigma^2$; and
iii. $((W_t - W_s) - \mu(t-s))/\sigma$ is a **normalized BM process**.

We close this section by noting that a $d$-dimensional random process $\{W_t = (W_t^1, ..., W_t^d)\}_{t\geq 0}$ is a $d$-dimensional SBM process if every $\{W_t^i\}_{t\geq 0}$ is a one-dimensional SBM process and the $\{W_t^i\}_{t\geq 0}$ are independent (or the $\sigma$-algebras $\sigma(W_t^i, t \geq 0)$ are independent), $i = 1, ..., d$. In this regard, if $\{W_t = (W_t^1, ..., W_t^d)\}_{t\geq 0}$ is a $d$-dimensional SBM process, then for $i, j = 1, ..., d$,

i. $E(W_t - W_s | \mathcal{A}_s) = 0, 0 \leq s < t < +\infty$;
ii. $E(W_t^i W_s^j) = min(t,s)\delta_{ij}, 1 \leq i, j \leq d$, where

$$\delta_{ij} = \begin{cases} 1, i \neq j; \\ 0, i \neq j \end{cases}$$

is the **Dirac delta function**.
iii. $E\left((W_t^i - W_s^i)(W_t^j - W_s^j)\right) = (t-s)\delta_{ij}, 0 \leq s < t.$

In addition, a $d$-dimensional SBM process is a continuous martingale. That is, let $\{X_t = (X_t^1, ..., X_t^d)\}_{t\geq 0}$ be a $d$-dimensional martingale with respect to the filtration $\{\mathcal{A}_t\}_{t\geq 0}$, with $X_0 = 0$ a.s., and having joint quadratic variations $(X^i, X^j)_t = t\delta_{ij}, 1 \leq i, j \leq d$. Then $\{X_t = (X_t^1, ..., X_t^d)\}_{t\geq 0}$ is a $d$-dimensional SBM process with respect to $\{\mathcal{A}_t\}_{t\geq 0}$.

### 3.5.2 BM as a Markov Process

Let $\{X_t\}_{t\geq 0}$ be a stochastic process defined on a probability space $(\Omega, \mathcal{A}, P)$. The $\sigma$-algebra $\mathcal{A}_t = \sigma(X_s, 0 \leq s < t)$, $t \geq 0$, is the **history** of the process up to and including time $t$. ($\mathcal{A}_t$ records the information available from our observation of $X_s$ for all times $0 \leq s < t$.) In addition, a real-valued, $\mathcal{A}_t$-adapted stochastic process $\{X_t\}_{t \geq 0}$ is called a **Markov process** if the **Markov property** $P(X_t \in B|\mathcal{A}_s) = P(X_t \in B|X_s)$ holds a.s. for all $0 \leq s \leq t < +\infty$ and all Borel sets $B \subset R$. So given $X_s$, one can predict the probabilities of future values $X_t$ just as well as if you knew the entire history of the process prior to time $s$. The process only knows $X_s$ and is not aware of how it got there so that the future depends on the past only through the present, that is, once the present is known, the past and future are independent. Looked at in another fashion, $\{X_t\}_{t \geq 0}$ is a Markov process if for any $0 \leq s < t$, the conditional distribution of $X_{t+s}|\mathcal{A}_t$ is the same as the conditional distribution of $X_{t+s}|$ $X_t$, that is,

$$P(X_{t+s} \leq y|\mathcal{A}_t) = P(X_{t+s} \leq y|X_t) \text{ a.s.}$$

The **transition probability** of the Markov process is a function $P(x, s; B, t)$ defined on $0 \leq s \leq t < +\infty, x \in R$, with the following properties:

1. For every $0 \leq s \leq t < +\infty, P(X_s, s; B, t) = P(X_t \in B|X_s)$.
2. For every $0 \leq s \leq t < +\infty$ and $x \in R, P(x, s; \cdot, t)$ is a probability measure on the family of Borel sets $\mathcal{B}$.
3. For every $0 \leq s \leq t < +\infty$ and $B \in \mathcal{B}, P(\cdot, s; B, t)$ is Borel measurable.
4. For every $0 \leq s \leq r \leq t < +\infty, x \in R,$ and $B \in \mathcal{B}$,

$$P(x, s; B, t) = \int_R P(y, r; B, t) P(x, s; dy, r).$$

[Chapman–Kolmogorov equation]

(That is, a single-step transition probability can be expressed in terms of a combination of two-step transition probabilities with respect to an arbitrary intermediate time $r$.)

In terms of the preceding transition probability, the Markov property becomes

$$P(X_t \in B|\mathcal{A}_s) = P(X_s, s; B, t);$$

and thus we can write

$$P(X_t \in B|X_s = x) = P(x, s; B, t),$$

the probability that the process will be in set $B$ at time $t$ given that the process was in state $x$ at time $s \leq t$.

A Markov process $\{X_t\}_{t \geq 0}$ is termed **homogeneous** (with respect to $t$) if its transition probability $P(x, s; B, t)$ is **stationary**, that is,

$$P(x, s + u; B, t + u) = P(x, s; B, t),$$

$0 \leq s \leq t < +\infty, x \in R,$ and $B \in \mathcal{B}$. In this circumstance, the transition probability is a function of $x$, $B$, and the difference $t - s$ since $P(x, s; A, t) = P(x, 0; B, t - s)$. Under this observation, we can simply write $P(x, 0; B, t) = P(x; B, t)$.

The process $\{X_t\}_{t\geq 0}$ is said to be a **strong Markov process** if the following **strong Markov property** holds: for any bounded Borel-measurable function $\xi : R \rightarrow R$, any finite $\mathcal{A}_t$ stopping time $\tau$, and $t \geq 0$,

$$E(\xi(X_{\tau+t})|\mathcal{A}_\tau) = E(\xi(X_{\tau+t})|X_\tau).$$

In terms of the transition probability, the strong Markov property can be written as

$$P(X_{\tau+t} \in B|\mathcal{A}_\tau) = P(X_\tau, \tau; B, \tau + t).$$

While a strong Markov process is a Markov process, the converse of this statement does not generally hold.

To relate the notion of a Markov process to our current discussion, we note the following: if $\{W_t\}_{t\geq 0}$ is an SBM process, then $\{W_t\}_{t\geq 0}$ is a Markov process with stationary transition probability

$$P(W_t \in B|W_s) = \frac{1}{\sqrt{2\pi(t-s)}} \int_B e^{\frac{-(x-W_s)^2}{2(t-s)}} dx$$

a.s., $0 \leq s \leq t < +\infty$, for Borel sets $B \subset R$. Moreover, this process is homogeneous with respect to its state space (since $W_0 = 0$) and time and, via Theorem 3.13 (the Kolmogorov continuity theorem) can be chosen so that it has continuous sample paths with probability 1. It is also true that $\{W_t\}$ is a strong Markov process.

### 3.5.3 Constructing BM

#### 3.5.3.1 BM Constructed from *N*(0, 1) Random Variables

Let us first consider the construction of an SBM process from countably many standard normal random variables (Ciesielski, 1961; Evans, 2013; Lévy, 1948; McKean, 1969). This construction will verify that an SBM process actually exists. (Section 3.5.3.2 offers a construction of an SBM process based upon the notion of the limiting behavior of a collection of symmetric random walks.)

It was determined earlier in Section 1.11.3 that, for functions $f(t)$, $g(t) \in H(= L^2[0,1])$, the norm and inner product were defined, respectively, as

$$\|f\| = \left( \int_0^1 f^2(t)dt \right)^{\frac{1}{2}} \text{ and}$$

$$(f,g) = \int_0^1 f(t)g(t)dt.$$

In addition, a countable set of functions $\Phi = \{\emptyset_1, \emptyset_2,...\}$ was defined as orthonormal if $\|\emptyset_i\| = 1$ for all $i$; and $(\emptyset_i, \emptyset_j) = 0$, $i \neq j$ Moreover, it can be demonstrated that $\{\emptyset_i\}$ is a *complete* orthonormal basis for $H$.

If $\{X_1, X_2, ...\}$ is a sequence of independent, identically distributed $N(0, 1)$ random variable defined on a probability space $(\Omega, \mathcal{A}, P)$, then for $n = 1, 2 ...,$ define

$$W_t^n = \sum_{i=1}^{n} X_i \int_0^t \emptyset_i(s)ds. \tag{3.8}$$

For each $t \geq 0$, $W_t^n$ is a Cauchy sequence in $L^2(\Omega, \mathcal{A}, P)$ whose limit $W_t$ is a $N(0, t)$ random variable. This being the case, the remaining issue of $W_t$ having continuous sample paths will now be addressed. While (3.8) holds for an *arbitrary* complete orthonormal basis, let us now be a bit more specific. To this end, let us define the **Haar functions** $\{h_0, h_{j,n}, j = 1, 2, ..., 2^{n-1}; n = 1, 2, ...\}$ as $h_0(t) = 1$ and with $k = 2j - 1$,

$$h_{j,n} = \begin{cases} 2^{(n-1)/2}, (k-1)/2^n \leq t \leq k/2^n; \\ -2^{(n-1)/2}, k/2^n < t \leq (k+1)/2^n; \\ 0, \text{elsewhere} \end{cases} \tag{3.9}$$

(Figure 3.2a). These Haar functions constitute a complete orthonormal basis in $L^2[0, 1]$. Moreover, when the $\emptyset_i$ in (3.8) are Haar functions, it follows that $W_t^n \to W_t$ uniformly a.s.

In this regard, let us define the **Schauder functions** $\{S_0, S_{n,j}\}$ as the indefinite integrals of the Haar functions or

$$S_{n,j} = \int_0^t h_{j,n}(s)\, ds \tag{3.10}$$

**Figure 3.2** (a) The Haar function $h_{j,n}$ and (b) the Schauder function $S_{n,j}(t)$.

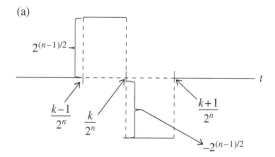

(a)

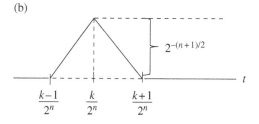

(b)

so that $S_0(t) = t$, and

$$
S_{n,j}(t) = \begin{cases} 2^{(n-1)/2}(t-(k-1)/2^n), (k-1)/2^n \le t \le k/2^n; \\ 2^{(n-1)/2}((k+1)/2^n - t), k/2^n \le t \le (k+1)/2^n \\ 0, \text{ elsewhere} \end{cases} \qquad (3.11)
$$

(Figure 3.2b).

Next, for $\{X_0, X_{n,j}, n = 1, 2, ...; j = 1, 2, ..., 2^{n-1}\}$ a collection of independent $N$ (0, 1) random variables defined on the probability space $(\Omega, \mathcal{A}, P)$, and for $t \in [0, 1]$, $N = 1, 2, ...$, and $\omega \in \Omega$, define (via (3.8)–(3.11))

$$
W_t^n(\omega) = X_0 S_0 + \sum_{n=1}^{N} Y_n(t, \omega), \qquad (3.12)
$$

where

$$
Y_n(t, \omega) = \sum_{j=1}^{2^{n-1}} X_{n,j}(\omega) S_{n,j}(t).
$$

Now, for each $N$ and $\omega \in \Omega$, the sample path function $t \mapsto W^N(t, \omega)$ is continuous. The justification that $W_t^N(\omega)$ has a limiting continuous-path process $W_t$ is provided by Theorem 3.15.

**Theorem 3.15** The sequence $\{W_t^N(\omega)\}$ defined by (3.12) converges uniformly a.s. as $N \to \infty$ to a stochastic process $W_t$ with continuous sample paths.

Moreover, $W_t$ is an SBM for $t \in [0, 1]$.

The gist of the preceding discussion is that, in general, we can construct an SBM process on a probability space on which there exists countably many independent standard normal random variables. That is, suppose we have countably many independent $N(0, 1)$ random variables $\{X_n\}_{n=1}^{\infty}$ defined on the probability space $(\Omega, \mathcal{A}, P)$. Then there exists a one-dimensional SBM process $W(t, \omega)$ defined for $t \ge 0, \omega \in \Omega$.

### 3.5.3.2 BM as the Limit of Symmetric Random Walks

What sort of random behavior can be used to generate (model) an SBM process? Our approach to framing the answer to this question, and to verify that SBM processes exists, is to look to the **Donsker's invariance principle** which states that an SBM process may be constructed as the limit of appropriately rescaled random walks. To see this, let $\{X_i\}_{i \ge 1}$ be a sequence of independent random variables defined on the probability space $(\Omega, \mathcal{A}, P)$ and which satisfy $P(X_i = 1) = P(X_i = -1) = 1/2$. In addition, for $n \ge 1$, let $Y_n = \sum_{i=1}^{n} X_i$ (our position after $n$ trials or steps); and for $n \geq 0$, let $\mathcal{A}_n = \sigma(Y_0, Y_1, ..., Y_n)$. Then the sequence $\{Y_n\}_{n \geq 0}, Y_0 = 0$, is a symmetric random walk on the set of integers. Next, the sequence of processes $\{Y_t^n\}_{t \in [0, 1]}, n \in N$, where

$$Y_t^n = \sqrt{n} \left[ \left( t - \frac{k}{n} \right) Y_{k+1} + \left( \frac{k+1}{n} - t \right) Y_K \right], \frac{k}{n} \leq t \leq \frac{k+1}{n},$$

is a stochastic process that represents a piecewise continuous linear interpolation of the rescaled discrete sequence $\{Y_n\}_{n \geq 0}$. Given this construction, we can now conclude our argument with Theorem 3.16.

**Theorem 3.16** The sequence $\{Y_t^n\}_{t \in [0, 1]}, n \in N$, converges in distribution (i.e., weakly) to an SBM process $\{W_t\}_{t \in [0, 1]}$ (Donsker, 1951).

### 3.5.4 White Noise Process

The time derivative of $W(t), \dot{W}(t) = dW(t)/dt = \xi(t)$, is termed (one-dimensional) **white noise**. As was noted earlier (see characteristic (5) of SBM processes), $\dot{W}(t)$ does not exist in the ordinary sense; it is a generalized or idealized derivative. What then are the properties of a white noise process $\xi(t)$? If $\{X_t\}_{t \geq 0}$ is a Gaussian stochastic process with $E(X_t^2) < +\infty, t \geq 0$, the **covariance function** of $\{X_t\}_{t \geq 0}$ is defined as $r(t, s) = E(X_t \cdot X_s)$, where $t, s \geq 0$. If $r(t, s) = h(t - s)$ for some real-valued function $h : R \to R$ and if $E(X_t) = E(X_s)$ for all $t, s \geq 0$, then $X_t$ is called **stationary in the wide sense**. Then a white noise process $\xi(t)$ is one that is Gaussian as well as wide-sense stationary, with $h(\cdot) = \delta_0$, where $\delta_0$ is the Dirac point mass[11] concentrated at $t = 0$.

# Appendix 3.A   Kolmogorov Existence Theorem: Another Look

A stochastic process $\{X_t\}_{t \in T}$ is usually described in terms of its family of finite-dimensional distributions. In this regard, for each $n$-tuple $(t_1, ..., t_n)$ of distinct elements of $T$, let the associated distribution of the random vector $(X(t_1), ..., X(t_n))$ over $R^n$ be denoted as $\mu_{t_1, ..., t_n}$, with

$$\mu_{t_1, ..., t_n}(A) = P((X(t_1), ..., X(t_n)) \in A), A \in R^n. \tag{3.A.1}$$

---

11  For $(\Omega, \mathcal{A})$ a measurable space with $x \in \Omega$, the function $\delta_x : \mathcal{A} \to \{0, 1\}$, defined for set $A \in \mathcal{A}$ by
$$\delta_x(A) = \begin{cases} 1, x \in A; \\ 0, x \notin A \end{cases}$$ is termed the **Dirac point mass function** at point $x$.

Then $\mu_{t_1,\ldots,t_n}$ is a probability measure on $R^n$ and is termed a **finite-dimensional distribution** of the process $\{X_t\}_{t\in T}$. We require that the family of probability measures or finite-dimensional distributions induced by $\{X_t\}_{t\in T}$ satisfy the following two consistency conditions; that is, if $t_1,\ldots,t_n \in T, \mathcal{B}(R)$ is the Borel $\sigma$-algebra on $R$, and if $\{\pi(1),\ldots,\pi(n)\}$ is a permutation of the numbers $1,\ldots,$ $n$, then

a. $\mu_{t_1,\ldots,t_n}(A_1 \times \cdots \times A_n) = \mu_{t_{\pi(1)},\ldots,t_{\pi(n)}}(A_{\pi(1)} \times \cdots \times A_{\pi(n)}), A_i \in \mathcal{B}(R), i = 1,\ldots,n;$

b. $\mu_{t_1,\ldots,t_n}(A_1 \times \cdots \times A_{n-1} \times R) = \mu_{t_1,\ldots,t_n}(A_1 \times \cdots \times A_{n-1}), A_i \in \mathcal{B}(R), i = 1,\ldots,n.$

$$(3.A.2)$$

(Note that (3.A.2a) holds because $(X(t_1),\ldots,X(t_n)) \in (A_1 \times \cdots \times A_n)$ and $(X(t_{\pi(1)}),\ldots,X(t_{\pi(n)})) \in (A_{\pi(1)} \times \cdots \times A_{\pi(n)})$ are the same events.)

The probability measures (3.A.1) generated by a process such as $\{X_t\}_{t\in T}$ must satisfy (3.A.2). Conversely, the Kolmogorov existence theorem states that if a given system of measures satisfies the consistency requirements (3.A.2), then there exists a stochastic process with these finite-dimensional distributions. In this regard, we have Theorem 3.A.1.

**Theorem 3.A.1** Kolmogorov Existence Theorem
Suppose for every $t_1,\ldots,t_n \in R^n$ we have a probability measure or system of distributions $\mu_{t_1,\ldots,t_n}$ on $R^n$ and that these measures satisfy the consistency conditions (3.A.2). Then there exists on some probability space $(\Omega, \mathcal{A}, P)$ a stochastic process $\{X_t\}_{t\in T}$ having $\mu_{t_1,\ldots,t_n}$ as its finite-dimensional distributions.

Hence, this theorem offers a justification for the existence of a stochastic process $\{X_t\}_{t\in T}$ having a family $\{\mu_{t_1,\ldots,t_n} | n \geq 1, t_i \in T\}$ of probability measures as its finite-dimensional distributions.

## Appendix 3.B   Nondifferentiability of BM

Our objective herein is to examine the proposition that, for any time $t$, almost all trajectories of BM are not differentiable. To this end, consider the interval from $t$ to $t + h$. For this interval, let us express the difference quotient for the BM process $\{W_t\}_{t\geq 0}$ as

$$Y_h = \frac{W(t+h) - W(t)}{h}. \qquad (3.B.1)$$

Here, $Y_h$ represents a normally distributed random variable with

$$E(Y_h) = \frac{1}{h}E(W(t+h) - W(t)) = 0;$$

$$V(Y_h) = \frac{1}{h^2}V(W(t+h) - W(t)) = \frac{1}{h^2}(h) = \frac{1}{h}.$$

Thus, the standard deviation of $Y_h$ is $\sqrt{V(Y_h)} = (1/\sqrt{h})$. From (3.6iii), we know that we can write $Y_h$ as $(1/\sqrt{h})Z, Z \sim N(0,1)$, and thus, for constant $k > 0$,

$$P(|Y_h| > k) = P\left(\left|\frac{Z}{\sqrt{h}}\right| > k\right).$$

Taking $k$ to be arbitrarily large, as $h \to 0, P(|Z/\sqrt{h}| > k) \to 1$ and thus $Y_h = Z/\sqrt{h} \to +\infty$ (in distribution) so that the rate of change in $W_t$ at time $t$ is infinite or the derivative of $W_t$ with respect to $t$ at time $t$ does not exist. In addition, with $t$ also arbitrary, the BM path is nowhere differentiable—one cannot determine at any time $t$ the immediate (local) direction of the BM path.

# 4

# Mathematical Foundations 4

Stochastic Integrals, Itô's Integral, Itô's Formula, and Martingale Representation

## 4.1  Introduction

Suppose we have an ordinary differential equation (ODE) depicting the growth in, say, population ($N$) over time or

$$\frac{dN(t)}{dt} = a(t)N(t), N(0) = N_0, \tag{4.1}$$

where the **relative growth rate** is $(dN(t)/dt)/N(t) = a(t)$. If $a(t)$ is deterministic or $a(t) = a = $ constant, then (1.4) has the solution

$$N(t) = N_0 e^{at}.^1 \tag{4.2}$$

However, what if $a(t)$ is subject to random fluctuations that, possibly, can be attributed to extraneous or chance factors such as the weather, location, and the general environment? In this circumstance, we can rewrite $a(t)$ as

$$a(t) = r(t) + j(t) \cdot (\textbf{random error})$$
$$= r(t) + j(t) \cdot \textbf{noise}, \tag{4.3}$$

where $r(t)$ and $j(t)$ are given functions of $t$ and the (one-dimensional) *noise* term follows a "known probability law."

In view of (4.3), Equation (4.1) can be rewritten as the stochastic differential equation (SDE)

$$\frac{dN(t)}{dt} = r(t)N(t) + j(t)N(t) \cdot \text{noise}, N(0) = N_0. \tag{4.1.1}$$

---

1  Rewrite (1.4) as $dN(t)/N(t) = a\,dt$. Then $ln N(t) = at + ln C$, where $ln C$ is the constant of integration, and $N(t) = Ce^{at}$. For $t = 0, N(0) = C$ so that (4.2) obtains.

*Stochastic Differential Equations: An Introduction with Applications in Population Dynamics Modeling*, First Edition. Michael J. Panik.
© 2017 John Wiley & Sons, Inc. Published 2017 by John Wiley & Sons, Inc.

Under integration, this expression becomes

$$N(t) = N_0 + \int_0^t r(s)N(s)ds + \int_0^t j(s)N(s) \cdot \text{noise} \cdot ds. \tag{4.4}$$

But how do we integrate

$$\int_0^t j(s)N(s) \cdot \text{noise} \cdot ds? \tag{4.5}$$

To answer this question, we must find a stochastic process that can be used to represent *noise*. An obvious choice is the *white noise* term $\dot{W}(t) = dW(t)/dt$—the time derivative of the Brownian motion (BM) process $\{W_t\}_{t \geq 0}$ (see Section 3.5.4). In this regard, a substitution of $\dot{W}(t)dt = dW(t) = \text{noise} \cdot dt$ into (4.5) yields

$$\int_0^t j(s)N(s) \cdot \text{noise} \cdot ds = \int_0^t j(s)N(s)dW(s). \tag{4.5.1}$$

What are the key issues surrounding the calculation of the integral given in (4.5.1)? For one thing, (4.5.1) is not an ordinary integral since BM is nowhere differential almost surely (a.s.) or with probability 1. For another, BM is a stochastic process whose paths are a.s. of infinite variation for almost every $\omega \in \Omega$. To circumvent these difficulties, we shall look to a specialized mathematical apparatus that directly exploits the random nature of BM. Specifically, we shall look to the Itô stochastic integral (Itô, 1946) defined with respect to BM.

## 4.2 Stochastic Integration: The Itô Integral

Let us now define the class of admissible functions or processes to which the Itô integral will be applied (Allen, 2010; Evans, 2013; Friedman, 2006; Mao, 2007; Øksendal, 2013). That is, our objective is to obtain a solution of the **stochastic integral**

$$\int_0^t f(s, \omega) \, dW(s, \omega), \omega \in \Omega, \text{ or, written more succinctly, of}$$

$$\int_0^t f(s) \, dW(s). \tag{4.6}$$

To this end, suppose $(\Omega, \mathcal{A}, P)$ is a complete probability space with a filtration $\mathcal{I} = \{\mathcal{A}_t\}_{t \geq 0}$ satisfying the *usual conditions* (see Section 3.1.3). In addition, let $\{W_t\}_{t \geq 0}$ be a one-dimensional BM process defined on the filtered probability space $(\Omega, \mathcal{A}, \mathcal{I}, P)$ that is adapted to the filtration $\mathcal{I}$. For $a, b \in R$, with $0 \leq a < b < +\infty$, let the interval $[a, b]$ be endowed with its Borel $\sigma$-algebra and let $m^2([a, b])$ denote the space of all real-valued processes $f = \{f(t)\}_{t \in [a, b]}$ on the product space $\Omega \times [a, b]$ such that

i.  process $f$ is measurable with respect to the product $\sigma$-algebra on $\Omega \times [a, b]^2$
ii. for each $t \in [a, b]$, the random variable $f(t)$ is $\mathcal{A}_t$-measurable

  i.e., the stochastic process $f$ is adapted to $\mathcal{F}$; and

iii. $\|f\|_2^2([a, b]) = E\left( \int_a^b |f(t)|^2 dt \right) < +\infty.$ \hfill (4.7)

Under the norm (4.7iii), $m^2([a, b])$ is complete and, for processes $\{f\}, \{h\} \in m^2([a, b])$, these processes are **equivalent** if $\|f - h\|_2^2([a, b]) = 0$. (The set of processes $m^2([a, b]) = \{f : \Omega \times [a, b] \to R | f \text{ satisfies } (4.7)\}$, which displays a Hilbert space structure, is termed the **natural domain of the Itô integral**.)

Given that $f(t) \in m^2([a, b])$, how should (4.6) be defined? A brief sketch of our approach to providing an answer to this question proceeds in three steps. For an integrand $f \in m^2([a, b])$, let the Itô stochastic integral be denoted as

$$I(f) = \int_a^b f(t) \, dW(t) \in L^2(P),$$ with $P$ a probability measure.

Step 1. If $\phi = \{\phi(t)\}_{t \in [a, b]} \in m_0^2([a, b])$ is an elementary or step process ($m_0^2([a, b])$ is the subspace of elementary functions in $m^2([a, b])$), define the stochastic integral of $\phi$ as

$$I(\phi) = \int_a^b \phi(t) dW(t) = \sum_{i=0}^{k-1} e_i [W(t_{i+1}) - W(t_i)].$$

Step 2. The Itô isometry on $m_0^2([a, b])$ holds, that is, for all elementary functions $\phi \in m_0^2([a, b])$, $\|\phi\|_{m_0^2([a, b])}^2 = \|I(\phi)\|_{L^2P}^2 = E\left( I(\phi)^2 \right)$ or, for $\phi \in m_0^2([a, b])$, the mapping $I$: $m_0^2([a, b]) \to L^2(P)$ is an isometry and is continuous.

Step 3. For every elementary function $\phi \in m_0^2([a, b])$, there exists a sequence of elementary processes $\phi_n \in m_0^2([a, b])$ such that

$$\|f - \phi_n\|_{m_0^2([a, b])}^2 \to 0 \text{ as } n \to \infty.$$

---

2 One way to determine measurability is: if $f(t)$ is adapted and $t \mapsto f(t)$ is a.s. right- (or left-) continuous in $t$, then $f$ is measurable.

Then define

$$I(f) = \lim_{n \to \infty} \int_a^b \phi_n dW(t) \text{ in } L^2(P),$$

that is, since each $f \in m_0^2([a, b])$ can be approximated by a step process, the Itô stochastic integral is the $L^2$-limit of integrals of elementary functions or $I(f)$ is the limit of $\int_a^b \phi(t)dW(t)$ as $\phi \to f \in m_0^2([a, b])$.

We now fill in some of the details.

We may view the notion of an elementary or step process in the following fashion. A real-valued stochastic process $\{\phi(t)\}_{t \in [a, b]}$ is an **elementary (step) process** if there exists a **partition** $a = t_0 < t_1 < \cdots < t_k = b$ of $[a, b]$, with **mesh size** $\max_{0 \le i \le k-1} |t_{i+1} - t_i|$, and bounded random variables $e_i, 0 \le i \le k-1$, such that $e_i$ is $\mathcal{A}_{t_i}$-measurable, $E(e_i^2) < +\infty$, and

$$\phi(t) = \sum_{i=0}^{k-1} e_i \chi_{(t_i, t_{i+1}]}(t), \tag{4.8}$$

where $\chi_{(t_i, t_{i+1}]}(t) = 1$ if $t \in (t_i, t_{i+1}]$ and 0 otherwise, and with the family of all such processes denoted as $m_0^2([a, b]) \subset m^2([a, b])$. In this regard, for an elementary processes $\phi(t) \in m_0^2([a, b])$, define the random variable

$$I(\phi) = \int_a^b \phi(t)dW(t) = \sum_{i=0}^{k-1} e_i[W(t_{i+1}) - W(t_i)] \tag{4.9}$$

as the **Itô stochastic integral** of $\phi$ with respect to the BM process $\{W_t\}_{t \ge 0}$.

Here, $\int_a^b \phi(t)dW(t)$ is $\mathcal{A}_b$-measurable, $e_i$ is $\mathcal{A}_{t_i}$-measurable, and $W(t_{i+1}) - W(t_i)$ is independent of $\mathcal{A}_{t_i}, i = 0, 1, ..., k-1$. Moreover, (4.9) is an element of $L^2(P)$ since, if $\phi \in m_0^2([a, b])$ (i.e., $\varnothing$ is elementary and bounded), then it is true that

    a. $E\left(\int_a^b \phi(t)dW(t)\right) = 0$; and

    b. $E\left(\left|\int_a^b \phi(t)dW(t)\right|^2\right) = E\left(\int_a^b |\phi(t)|^2 dt\right)^3$     (4.10)

    [the Itô isometry]

---

3 In general, think of an **isometry** as involving a function from one metric space to another metric space that displays the property that the distance between two points in the first space is equal to the distance between the image points in the second space. We thus have a function that preserves a metric or preserves length (e.g., preserves norms).

Moreover, if both $\phi_1, \phi_2 \in m_0^2([a, b])$, with $c_1, c_2 \in R$, then $c_1\phi_1 + c_2\phi_2 \in m_0^2([a, b])$ and

$$\int_a^b \left( c_1\phi_1(t) + c_2\phi_2(t) \right) dW(t) = c_1 \int_a^b \phi_1(t) dW(t) + c_2 \int_a^b \phi_2(t) dW(t) dt.$$

$$(4.11)$$

Given (4.10b) and (4.11), we can now broaden the definition of the integral based upon elementary processes to also include processes in $m^2([a, b])$. The justification for this extension is Theorem 4.1, an important approximation result:

**Theorem 4.1** For any $f \in m^2([a, b])$, there exists a sequence $\{\phi_n(t)\}_{n=1}^\infty$ of elementary processes such that

$$\lim_{n \to \infty} E\left( \int_a^b |f(t) - \phi_n(t)|^2 dt \right) = 0.$$

$$(4.12)$$

As promised, we can now define the Itô stochastic integral for a process $f \in m^2([a, b])$. To this end, via (4.12) (4.11), and (4.10b),

$$\lim_{n,m \to \infty} E\left( \left| \int_a^b \phi_n(t) dW(t) - \int_a^b \phi_m(t) dW(t) \right|^2 \right) =$$

$$\lim_{n,m \to \infty} E\left( \left| \int_a^b (\phi_n(t) - \phi_m(t)) dW(t) \right|^2 \right) =$$

$$\lim_{n,m \to \infty} E\left( \int_a^b |\phi_n(t) - \phi_m(t)|^2 dt \right) = 0.$$

Hence, $\left\{ \int_a^b \phi_n(t) dW(t) \right\}$ is a Cauchy sequence in $L^2(P)$ so that its limit exists in $L^2(P)$. In addition, we will define this limit as the Itô stochastic integral. To summarize, the **Itô stochastic integral of $f$ with respect to** $\{W_t\}_{t \geq 0}$ is

$$\int_a^b f(t) dW(t) = \lim_{n \to \infty} \int_a^b \phi_n(t) dW(t) \operatorname{in} L^2(P),$$

$$(4.13)$$

where $\{\phi_n\}$ is a sequence of elementary processes satisfying (4.12). (Note that the limit in (4.13) does not depend upon the choice of $\{\phi_n\}$ so long as (4.12) holds.)

Given that $f,g \in m^2([a,b])$ and $\alpha,\beta \in R$, this stochastic integral has the following properties:

1. (**Measurability**) $\int_a^b f(t)dW(t)$ is $\mathcal{A}_b$-measurable.

2. $E\left(\int_a^b f(t)dW(t)\right) = 0$ and $E\left(\int_a^b f(t)dW(t)|\mathcal{A}_a\right) = 0$.

3. $E\left(\left|\int_a^b f(t)dW(t)\right|^2\right) = E\left(\int_a^b |f(t)|^2 dt\right)$ and

$$E\left(\left|\int_a^b f(t)dW(t)\right|^2 |\mathcal{A}_a\right) = E\left(\int_a^b |f(t)|^2 dW(t)|\mathcal{A}_a\right)$$

$$= \int_a^b E\left(|f(t)|^2|\mathcal{A}_a\right)dt.$$

4. (**Linearity**) $\int_a^b (\alpha f(t) + \beta g(t))\, dW(t) = \alpha \int_a^b f(t)\, dW(t) + \beta \int_a^b g(t)\, dW(t).$

5. $E\left(\int_a^b f(t)\, dW(t) \int_a^b g(t)\, dW(t)\right) = E\left(\int_a^b f(t)g(t)\, dW(t)\right).$

6. If $Z$ is a real – valued, bounded $\mathcal{A}_b$-measurable random variable, then

$$Zf \in m^2([a,b])$$

$$\int_a^b Zf(t)\, dW(t) = Z\int_a^b f(t)\, dW(t).$$

7. (**Gaussian**) if the integral $f$ is deterministic (i.e., $f$ is independent of $\omega \in \Omega$), then the Itô stochastic integral $\int_a^b f(t)\, dW(t)$ is $N\left(0, \int_a^b f(t)^2 dt\right).$

8. (**Additivity**) $\int_a^b f(t)\, dW(t) = \int_a^c f(t)\, dW(t) + \int_c^b f(t)\, dW(t)$ for $a < c < b.$

9. For a set $A \subset [a,b]$, $\int_A f(t)\, dW(t) = \int_a^b f(t)\chi_A\, dW(t).$

$$(4.14)$$

We next define, for $f \in m^2([0, T])$, the **indefinite Itô stochastic integral** as

$$I(t) = \int_0^t f(s) \, dW(s), t \in [0, T], \tag{4.15}$$

where $I(0) = 0$ and $\{I(t)\}_{t \in [0, T]}$ is $\mathcal{A}_t$-adapted and has the following martingale property. That is, if $f \in m^2([0, T])$, then $\{I(t)\}_{t \in [0, T]}$ is a square integrable martingale with respect to the filtration $\mathcal{I} = \{\mathcal{A}_t\}_{t \geq 0}$ so that, via (4.14.2), $\{I(t)\}_{t \in [0, T]}$ has the martingale property

$$E(I(t) | \mathcal{A}_s) = E(I(s) | \mathcal{A}_s) + E\left( \int_s^t f(r) dW(r) | \mathcal{A}_s \right) = I(s),$$

$0 \leq s < t \leq T$. Specifically, Doob's martingale property (3.5b) holds or

$$E\left( \max_{0 \leq t \leq T} \left| \int_0^t f(s) \, dW(s) \right|^2 \right) \leq 4E\left( \int_0^T |f(s)|^2 ds \right).$$

Moreover, $\{I(t)\}_{t \in [0, T]}$ has a **continuous version** a.s. That is, for $f \in m^2([0, T])$, there exists a **$t$-continuous** version of (4.15) in that there exists a $t$-continuous stochastic process $\{J_t\}$ on $(\Omega, \mathcal{A}, P)$ with continuous sample paths such that $P\left( J_t = \int_0^t f dW \right) = 1, t \in [0, T]$. (From this point on, when we speak of an indefinite Itô integral (4.15), we mean its $t$-continuous version.) In sum, for $f \in m^2([0, T])$, the indefinite Itô integral $\{I(t)\}_{t \in [0, T]}$ is a square-integrable continuous martingale with quadratic variation.

$$(I(t), I(t))_t = \int_0^t |f(s)|^2 ds, t \in [0, T].$$

Given that the Itô stochastic integral (4.15) is a one-dimensional stochastic process, we write $X_t = \int_0^t f(s) \, dW(s), t \in [0, T]$. Then $X_t - X_s = \int_s^t f(u) \, dW(u)$, $0 \leq s \leq t \leq T$. In addition,

a. $E(X_t) = 0$ and $V(X_t) = E(X_t^2), t \in [0, T]$;

b. $COV(X_t, X_s) = E(X_t X_s) = \int_0^{min(s, t)} E(f(u)^2) \, du, 0 \leq s \leq t \leq T$;

c. $E(|X_t|^2) = \int_0^t E(|f(u)|^2) \, du, t \in [0, T]$; and

d. the process $X_t = \int_0^t f(s) \, dW(s), t \in [0, T]$, has **orthogonal increments**, that is, for $0 \leq r \leq s \leq t \leq T, E[(X_u - X_t)(X_s - X_r)] = 0$. (Note that although the Itô stochastic integral has orthogonal, and thus uncorrelated, increments, these increments are not, in general, independent. However, if $f$ is independent of $\omega$, then property (4.14.7) obtains. In this instance, the uncorrelatedness of a set of normally distributed random variables implies their independence so that $X_t$ is a random process with independent increments. This said, we can thus conclude that a normal process $X_t$ that has $E(X_t) = 0$ and $X(0) = 0$ has independent increments in the interval and thus can be represented by an Itô stochastic integral of the form (4.15).)

It is important to note that, in our subsequent discussions, all (three) of the following notations are equivalent and will be used interchangeably:

$$\int_0^t f dW = \int_0^t f(s) \, dW(s) = \int_0^t f(s, \omega) \, d(s, \omega).$$

## 4.3 One-Dimensional Itô Formula

Let $\{W_t\}_{t \geq 0}$, $W_0 = 0$, be a one-dimensional BM on a probability space $(\Omega, \mathcal{A}, P)$ that is adapted to the filtration $\mathcal{F} = \{\mathcal{A}_t\}_{t \geq 0}$. What is the solution of the integral $\int_0^t W(s) \, dW(s)$? Under ordinary (Riemann) integration,

$$\int_0^t W(s) \, dW(s) = \frac{1}{2} W(t)^2.$$

However, as will be determined later, the desired integral is actually

$$\int_0^t W(s) \, dW(s) = \frac{1}{2} W(t)^2 - \frac{1}{2} t,$$

that is, the extra term $-(1/2)t$ demonstrates that the Itô stochastic integral cannot be evaluated as one would an ordinary integral. To correctly evaluate Itô stochastic integrals, we shall use what is called the Itô chain rule or Itô formula.

To this end, let $\{W_t\}_{t \geq 0}$ be a one-dimensional BM defined on a complete probability space $(\Omega, \mathcal{A}, P)$ and adapted to the filtration $\mathcal{F} = \{\mathcal{A}_t\}_{t \geq 0}$. In addition, let $f(t) \in L_{R+}^1$ and $g(t) \in L_{R+}^2$. Then a one-dimensional continuous adapted process $\{X_t\}_{t \geq 0}$ of the form

$$X(t) = X_0 + \int_0^t f(s) \, ds + \int_0^t g(s) \, dW(s) \tag{4.16}$$

is called an **Itô process** with **stochastic differential**

$$dX(t) = f(t)dt + g(t)dW(t). \tag{4.17}$$

Next, suppose $V = V(X,t):R \times R_+ \to R$ is twice continuously differentiable in $X$ and once in $t$. Then under (4.16) and (4.17), $Y(t) = V(X(t),t)$ is also an Itô process with stochastic differential

$$dY(t) = \left[ V_t(X(t),t) + V_X(X(t),t)f(t) + \frac{1}{2}V_{XX}(X(t),t)g(t)^2 \right] dt$$
$$+ V_X(X(t),t)g(t)dW(t) \text{ a.s.,}$$

(4.18)

where $V_t = \partial V/\partial t, V_X = \partial V/\partial X$, and $V_{XX} = \partial^2 V/\partial X^2$. This expression is commonly known as the one-dimensional **Itô chain rule** or **Itô formula**.[4] Moreover, $(X(t), t)$ is the argument of $V_t$, $V_X$, and $V_{XX}$ while for all times $t \geq 0$, (4.18) is interpreted as

$$Y(t) - Y(0) = V(X(t),t) - V(X(0),0)$$
$$= \int_0^t \left( V_t(X(s),s) + V_X(X(s),s)f(s) + \frac{1}{2}V_{XX}(X(s),s)g(s)^2 \right) ds$$
$$+ \int_0^t V_X(X(s),s) g(s)dW \text{ a.s.}$$

(4.19)

A convenient computational device for evaluating certain Itô integrals is Theorem 4.2.

**Theorem 4.2**  Integration-by-Parts Formula
Let $f(s)$ depend only on $s$ ($w$ must not be an argument of $f$) with $f$ continuous and of bounded variation on $[0, t]$. Then

$$\int_0^t f(s) dW(s) = f(t)W(t) - \int_0^t W(s) df(s).$$

(4.20)

One final point is in order. Suppose $\{X_t\}_{t \geq 0}$ is an Itô process satisfying $dX(t) = fdt + gdW(t)$ and let both $f$ and $g$ be *deterministic* functions of $t$, with $X(0) = X_0 = $ constant, that is,

$$X_t = X_0 + \int_0^t f(s) ds + \int_0^t g(s) dW(s).$$

Then $\{X_t\}_{t \geq 0}$ is a

$$N[E(X(t)), V(X(t))] = N\left( X_0 + \int_0^t f(s) ds, \int_0^t g(s)^2 ds \right)$$

process with independent increments.

---

4  A rationalization of Itô's formula is provided by Appendix 4.A.

**Example 4.1** Let us calculate $\int_0^t W(s)\, dW(s)$. Two approaches will be offered: (1) direct use of Itô's formula and (2) a modification thereof.

1. From experience with classical theory, one might surmise that the solution should include a term of the form $(1/2)W(t)^2$. Hence, we shall apply Itô's formula to $Y = V(X(t),t) = (1/2)X(t)^2$ or, for $X(t) = W(t)$, to $V = (1/2)W(t)^2$, that is, since $V_X = W(t), V_{XX} = 1$, and (4.17) becomes $dX(t) = (0)dt + (1)dW(t)$, (4.18) yields

$$d\left(\frac{W(t)^2}{2}\right) = \left[\underset{(=0)}{V_t} + \underset{(=0)}{V_X f(t)} + \frac{1}{2}\underset{(=1/2)}{V_{XX} g(t)^2}\right] dt + \underset{(=W(t))}{V_X g(t) dW(t)}$$

$$= \frac{1}{2}dt + W(t)dW(t).$$

Thus, the solution in *differential form* is

$$d\left(\frac{W(t)^2}{2}\right) = \frac{1}{2}dt + W(t)dW(t).$$

In integral form, the solution is, from (4.19),

$$\frac{W(t)^2}{2} = \frac{1}{2}t + \int_0^t W(s)\, dW(s) \text{ or}$$

$$\int_0^t W(s)\, dW(s) = \frac{W(t)^2}{2} - \frac{1}{2}t.$$

2. Since $V(X(t), t)$ might involve only $X(t)(= W(t))$ and not explicitly include a "$t$" term (the present case), let us examine the resulting differential and integral forms of the solution by supposing that $V = V(X(t))$ is twice differentiable with a continuous second derivative. Then for $X(t) = W(t)$, a second-order Taylor expansion of $V = V(W(t))$ renders Itô's formula in *differential form* or

$$dV(W(t)) = V'(W(t))dW(t) + \frac{1}{2}V''(W(t))(dW(t))^2$$

$$= V'(W(t))dW(t) + \frac{1}{2}V''(W(t))dt$$

(4.21)

via Table 4.A.1. The integral form of Itô's formula then appears as

$$V(W(t)) = V(W(0)) + \int_0^t V'(W(s))\, dW(s) + \frac{1}{2}\int_0^t V''(W(s))\, ds.$$

(4.22)

So given $V = V(W(t)) = (1/2)W(t)^2$, (4.21) yields

$$d\left(\frac{W(t)^2}{2}\right) = W(t)\,dW(t) + \frac{1}{2}(1)\,dt$$

$$= W(t)\,dW(t) + \frac{1}{2}\,dt;$$

and from (4.22)

$$\frac{W(t)^2}{2} = 0 + \int_0^t W(s)\,dW(s) + \frac{1}{2}\int_0^t ds \text{ or}$$

$$\int_0^t W(s)\,dW(s) = \frac{W(t)^2}{2} - \frac{1}{2}t.\,\blacksquare$$

**Example 4.2** Find $\int_0^t s\,dW(s)$. From classical analysis, we might guess that the solution should have a term of the form $tW(t)$. In this regard, set $Y = V(X(t), t) = tX(t)$ or, for $X(t) = W(t)$, $V = tW(t)$ so that (4.17) becomes $dX(t) = (0)\,dt + (1)\,dW(t)$ and Itô's formula (4.18) yields

$$d(tW(t)) = \left[\underset{(=W(t))}{V_t} + \underset{(=0)}{V_x f(t)} + \frac{1}{2}\underset{(=0)}{V_{xx}g(t)^2}\right]dt + \underset{(=t)}{V_x g(t)}dW(t).$$

Hence, the solution in *differential form* is

$$d(tW(t)) = W(t)dt + t\,dW(t);$$

and the solution in *integral form* is

$$tW(t) = \int_0^t W(s)\,ds + \int_0^t s\,dW(s) \text{ or}$$

$$\int_0^t s\,dW(s) = tW(t) - \int_0^t W(s)\,ds.\,\blacksquare$$

**Example 4.3** Evaluate $\int_0^t e^{W(s)}dW(s)$. For $X(t) = W(t)$, (4.17) becomes $dX(t) = (0)dt + (1)dW(t)$. Then set $V(X(t)) = V(W(t)) = e^{W(t)}$ (since deterministic theory supports the solution having a term of the form $e^{W(t)}$). Then from (4.21), the *differential form* of the solution is

$$d\left(e^{W(t)}\right) = e^{W(t)}dW(t) + \frac{1}{2}e^{W(t)}dt;$$

and the *integral form* is

$$e^{W(t)} = \int_0^t e^{W(s)} dW(s) + \frac{1}{2} \int_0^t e^{W(s)} ds + V(W(0)) \text{ or}$$

$$\int_0^t e^{W(s)} dW(s) = e^{W(t)} - 1 - \frac{1}{2} \int_0^t e^{W(s)} ds. \quad \blacksquare$$

**Example 4.4**  Find $\int_0^t e^{\mu s + \sigma W(s)} dW(s)$. Set $X(t) = W(t)$. Then from (4.17),

$dX(t) = (0)dt + (1)dW(t)$, and thus $V = V(X(t),t) = V(W(t),t) = e^{\mu t + \sigma W(t)}$.
Then by Itô's formula (4.18),

$$dV = \left[ V_t + V_X f(t) + \frac{1}{2} V_{XX} g(t)^2 \right] dt + V_X g(t) dW(t)$$

$$(=\mu V)(=0) \quad \left( = \left( \frac{1}{2} \right) \sigma^2 V \right) \quad (=\sigma V)$$

so that the solution in *differential form* is

$$dV = \left( \mu V + \frac{1}{2} \sigma^2 V \right) dt + \sigma V dW(t)$$

$$= \left( \mu + \frac{1}{2} \sigma^2 \right) V dt + \sigma V dW(t).$$

The solution in *integral form* is, for $V(W(0),0) = 1$,

$$e^{\mu t + \sigma W(t)} = 1 + \left( \mu + \frac{\sigma^2}{2} \right) \int_0^t e^{\mu s + \sigma W(s)} ds + \sigma \int_0^t e^{\mu s + \sigma W(s)} dW(s) \text{ or}$$

$$\int_0^t e^{\mu s + \sigma W(s)} dW(s) = \frac{1}{\sigma} e^{\mu t + \sigma W(t)} - \frac{1}{\sigma} - \left( \frac{\mu}{\sigma} + \frac{\sigma}{2} \right) \int_0^t e^{\mu s + \sigma W(s)} ds.$$

(Note that if $\mu = 0$ and $\sigma = 1$, then the result of Example 4.3 emerges.) $\blacksquare$

**Example 4.5**  Determine $\int_0^t W(s)^2 dW(s)$. Classical analysis provides a hint
that the solution should contain a term of the form $(1/3)W(t)^3$. Hence, we shall
apply Itô's formula to $Y = V(X(t)) = (1/3)X(t)^3$ or, for $X(t) = W(t)$, to $V =
(1/3)W(t)^3$. In this regard, since $V' = W(t)^2$ and $V'' = 2W(t)$, Itô's formula
(4.21) yields the solution in *differential form*

$$d \left( \frac{1}{3} W(t)^3 \right) = W(t)^2 dW(t) + W(t) dt$$

or, in *integral form,*

$$\frac{1}{3}W(t)^3 = \int_0^t W(s)^2 dW(s) + \int_0^t W(s)\, ds \text{ or}$$

$$\int_0^t W(s)^2\, dW(s) = \frac{1}{3}W(t)^3 - \int_0^t W(s)\, ds. \blacksquare$$

For the convenience of the reader, Table 4.1 of some common stochastic integrals and their solutions now follows.

**Table 4.1** Common stochastic integrals (for real $a < b; t > 0$).

| | |
|---|---|
| 1. | $\int_0^t dW(s) = W(t)$ |
| 2. | $\int_0^t W(s)\, dW(s) = \frac{1}{2}W(t)^2 - \frac{1}{2}t$ |
| 3. | $\int_a^b c\, dW(s) = c(W(b) - W(a)), c$ a constant |
| 4. | $\int_a^b W(s)\, dW(s) = \frac{1}{2}\left(W(b)^2 - W(a)^2\right) - \frac{1}{2}(b-a)$ |
| 5. | $\int_0^t s\, dW(s) = tW(t) - \int_0^t W(s)\, ds$ |
| 6. | $\int_0^t W(s)^2\, dW(s) = \frac{1}{3}W(t)^3 - \int_0^t W(s)\, ds$ |
| 7. | $\int_0^t e^{W(s)} dW(s) = e^{W(t)} - 1 - \frac{1}{2}\int_0^t e^{W(s)} ds$ |
| 8. | $\int_0^t W(s) e^{W(s)} dW(s) = 1 + W(t)e^{W(t)} - e^{W(t)} - \frac{1}{2}\int_0^t e^{W(s)}(1 + W(s)) dW(s)$ |
| 9. | $\int_0^t s W(s)\, dW(s) = \frac{t}{2}\left(W(t)^2 - \frac{t}{2}\right) - \frac{1}{2}\int_0^t W(s)^2 ds$ |
| 10. | $\int_0^t \left(W(s)^2 - s\right) dW(s) = \frac{1}{3}W(t)^3 - tW(t)$ |
| 11. | $\int_0^t e^{-\frac{s}{2} + W(s)} dW(s) = e^{-\frac{t}{2} + W(t)} - 1$ |
| 12. | $\int_0^t \sin W(s)\, dW(s) = 1 - \cos W(t) - \frac{1}{2}\int_0^t \cos W(s)\, ds$ |

*(Continued overleaf)*

**Table 4.1** (*Continued*)

| | |
|---|---|
| 13. | $\int_0^t \cos W(s)\,dW(s) = \sin W(s) + \dfrac{1}{2}\int_0^t \sin W(s)\,ds$ |
| 14. | $E\left(\int_0^t dW(s)\right) = 0$ |
| 15. | $E\left(\int_0^t W(s)dW(s)\right) = 0$ |
| 16. | $V\left(\int_0^t W(s)dW(s)\right) = \dfrac{t^2}{2}$ |

## 4.4 Martingale Representation Theorem

It was mentioned earlier in Section 4.2 that the Itô integral is a martingale with respect to $A_t$. The converse of this statement is also true—any $A_t$-martingale $\{X_t\}_{t\geq 0}$ with $E(X_t^2) < +\infty$ can be written as a stochastic integral. In what follows, we shall argue that *any* square-integrable random variable that is measurable with respect to a BM is representable as a stochastic integral in terms of this BM. In particular, any martingale that is adapted to a BM filtration can be written as a stochastic integral.

To set the stage for this discussion, suppose $\{W_t\}_{t\geq 0}$ is a BM. If $f \in L^2(P)$, then the random variable $V = \int_0^t f(s)\,dW(s), t \in [0, T]$, is $A_t$-measurable and, by the Itô isometry, $E(V^2) = E\left(\int_0^t f(s)^2 ds\right) < +\infty$ so that $V \in L^2(P)$. Theorem 4.3, the Itô representation theorem, informs us that any $f \in L^2(P)$ can be represented in this fashion. That is,

**Theorem 4.3** Itô Representation Theorem
Let $\{W_t\}_{t\geq 0}$ be a BM. For every function $f \in L^2(P)$ and adapted to the filtration $\mathcal{I} = \{A_t\}_{t\geq 0}$, there is a unique stochastic process $\{U_t\}_{t\geq 0}$ such that $E\left(\int_0^t U(s)^2 ds\right) < +\infty$.

The importance of this theorem is that it enables us to obtain the following characterization of the square-integrable martingales adapted to the filtration $\mathcal{I}$. Specifically, Theorem 4.4 describes the martingale representation theorem.

**Theorem 4.4** Martingale Representation Theorem

Let $\{W_t\}_{t \geq 0}$ be a BM with $\{X_t\}_{t \geq 0}$ a square-integrable martingale adapted to the filtration $\mathfrak{F} = \{\mathcal{A}_t\}_{t \geq 0}$. Then there is a unique stochastic process $\{U_t\}_{t \geq 0}$ such that, for every $t \geq 0$, $E\left(\int_0^t U(s)^2 ds\right) < +\infty$.

In this regard, any square-integrable martingale adapted to the filtration $\mathfrak{F}$ admits a continuous version.

## 4.5 Multidimensional Itô Formula

Let us now extend the one-dimensional Itô formula (Section 4.3) to the multidimensional case. To this end, suppose $\{W(t)\}_{t \geq 0} = (W_1(t), \ldots, W_m(t))'$ (a *prime* "'" denotes matrix transposition) is an $m$-dimensional BM defined on the complete probability space $(\Omega, \mathcal{A}, P)$ and adapted to the filtration $\mathfrak{F} = \{\mathcal{A}_t\}_{t \geq 0}$, where the components $W_r(t)$ and $W_s(t)$ are independent if $r \neq s$. In addition, let $\boldsymbol{f}(t) = (f_1(t)), \ldots, f_d(t)' \in L^1_{R_+}(R^d)$ and $\boldsymbol{g}(t) = [g_{ij}]_{dxm} \in L^2_{R_+}(R^{dxm})$. Then a $R^d$-valued continuous adapted process $\{X(t)\}_{t \geq 0} = (X_1(t), \ldots, X_d(t))'$ of the form

$$X(t) = X_0 + \int_0^t \boldsymbol{f}(s)ds + \int_0^t \boldsymbol{g}(s)d\boldsymbol{W}(s) \tag{4.23}$$

is termed a **$d$-dimensional Itô process** with **stochastic differential**

$$dX(t) = \boldsymbol{f}(t)dt + \boldsymbol{g}(t)d\boldsymbol{W}(t), \tag{4.24}$$

where

$$\underset{(d \times 1)}{X(t)} = \begin{bmatrix} X_1(t) \\ \vdots \\ X_d(t) \end{bmatrix}, \quad \underset{(d \times 1)}{\boldsymbol{f}} = \begin{bmatrix} f_1(t) \\ \vdots \\ f_d(t) \end{bmatrix}, \quad \underset{(d \times m)}{\boldsymbol{g}(t)} = \begin{bmatrix} g_{11} \cdots g_{1m} \\ \vdots \quad \vdots \\ g_{d1} \cdots g_{dm} \end{bmatrix},$$

$$\text{and } \underset{(m \times 1)}{d\boldsymbol{W}(t)} = \begin{bmatrix} dW_1(t) \\ \vdots \\ dW_m(t) \end{bmatrix}.$$

Here, the $i$th component of (4.23) is

$$X_i(t) = X_i(0) + \int_0^t f_i(s)ds + \int_0^t \sum_{j=1}^m g_{ij}(s)dW_j(s), i = 1, \ldots, d; \tag{4.23.1}$$

and the $i$th component of (4.24) is

$$dX_i(t) = f_i(t)dt + \sum_{j=1}^{m} g_{ij}(t)dW_j(t), i = 1, \ldots, d. \tag{4.24.1}$$

Now, suppose $V = V(X(t), t) : R^d \times R_+ \rightarrow R$ is twice continuously differentiable in $X$ and once in $t$. Then under (4.23) and (4.24), $Y(t) = V(X(t), t)$ is also an Itô process with stochastic differential

$$dY(t) = \left[ V_t(X(t), t) + V_X'(X(t), t)f(t) + \frac{1}{2}tr(g'(t)V_{XX}(X(t), t)g(t)) \right] dt$$
$$+ V_X'(X(t), t)g(t)dW(t) \text{ a.s.,}$$

$$\tag{4.25}$$

or

$$dY(t) = \left[ V_t + \sum_{i=1}^{d} V_{X_i}f_i(t) + \frac{1}{2}\sum_{i=1}^{d}\sum_{j=1}^{d}\sum_{k=1}^{m} V_{X_iX_j}g_{ik}(t)g_{jk}(t) \right] dt$$
$$+ \sum_{i=1}^{d}\sum_{j=1}^{m} V_{X_i}g_{ij}(t)dW_j(t) \text{ a.s.,}^5 \tag{4.25.1}$$

where $V_t$, $V_{X_i}$, and $V_{X_iX_j}$ are all evaluated at the point $(X(t), t)$, $i = 1, \ldots, d; j = 1, \ldots, m$.

**Example 4.6** Suppose $d = 1$ and $m = 3$. Also, $X(t) = X_1(t)$, $f(t) = f_1 = 0$, $g(t) = [g_{11}(t), \ldots, g_{13}(t)] = [a, b, c]$, with $a, b, c \in R, \{W(t)\}_{t \geq 0} = (W_1(t), \ldots, W_3(t))'$, and $Y - V(X(t), t) = V(X_1, t) = 2X_1^2(t)$. Then from (4.24.1),

$$dX_1(t) = f_1 dt + g_{11}dW_1(t) + g_{12}dW_2(t) + g_{13}dW_3(t)$$
$$= adW_1(t) + bdW_2(t) + cdW_3(t)$$

and, from (4.25.1),

$$dY(t) = \left[ V_t + V_{X_1}f_1(t) + \frac{1}{2}\sum_{k=1}^{3} V_{X_1X_1}g_{1k}(t) \right] dt$$
$$+ \sum_{j=1}^{m} V_{X_1}g_{1j}(t)dW_j(t)$$
$$= \left[ V_t + V_{X_1}f_1(t) + \frac{1}{2}V_{X_1X_1}(g_{11}(t)g_{11}(t) + g_{12}(t)g_{12}(t) + g_{13}(t)g_{13}(t)) \right] dt$$
$$+ V_{X_1}(g_{11}(t)dW_1(t) + g_{12}(t)dW_2(t) + g_{13}(t)dW_3(t))$$
$$= 2(a^2 + b^2 + c^2)dt + 4X_1(adW_1(t) + bdW_2(t) + cdW_3(t)). \blacksquare$$

---

5 A rationalization of this expression is provided by Appendix 4.B.

## Appendix 4.A    Itô's Formula

Let $\{W_t\}_{t \geq 0}$ be a one-dimensional continuous BM defined on the filtered probability space $(\Omega, \mathcal{A}, \mathcal{I}, P)$ that is adapted to the filtration $\mathcal{I} = \{\mathcal{A}_t\}_{t \geq 0}$. Suppose $[t, t + dt]$ is an infinitesimal time interval, where both $dt$ and $(dt)^{\frac{1}{2}}$ are positive and $(dt)^i, i = 2, 3, \ldots$, is 0. For $dW_t = W_{t+dt} - W_t$, we know that

$$dW_t \sim N[E(dW_t), V(dW_t)] = N(0, dt)$$

so that $dW_t$ is a differential of size $(dt)^{1/2}$. Moreover,

     a.   $E\big((dW_t)^2\big) = dt$; and

     b.   $V\big((dW_t)^2\big) = (dt)^2 = 0 = \text{constant}$          (4.A.1)

so that $(dW_t)^2 = dt$, that is, under (4.A.1b), $(dW_t)^2$ must equal its mean value of $dt$.

In a similar vein,

     a.   $E(dt \cdot dW_t) = dt \, E(dW) = 0$; and

     b.   $V(dt \cdot dW_t) = (dt)^2 V(dW_t) = 0 = \text{constant}$          (4.A.2)

so that $dt \cdot dW_t = 0$ (since (4.A.2b) implies that $dt \cdot dW_t$ must equal its mean value of 0).

To summarize,

$$(dW_t)^2 = dt, (dt)^2 = 0, dt \cdot dW_t = dW_t \cdot dt = 0. \qquad (4.A.3)$$

An aid to remembering these results is provided by the **Itô multiplication table** (Table 4.A.1).

The second-order Taylor expansion of $V(X(t), t)$ *generally* appears as

$$dV(X, t) = V_t \, dt + V_X \, dX + \frac{1}{2}\Big[V_{XX}(dX)^2 + 2V_{Xt} \, dX dt + V_{tt}(dt)^2\Big]$$

$$= V_t \, dt + V_X(f dt + g dW)$$

$$+ \frac{1}{2}\Big[V_{XX}(f dt + g dW)^2 + 2V_{Xt}(f dt + g dW)dt + V_{tt}(dt)^2\Big].$$

$$(4.A.4)$$

Then from the assumptions on $V(X, t)$ and Table 4.A.1, Equation (4.A.4) simplifies to

$$dV(X, t) = \left(V_t + V_X f + \frac{1}{2} V_{XX} g^2\right) dt + V_X g \, dW \qquad (4.A.5)$$

or Equation (4.18).

**Table 4.A.1** Itô multiplication table.

| • | $dt$ | $dW_t$ |
|---|---|---|
| $dt$ | 0 | 0 |
| $dW_t$ | 0 | $dt$ |

Note that Itô's chain rule (4.A.5) reduces to the chain rule of ordinary calculus in the deterministic case where $g \equiv 0$. The stochastic version of the chain rule has an additional term involving $V_{XX}$.

Table 4.A.1 can also be used to develop Itô's product formula. To this end, suppose $X_t$ and $Y_t$ are one-dimensional continuous adapted processes of the form (4.16), that is,

$$X_t = X_0 + \int_0^t a(s)\,ds + \int_0^t b(s)\,dW(s),$$

$$Y_t = Y_0 + \int_0^t \alpha(s)\,ds + \int_0^t \beta(s)\,dW(s),$$

and let $V(x,y) = xy$. Then, via Itô's formula,

$$
\begin{aligned}
d(X_t Y_t) = dV(X_t, Y_t) &= V_x\,dX_t + V_y\,dY_t \\
&\quad + \frac{1}{2}V_{xx}(dX_t)^2 + V_{xy}\,dX_t dY_t + \frac{1}{2}V_{yy}(dY_t)^2 \qquad (4.A.6) \\
&= Y_t\,dX_t + X_t\,dY_t + dX_t dY_t,
\end{aligned}
$$

the **Itô's product formula**. (Note that each of the partial derivatives in (4.A.6) is evaluated at the point $(X_t, Y_t)$.) From this expression, one can readily obtain the general **integration by parts formula**

$$
\begin{aligned}
X_t Y_t &= X_0 Y_0 + \int_0^t Y_s\,dX_s + \int_0^t X_s\,dY_s + \int_0^t dX_s\,dY_s \\
&= X_0 Y_0 + \int_0^t Y_s\,dX_s + \int_0^t X_s\,dY_s + \int_0^t b(s)\,\beta(s)\,ds.
\end{aligned}
\qquad (4.A.7)
$$

## Appendix 4.B   Multidimensional Itô Formula

Our objective here is to provide some guidance for the development of Equation (4.25.1). We shall first collect some of our major results from Section 4.5. To this end, set $Y(t) = V(X(t), t)$. Let us write the $i$th component of

$$dX(t) = f(t)dt + g(t)dW(t) \qquad (4.B.1)$$

**Table 4.B.1** Itô multiplication table ($i \neq j$).

| $\cdot$ | $dt$ | $dW_i$ | $dW_j$ |
|---|---|---|---|
| $dt$ | 0 | 0 | 0 |
| $dW_i$ | 0 | $dt$ | 0 |
| $dW_j$ | 0 | 0 | $dt$ |

as

$$dX_i(t) = f_i(t)dt + \sum_{j=1}^{m} g_{ij}(t)dW_j(t), i = 1,\ldots,d. \qquad (4.B.2)$$

Next, the second-order Taylor expansion of $Y(t)$ can be written as

$$dY(t) = V_t dt + \sum_{j=1}^{d} V_{X_i} dX_i + \frac{1}{2}\sum_{i=1}^{d}\sum_{j=1}^{m} V_{X_i X_j} dX_i dX_j \qquad (4.B.3)$$

(here, all partial derivatives are evaluated at the point $(X(t), t)$, and a substitution of (4.B.2) into (4.B.3) and simplifying (by "brute force") yields Equation (4.25.1). An aid for executing the said simplification is the following set of relationships among the various differentials:

$$(dt)^2 = 0;$$
$$dW_i \cdot dt = dt \cdot dW_i = 0, i = 1,\ldots,d;$$
$$dW_j \cdot dt = dt \cdot dW_j = 0, j = 1,\ldots,m; \text{ and}$$
$$dW_i \cdot dW_j = \delta_{ij}dt, \text{ where } \delta_{ij} = \begin{cases} 1, i = j; \\ 0, i \neq j. \end{cases}$$

These results are summarized in the **Itô multiplication table** (Table 4.B.1).

# 5

# Stochastic Differential Equations

## 5.1 Introduction

Our objective in this chapter is to examine the solution to a **stochastic differential equation** (SDE) of the form

$$dX(t) = f(X(t),t)dt + g(X(t),t)dW(t), t \in [t_0, T], T > 0, \tag{5.1}$$

with initial condition $X(t_0) = X_0$, where $X_0$ is taken to be a random variable independent of $W(t) - W(t_0)$ and $E(|X_0|^2) < +\infty$. Here, $f$ and $g$ are termed the **drift** and **diffusion coefficient functions**, respectively. In particular, we can view $f$ as measuring short-term growth and $g$ as depicting short-term variability. (If $g(X(t),t) \equiv 0$ in (5.1), then the resulting expression is simply the ordinary differential equation (ODE) $dX(t)/dt = f(X(t),t), t \in [t_0, T]$.) Looking to specifics, we shall endeavor to answer questions such as the following:

Under what conditions does a solution *exist*?
If a solution does exist, is it *unique*?
How, and under what circumstances, can we determine an *exact solution*?
If an exact solution is found, in what sense is it *stable*?

To address these issues, let $(\Omega, \mathcal{A}, P)$ be a complete probability space with filtration $\mathfrak{I} = \{\mathcal{A}_t\}_{t \geq 0}$ satisfying the *usual conditions* (see Section 3.1.3). In addition, let $\{W_t\}_{t \geq 0}$ be a one-dimensional Brownian motion (BM) defined on the filtered probability space $(\Omega, \mathcal{A}, \mathfrak{I}, P)$. For $0 \leq t_0 < T < +\infty$, let $X_0$ be a $\mathcal{A}_{t_0}$-measurable random variable such that $E(|X_0|^2)$ is finite. Also, suppose $f : R \times [t_0, T] \rightarrow R$ and $g : R \times [t_0, T] \rightarrow R$ are each Borel-measurable functions. Then the solution to the SDE (5.1) in integral form is

$$X(t) = X_0 + \int_{t_0}^{t} f(X(s),s)\, ds + \int_{t_0}^{t} g(X(s),s)\, dW(s), t \in [t_0, T]. \tag{5.2}$$

*Stochastic Differential Equations: An Introduction with Applications in Population Dynamics Modeling*, First Edition. Michael J. Panik.
© 2017 John Wiley & Sons, Inc. Published 2017 by John Wiley & Sons, Inc.

More to the point, a one-dimensional stochastic process $\{X(t)\}_{t\in[t_0,T]}$ is termed a **solution** of (5.1) if

   i. process $\{X(t)\}$ is continuous and $\mathcal{A}_t$-adapted;

   ii. process $\{f(X(t),t)\} \in L^1([t_0,T])$ and process $\{g(X(t),t)\} \in L^2([t_0,T])$; and

   iii. Equation (5.2) holds for every $t \in [t_0,T]$ almost surely (a.s.).

Moreover, a solution $\{X(t)\}_{t\in[t_0,T]}$ is **unique** if any other solution $\{\tilde{X}(t)\}_{t\in[t_0,T]}$ is indistinguishable from $\{X(t)\}_{t\in[t_0,T]}$ or $P\big(X(t) = \tilde{X}(t), t \subset [t_0,T]\big) = 1$.

## 5.2 Existence and Uniqueness of Solutions

Under what circumstances will a unique solution to (5.1) exist? The answer to this question is provided by Theorem 5.1 (Allen, 2010; Arnold, 1974; Evans, 2013; Gard, 1988; Mao, 2007; Øksendal, 2013; Steele, 2010).

**Theorem 5.1**    Existence and Uniqueness Theorem

If the coefficient functions $f$, $g$ of SDE (5.1) satisfy the conditions

$$|f(x,t) - f(y,t)| + |g(x,t) - g(y,t)| \le k|x-y| \tag{5.3}$$
$$\text{[uniform Lipschitz condition]}$$

for some constant $k$ and all $t \in [t_0,T]$, $T > 0$; and

$$|f(x,t)|^2 + |g(x,t)|^2 \le k^2\big(1 + |x|^2\big) \tag{5.4}$$
$$\text{[linear growth condition]}$$

for some constant $k^2$ and all $t \in [t_0,T]$, $T > 0$, then there exists a continuously adapted solution $X_t$ of (5.1) such that $X(t) = X_0$ and $sup_{t_0 \le t \le T} E\big(|X_t|^2\big) < +\infty$ (i.e., $X_t$ is uniformly bounded in $\mathcal{m}^2([t_0,T])$). Moreover, if $X_t$ and $\tilde{X}_t$ are both continuous $\mathcal{m}^2([t_0,T])$ bounded solutions of (5.1), then

$$P\left( \sup_{0 \le t \le T} |X_t - \tilde{X}_t| = 0 \right) = 1, \tag{5.5}$$

and thus the solution $X_t$ is unique.

Here, the uniform Lipschitz condition (5.3) guarantees that the functions $f$ and $g$ do not change faster than a linear function of $X$ relative to changes in $X$, thus implying the Lipschitz continuity of $f(\cdot,t)$ and $g(\cdot,t)$ for all $t \in [t_0,T]$. Condition (5.4) serves to bound $f$ and $g$ uniformly with respect to $t \in [t_0,T]$; it allows at most linear increases in these functions with respect to $X$. This restriction on the growth of $f$ and $g$ guarantees that, a.s., the solution $X_t$ does not "explode" (tend to $+\infty$) in the interval $[t_0, T]$ no matter what $X_0$ happens to be.

The solution of (5.1) just described is termed a **strong solution** in that it has a **strongly unique** sample path and because the filtered probability space, the BM, and the coefficient functions $f$ and $g$ are all specified in advance. However, if only $f$ and $g$ are specified in advance and the pair of processes $(\tilde{X}_t, \tilde{W}_t)$ are defined on a suitably filtered probability space, then, provided (5.1) holds, $\tilde{X}_t$ is called a **weak solution**. If two weak solutions determined under these conditions are indistinguishable, then **path uniqueness** holds for (5.1). In addition, two solutions to (5.1) (either strong or weak) are termed **weakly unique** if they have the same finite-dimensional probability distribution (even though they have different sample paths). While a strong solution to (5.1) is also a weak solution, the converse is not generally true.

The uniform Lipschitz condition (5.3) and the linear growth condition (5.4) tend to be fairly restrictive requirements for the existence and uniqueness of solutions to the SDE (5.1). To overcome this characteristic, let us broaden the class of functions that can serve as the coefficient functions $f$ and $g$ in (5.1). To this end, let us relax the uniform Lipschitz condition to a local one: for every integer $n \geq 1$, there exists a positive constant $k_n$ such that, for all $x, y \in R, t \in [t_0, T], T > 0$, and with $max\{|x|, |y|\} \leq n$,

$$|f(x,t) - f(y,t)| + |g(x,t) - g(y,t)| \leq k_n |x - y|. \tag{5.3.1}$$
[local Lipschitz condition]

Now, if the linear growth condition (5.4) holds but the uniform Lipschitz condition (5.3) is replaced by the local Lipschitz condition (5.3.1), then there exists a unique solution $X_t$ to (5.1) with $X_t \in m^2([t_0, T])$. (Note that (5.3.1) will be in force when $f$ and $g$ are continuously differentiable in $x$ on $R \times [t_0, T]$.)

Similarly, let us replace the linear growth condition by a monotone condition: suppose there exists a positive constant $k'$ such that, for all $(X(t), t) \in R \times [t_0, T], T > 0$,

$$xf(x,t) + \frac{1}{2}|g(x,t)|^2 \leq k'\left(1 + |x|^2\right). \tag{5.4.1}$$
[monotone condition]

If the local Lipschitz condition (5.3.1) holds but the linear growth restriction (5.4) is replaced by the monotone condition (5.4.1), then there exists a unique solution $X_t$ to (5.1) with $X_t \in m^2([t_0, T])$. The role of the monotone condition is that it ensures the existence of a solution on the entirety of the interval $[t_0, T]$. (Note that if the monotone condition (5.4.1) holds, then the linear growth restriction (5.4) may or may not hold. However, if (5.4) is satisfied, then (5.4.1) is satisfied as well.)

A couple of final points are in order. First, suppose that the coefficient functions $f$ and $g$ are defined on $R \times [t_0, T]$ and the assumptions of Theorem 5.1 hold on every finite subinterval $[t_0, T] \subset [t_0, +\infty)$. Then the SDE $dX(t) = f(X(t), t)\,dt + g(X(t), t)\,dW(t), t \in [t_0, +\infty], X(t_0) = X_0$, has a unique solution defined on the entirety of $[t_0, +\infty)$. In this circumstance, $X_t$ is called a **global solution** to this SDE.

Next, suppose we have an **autonomous SDE** of the form

$$dX(t) = f(X(t))\,dt + g(X(t))\,dW(t), t \in [t_0, T], T > 0, \tag{5.6}$$

with $X(t_0) = X_0$ (here both $f$ and $g$ are independent of $t$ so that $f(X(t), t) \equiv f(X(t)), g(X(t), t) \equiv g(X(t))$. If $X_0$ is a random variable independent of $W(t) - W(t_0)$, then, provided that the assumptions underlying Theorem 5.1 are in effect, this SDE has a unique, continuous global solution $X_t$ on $[t_0, +\infty)$ such that $X(t_0) = X_0$ given that the Lipschitz condition

$$|f(x) - f(y)| + |g(x) - g(y)| \le k|x - y| \tag{5.3.2}$$

[Lipschitz condition]

is satisfied; and, for a fixed $y$ value, the linear growth condition

$$|f(x)|^2 + |g(x)|^2 \le k^2 \left(1 + |x|^2\right) \tag{5.4.2}$$

[linear growth condition]

holds.

## 5.3 Linear SDEs

In the SDE (5.1) (repeated here as $dX(t) = f(X(t), t)dt + g(X(t), t)dW(t)$, $t \in [t_0, T], T > 0$), the coefficient functions $f$ and $g$ are typically nonlinear in form. When nonlinearity does hold, approximation methods are used to generate a solution $X_t$ to (5.1) since, in general, nonlinear SDEs do not have explicit or exact solutions. However, it is possible to determine an explicit solution to a linear SDE, that is, a more thorough treatment of SDEs can be undertaken when the coefficient functions $f$ and $g$ are linear functions of $X$ (and especially when the coefficient functions are independent of $X$).

More specifically, the SDE (5.1) is linear if the functions $f$ and $g$ are each linear functions of $X$ on $R \times [t_0, T]$, that is, if

$$
\begin{aligned}
f(X(t), t) &= a(t) + A(t)X(t); \\
g(X(t), t) &= b(t) + B(t)X(t),
\end{aligned} \tag{5.7}
$$

where $a(t)$, $b(t)$, $A(t)$, and $B(t)$ are all mappings from $[t_0, T]$ to $R$. Then, under (5.7), our **linear SDE** appears as

$$dX(t) = (a(t) + A(t)X(t)) \, dt + (b(t) + B(t)X(t)) \, dW(t),$$
$$t \in [t_0, T], T > 0, X(t_0) = X_0. \tag{5.8}$$

Moreover, again assuming that $X(t_0) = X_0$, a linear SDE is said to be **homogeneous** if $a(t) = b(t) \equiv 0, t \in [t_0, T]$, or

$$dX(t) = A(t)X(t)dt + B(t)X(t)dW(t), t \in [t_0, T], T > 0; \tag{5.9}$$

and it is termed **linear in the narrow sense** if $B(t) \equiv 0$
or

$$dX(t) = (a(t) + A(t)X(t)) \, dt + b(t) \, dW(t), t \in [t_0, T], T > 0. \tag{5.10}$$

What can be said about the existence and uniqueness of a solution to a linear SDE? To answer this, suppose that for the linear SDE (5.8), the functions $a(t)$, $A(t)$, $b(t)$, and $B(t)$ are all measurable and bounded on $[t_0, T], T > 0$, that is,

$$\sup_{t_0 \le t \le T} (|a(t)| + |A(t)| + |b(t)| + |B(t)|) < +\infty.$$

Then $f$ and $g$ in (5.7) satisfy the hypotheses of the existence and uniqueness Theorem 5.1. Hence, (5.8) has a unique, continuous solution $X_t$ over $[t_0, T]$ provided $E(|X_0|^2) < +\infty$ and $X_0$ is independent of $W(t) - W(t_0)$, $t \ge t_0$. If these conditions hold in every subset of $[t_0, +\infty)$, then $X_t$ is a unique, continuous global solution to (5.8).

If the functions $a(t)$, $A(t)$, $b(t)$, and $B(t)$ are all independent of $t$, then (5.8) becomes an **autonomous linear SDE** or

$$dX(t) = (a + AX(t))dt + (b + BX(t))dW(t), t \in [t_0, T], t > 0, \tag{5.11}$$

with $a$, $A$, $b$, and $B$ real constants. For this type of SDE, a unique, continuous global solution always exists.

### 5.3.1 Strong Solutions to Linear SDEs

Consider the stochastic process defined by $Y_t = V(X_t, t)$ (Arnold, 1974; Gard, 1988; Steele, 2010). This process can also be written as a function of the BM process $Y_t = h(W_t, t)$ (e.g., $X_t$ may be a smooth function of $W_t$ and $t$). In this regard, if we are looking for a solution to (5.1) of the form $X_t = h(W_t, t)$, then, to determine any such solution, we need only apply Itô's formula to this expression by (1) writing $X_t = h(W_t, t)$ in terms of a second-order Taylor expansion and (2) invoking the products displayed within the Itô multiplication table (Table 4.A.1). To this end, let

$$dX_t = h_t \, dt + h_W \, dW_t + \frac{1}{2}\left[h_{WW}(dW_t)^2 + 2h_{tW} \, dW_t dt + h_{tt}(dt)^2\right]$$

$$= h_t \, dt + h_W \, dW_t + \frac{1}{2}h_{WW} \, dt \tag{5.12}$$

$$= \left(h_t + \frac{1}{2}h_{WW}\right)dt + h_W \, dW_t.$$

But since $h(W_t,t) = V(X_t,t)$, (5.12) can be rewritten (via the chain rule) as

$$dX_t = \left(h_t(W_t,t) + \frac{1}{2}h_{XX}(W_t,t)\right)dt + h_X(W_t,t)\,dW_t. \tag{5.13}$$

So to solve (5.1), we must match the coefficients on $dt$ and $dW_t$ in (5.13) to those on the respective terms in (5.1). That is, we need to find a function $h(x, t)$ that satisfies the two equations

$$f\big(h(x,t),t\big) = h_t(x,t) + \frac{1}{2}h_{xx}(x,t), \ g\big(h(x,t),t\big) = h_x(x,t). \tag{5.14}$$

[Itô coefficient matching]

This *matching procedure* will be termed **Itô coefficient matching**.

**Example 5.1** (Geometric BM)
Suppose we choose to solve the geometric BM SDE

$$dX_t = \mu X_t \, dt + \sigma X_t \, dW_t, t \in [0, +\infty), X(0) = X_0, \tag{5.15}$$

where $-\infty < \mu < +\infty$ measures short-term growth (drift) and $\sigma > 0$ gauges short-term variability. More specifically, in structure, $X_t$ exhibits both short-term growth and short-term variability in proportion to the level of the process.

To find a function $h(x, t)$ that satisfies (5.14), set

a. $\mu h(x,t) = h_t(x,t) + \frac{1}{2}h_{xx}(x,t),$

b. $\sigma h(x,t) = h_x(x,t).$ $\tag{5.16}$

Then solving the partial differential equation system (5.16) involves the following three steps:

1. Integrate (5.16b) partially with respect to $x$ to obtain $h(x, t)$ up to an additive (arbitrary) function of integration $j(t)$.
2. Substitute the function $h(x, t)$ into (5.16b) in order to determine $j(t)$.
3. The final solution is of the form $X_t = h(W_t,t) + c$, where c is a constant that is determined from the initial condition $X_0$ on $X_t$.

From (5.16b), $h_x/h = \sigma, \ln h = \sigma x + j(t)$, and thus

$$h(x,t) = e^{\sigma x + j(t)}, j(t) \text{ arbitrary.} \tag{5.17}$$

Then a substitution of (5.17) into (5.16a) gives

$$\mu e^{\sigma x + j(t)} = j'(t)e^{\sigma x + j(t)} + \frac{1}{2}\sigma^2 e^{\sigma x + j(t)}$$

or $j'(t) = \mu - (1/2)\sigma^2$. Hence, $j(t) = \mu t + (1/2)\sigma^2 t + \ln c, c = \text{constant}$, and thus a particular solution to (5.15) is

$$X_t = X_0 e^{\sigma W_t + \left(\mu - \frac{1}{2}\sigma^2\right)t}. \tag{5.18}$$
[Geometric BM process]

It is important to note that $X_t$ reflects the fact that the log of the process is BM so that, at each point in time, the distribution of the process is log-normal.

Looking to the long-term dynamics of (5.18), we may conclude the following:

i. If $\mu > (1/2)\sigma^2$, then $X_t \to +\infty$ as $t \to +\infty$.
ii. If $\mu < (1/2)\sigma^2$, then $X_t \to 0$ as $t \to +\infty$.
iii. If $\mu = (1/2)\sigma^2$, then $X_t$ fluctuates between arbitrarily large and arbitrarily small values as $t \to +\infty$. ∎

How do we know if (5.15) has been solved correctly? A direct check is the following. Let $X = k(y,t) = X_0 e^Y$, where $Y = (\mu_Y - (1/2)\sigma^2)t + \sigma W_t$. By Itô's formula (4.18), $dX = X_0 e^Y dY + (1/2)X_0 e^Y dY \cdot dY$, where $dY = \left(\mu - (1/2)\sigma^2\right)dt + \sigma dW_t$ and $dY \cdot dY = \sigma^2 dt$ (see Table 4.A.1). Then

$$dX = X_0 e^Y \left[\left(\mu - \frac{1}{2}\sigma^2\right)dt + \sigma dW_t\right] + \frac{1}{2}X_0 e^Y \sigma^2 dt$$

$$= X_0 e^Y \mu dt + X_0 e^Y \sigma dW_t = \mu X dt + \sigma X dW_t.$$

A specific class of SDEs that are particularly well suited for solution by Itô coefficient matching are those that may be characterized as **exact**. (A review of how to handle exact ODEs is provided by Appendix D.) A test for exactness appears as Theorem 5.2.

**Theorem 5.2** Test for Exactness
If the SDE (5.1) (or the SDE $dX_t = f(W_t, t) + g(W_t, t)dW_t$) is exact, then the coefficient functions $f(x, t)$ and $g(x, t)$ satisfy the condition

$$f_x = g_t + \frac{1}{2}g_{xx}.$$
(Exactness Criterion)

To see this, let us note that if the SDE is exact, then there is a function $h(x, t)$ such that system (5.14) holds. Differentiating the first equation of this pair with respect to $x$ yields

$$f_x = \frac{\partial}{\partial x} h_t + \frac{1}{2} \frac{\partial}{\partial x} h_{xx}.$$

Then substituting $g = h_x$ renders

$$f_x = h_{xt} + \frac{1}{2} \frac{\partial}{\partial x} g_x \text{ (since } h_{xt} = h_{tx})$$

$$= g_t + \frac{1}{2} g_{xx}.$$

For instance, is the SDE $dX_t = e^t \left(1 + 2W_t^2\right)dt + \left(1 + 4e^t W_t\right)dW_t, X_0 = 0$, exact? Here,

$$f(x,t) = e^t(1 + 2x^2),$$
$$g(x,t) = 1 + 4e^t x.$$

Since $f_x = 4xe^t$ and $g_t + \frac{1}{2}g_{xx} = 4e^t x + \frac{1}{2}(0) = 4e^t x$ are equal, we can legitimately conclude that this SDE is exact and thus may be solved via Itô coefficient matching.

**Example 5.2**  It is instructive to note that the SDE (5.15) specified in Example 5.1 can (alternatively) be solved by utilizing Equation (4.18) (Itô's formula) directly. That is, if we rewrite (5.15) as

$$\frac{dX_t}{X_t} = \mu dt + \sigma dW_t,$$

then we see that the term $\ln X_t$ should appear in the solution. In this regard, set $V(X_t, t) = \ln X_t, f = \mu X_t$, and $g = \sigma X_t$ in (4.18) so that

$$dV = \left(0 + \mu X_t \left(\frac{1}{X_t}\right) + \frac{1}{2}\sigma^2 X_t^2 \left(\frac{1}{X_t^2}\right)\right)dt + \sigma X_t \left(\frac{1}{X_t}\right)dW_t$$

or

$$d\ln X_t = \left(\mu - \frac{1}{2}\sigma^2\right)dt + \sigma dW_t,$$

$$\int_0^t d\ln X_u = \left(\mu - \frac{1}{2}\sigma^2\right)\int_0^t du + \sigma \int_0^t dW_u, \text{ and}$$

$$\ln X_t - \ln X_0 = \left(\mu - \frac{1}{2}\sigma^2\right)t + \sigma(W_t - W_0).$$

Then, for $W_0 = 0$,

$$\ln\left(\frac{X_t}{X_0}\right) = \left(\mu - \frac{1}{2}\sigma^2\right)t + \sigma W_t.$$

Thus,

$$X_t = X_0 e^{\left(\mu - \frac{1}{2}\sigma^2\right)t + \sigma W_t}.$$

More generally, if

$$dX_t = u(t)X_t\, dt + v(t)X_t\, dW_t, X(0) = X_0, \tag{5.19}$$

then Itô's formula (4.18) can be applied to $V(X_t,t) = \ln X_t$ (with $f = u(t)X_t, g = v(t)X_t$) to yield the particular solution

$$X_t = X_0 \exp\left(\int_0^t \left(u(s) - \frac{1}{2}v(s)^2\right)ds + \int_0^t v(s)dW_s\right). \blacksquare \tag{5.20}$$

**Example 5.3**  Use the method of Itô coefficient matching to solve the SDE

$$dX_t = -\frac{1}{2}e^{-2X_t}dt + e^{-X_t}dW_t, t \in [0, +\infty), X(0) = X_0. \tag{5.21}$$

Our objective is to find a function $h(x, t)$ that satisfies (5.14).

a.  $-\frac{1}{2}e^{-2h(x,t)} = h_t(x,t) + \frac{1}{2}h_{xx}(x,t);$

b.  $e^{-h(x,t)} = h_x(x,t).$
$$\tag{5.22}$$

From (5.22b),

$$x + j(t) = e^h \text{ or } h = \ln(x + j(t)).$$

A substitution of this expression for $h$ into (5.22a) gives

$$-\frac{1}{2}e^{-2\ln(x+j(t))} = \frac{j'(t)}{x+j(t)} - \frac{1}{2}\frac{1}{(x+j(t))^2}.$$

Upon simplifying this latter result, we obtain $j'(t)/(x+j(t)) = 0$ or $j'(t) = 0$. Clearly, $j(t) = \text{constant} = c$. Hence, $h(x,t) = \ln(x+j(t)) = \ln(x+c)$ or

$$X_t = \ln(W_t + c).$$

With $W_0 = 0, X_0 = \ln c$ or $c = e^{X_0}$. Hence, a particular solution to (5.12) is

$$X_t = \ln\left(W_t + e^{X_0}\right). \tag{5.23}$$

Given that $e^{X_0}$ is constant, after a finite random time, $X_t \to +\infty$. More concretely, suppose the SDE in (5.21) holds for $t \in [t_0, T)$. Then

$$X_t = ln\big(W(t) - W(t_0) + e^{X_0}\big) \tag{5.23.1}$$

so that this expression "explodes" when $T = inf\{t | W(t) - W(t_0) = -e^{X_0}\}$. ■

**Example 5.4**   To solve the SDE

$$dX_t = X_t\, dt + dW_t, t \geq 0, X(0) = X_0, \tag{5.24}$$

let us set $V(X_t, t) = e^{-t}X_t$ and apply Itô's formula (4.18) to this expression (here $f = X_t, g = 1$).

$$d(e^{-t}X_t) = \left( -e^{-t}X_t + X_t e^{-t} + \frac{1}{2}(0) \right) dt + e^{-t}dW_t$$

$$= e^{-t}dW_t.$$

Then

$$\int_0^t d(e^{-s}X_s) = \int_0^t e^{-s}dW_s \text{ or}$$

$$e^{-t}X_t - X_0 = \int_0^t e^{-s}dW_s$$

so that a particular solution to (5.24) is

$$X_t = X_0 e^t + \int_0^t e^{t-s}dW_s. \tag{5.25}$$

More generally, if

$$dX_t = \mu X_t\, dt + \sigma\, dW_t, X(0) = X_0, \tag{5.26}$$

with $\alpha$ and $\sigma$ constants, set $V(X_t, t) = e^{-\mu t}X_t$ and apply Itô's formula (4.18). The resulting particular solution has the form

$$X_t = X_0 e^{\mu t} + \int_0^t e^{\mu(t-s)}dW_s. \tag{5.27}$$

(For solution details and a special case of (5.26), see Example 5.6.) ■

**Example 5.5**   Let's take a look at a slightly different type of SDE—one with coefficients that are not constants but that are deterministic functions indexed by time. Specifically,

$$dX_t = \big(a_t + b_t X_t\big)\, dt + \sigma_t\, dW_t, X(0) = X_0, \tag{5.28}$$

where $a_t$, $b_t$, and $\sigma_t$ are all continuous, deterministic functions of $t$. If we rewrite (5.28) as

$$dX_t - b_t X_t \, dt = a_t \, dt + \sigma_t \, dW_t \tag{5.29}$$

and concentrate on the left-hand side of this expression, we see that

$$\frac{d}{dt}\left( X_t e^{-\int_0^t b_u du} \right) = \frac{dX_t}{dt} e^{-\int_0^t b_u du} - X_t b_t e^{-\int_0^t b_u du} \quad \text{or}$$

$$d\left( X_t e^{-\int_0^t b_u du} \right) = e^{-\int_0^t b_u du} (dX_t - X_t b_t \, dt).^1 \tag{5.30}$$

Hence, (5.30) implies that $e^{-\int_0^t b_u du}$ is an *integrating factor* for (5.29) so that

$$d\left( X_t e^{-\int_0^t b_u du} \right) = a_t e^{-\int_0^t b_u du} dt + \sigma_t e^{-\int_0^t b_u du} dW_t. \tag{5.31}$$

Since the left-hand side of (5.31) integrates to

$$\int_0^t d\left( X_s e^{-\int_0^t b_u du} \right) = X_t e^{-\int_0^t b_u du} - X_0,$$

it follows that a particular solution to (5.28) is

$$X_t = X_0 \, e^{\int_0^t b_u du} + \int_0^t a_s \, e^{\int_s^t b_u du} ds + \int_0^t \sigma_s \, e^{\int_s^t b_u du} dW_s.^2 \tag{5.32}$$

---

1 Let $f(t)$ be continuous, with $a \le t \le b$, and define $F(x) = \int_a^x f(t)dt$. Then $F(x)$ is differentiable with derivative $F'(x) = f(x)$.

2 To evaluate (simplify) expressions such as these, remember that

i. $\int_a^a f(t) \, dt = 0$

ii. $\int_a^b f(t) \, dt = \int_0^b f(t) \, dt - \int_0^a f(t) \, dt$; and

iii. if $a < b$, then $\int_b^a f(t) \, dt = -\int_a^b f(t) \, dt$.

If $b_t$ is replaced by $b$ = constant in Equation (5.28), then the preceding solution becomes

$$X_t = X_0 e^{bt} + \int_0^t e^{b(t-s)} (a_s \, ds + \sigma_s \, dW_s). \; \blacksquare \tag{5.32.1}$$

For a more general discussion on integrating factors, see Appendix 5.B. (Also included therein is the technique of variation of parameters for solving certain SDEs.)

**Example 5.6**   (Ornstein–Uhlenbeck (OU) processes)

To obtain a solution to the OU SDE

$$dX_t = -\alpha X_t \, dt + \sigma \, dW_t, t \geq 0, X(0) = X_0, \tag{5.33}$$

with $\alpha$ and $\sigma$ positive constants, let us apply Itô's formula (4.18) to $V(X_t, t) = e^{\alpha t} X_t$ (here $f = -\alpha X_t, g = \sigma$):

$$d(e^{\alpha t} X_t) = \left( \alpha e^{\alpha t} X_t - \alpha X_t e^{\alpha t} + \frac{1}{2}(0) \right) dt + \sigma e^{\alpha t} dW_t \; \text{ or}$$

$$d(e^{\alpha t} X_t) = \sigma e^{\alpha t} dW_t.$$

Then

$$\int_0^t d(e^{\alpha s} X_s) = \sigma \int_0^t e^{\alpha s} dW_s \; \text{ or}$$

$$e^{\alpha t} X_t - X_0 = \sigma \int_0^t e^{\alpha s} dW_s.$$

Hence, a particular solution to (5.23) has the form

$$X_t = X_0 e^{-\alpha t} + \sigma \int_0^t e^{-\alpha(t-s)} dW_s. \tag{5.34}$$

[Ornstein – Uhlenbeck process, mean = 0]

Equation (5.33) has a very interesting property. Specifically, the drift term $-\alpha X_t \, dt$ is negative (resp. positive) when $X_t > 0$ (resp. $X_t < 0$). Given that the OU process always experiences random fluctuations, $X_t$ is **stationary**, that is, it is drawn back to its mean of zero whenever it drifts away from zero. Hence, the OU process is said to experience **mean reversion**. Moreover, since the short-term variability term $g(X_t, t) = \sigma$ = constant, $X_t$ tends to fluctuate robustly even when it is near its average value of zero. In this regard, $X_t$ crosses the zero level rather frequently. A slightly more general mean-reverting OU process involves modeling fluctuations that occur about a nonzero mean (or equilibrium) value $\mu(\neq 0)$. The OU-type SDE that captures such behavior appears as

$$dX_t = -\alpha(X_t - \mu)dt + \sigma dW_t, t \geq 0, X(0) = X_0, \tag{5.35}$$

with $\alpha$, $\mu$, and $\sigma$ all constant. A particular solution to (5.35) is

$$X_t = X_0 e^{-\alpha t} + \mu(1 - e^{-\alpha t}) + \sigma \int_0^t e^{-\alpha(t-s)} dW_s. \ \blacksquare \tag{5.36}$$

[Ornstein – Uhlenbeck process, mean = $\mu$]

The aforementioned mean-reversion characteristic can be applied to many varieties of stochastic processes. For instance, consider the SDE for the **geometric mean-reversion process** (Tvedt, 1995)

$$dX_t = \kappa(\alpha - \ln X_t)X_t \, dt + \sigma X_t \, dW_t, X(0) = X_0 > 0,$$

with $\kappa$, $\alpha$, $\sigma$, and $X_0$ all positive constants. The solution to this expression is

$$X_t = \exp\left\{ e^{-\kappa t} \ln X_0 + (\alpha - \sigma^2/2\kappa)(1 - e^{-\kappa t}) + \sigma e^{-\kappa t} \int_0^t e^{\kappa s} dW_s \right\}.$$

Up to this point in our discussion about generating solutions to linear SDEs, we considered two (related) approaches: Itô coefficient matching; and the application of Itô's formula (4.18) proper. Other possibilities exist, for example, we may also express solutions in, for example, what is called *product form* (see Appendix 5.A).

Let us, at this point, summarize some of our results pertaining to the solution of linear SDEs (Table 5.1). The principal structures are included therein.

**Example 5.7** (Brownian bridge processes)
Think of the Brownian bridge process as a BM process, defined over $t \in [0,1)$, that returns to $t = 0$ at time $t = 1$, that is, we have a "bridge" between the origin at time $t = 0$ and at time $t = 1$. Let us write the Brownian bridge SDE as

$$dX_t = -\frac{X_t}{1-t} dt + dW_t, X(0) = X_0 = 0. \tag{5.37}$$

To solve this SDE, we need only look to Table 5.1. Equation (5.37) appears to be of the linear narrow sense (LNS) form with $a_t \equiv 0$, $b_t = -(1/1-t)$ and $c_t \equiv c$. Then

**Table 5.1** Taxonomy of SDEs and their solutions.

---

STRUCTURES

(G) **General form:** $dX_t = f(X_t, t)\, dt + g(X_t, t) dW_t, 0 \le t \le T.$

(L) **Linear:** $f(X_t, t) = a_t + b_t X_t; g(X_t, t) = c_t + e_t X_t$ so that
$$dX_t = (a_t + b_t X_t)\, dt + (c_t + e_t X_t) dW_t.$$

(LH) **Homogeneous:** $a_t = c_t \equiv 0$ so that
$$dX_t = b_t X_t\, dt + e_t X_t\, dW_t.$$

(LNS) **Narrow sense:** $e_t \equiv 0$ so that
$$dX_t = (a_t + b_t X_t)\, dt + c_t\, dW_t.$$

(A) **Autonomous:** $f(X_t, t) \equiv f(X_t); g(X_t, t) \equiv g(X_t)$ so that
$$dX_t = f(X_t)\, dt + g(X_t)\, dW_t.$$

(AL) **Linear:** $f(X_t) = a + bX_t; g(X_t) = c + eX_t$ so that
$$dX_t = (a + bX_t)\, dt + (c + eX_t)\, dW_t.$$

(ALH) **Homogeneous:** $a = c \equiv 0$ so that
$$dX_t = bX_t\, dt + eX_t\, dW_t.$$

(ALNS) **Narrow sense:** $e \equiv 0$ so that
$$dX_t = (a + bX_t)dt + c\, dW_t.$$

SOLUTIONS

(L) See Equation (5.A.20).

(LH) $X_t = X_0 \exp\left\{ \int_0^t \left( b_s - \frac{1}{2} e_s^2 \right) ds + \int_0^t e_s\, dW_s \right\}.$

(LNS) $X_t = e^{\int_0^t b_s ds} \left\{ X_0 + \int_0^t e^{-\int_0^s b_u du} a_s\, ds + \int_0^t e^{-\int_0^s b_u du} c_s\, dW_s \right\}.$

(AL) $X_t = \exp\left\{ e \int_0^t dW_s + \left( b - \frac{1}{2} e^2 \right) \int_0^t ds \right\} \times \left\{ X_0 + (a - ec) \int_0^t \Phi_s^{-1} ds + c \int_0^t \Phi_s^{-1} dW_s \right\},$

where

$$\Phi_s^{-1} = \exp\left\{ -e \int_0^s dw_u - \left( b - \frac{1}{2} e^2 \right) \int_0^s du \right\}.$$

(ALH) See Equation (5.A.9).

(ALNS) $X_t = X_0 e^{bt} + a \int_0^t e^{b(t-s)} ds + c \int_0^t e^{b(t-s)} dW_s.$

---

$$X_t = e^{-\int_0^t \frac{1}{1-s}ds}\left\{0+0+\int_0^t e^{\int_0^s \frac{1}{1-u}du}dW_s\right\}$$

$$= e^{ln(1-t)}\left\{\int_0^t e^{-ln(1-s)}dW_s\right\}$$

$$= (1-t)\int_0^t \frac{1}{1-s}dW_s, t \in [0,1).$$

**[Brownian bridge process]** $\qquad\qquad(5.38)$

A more general version of (5.37), for $a$ and $b$ constants, is

$$dX_t = \frac{b-X_t}{1-t}dt + dW_t, t \in [0,1), X(0) = X_0 = a. \qquad (5.39)$$

The solution to this expression is

$$X_t = a(1-t) + bt + (1-t)\int_0^t \frac{dW_s}{1-s}, t \in [0,1). \blacksquare \qquad (5.40)$$

### 5.3.2 Properties of Solutions

Section 5.3.1 addressed the issue of determining strong solutions to linear SDEs. Let us now examine some of the properties of such solutions. In particular, we shall consider items such as boundedness, continuity, the specification of exact moments of a solution, and the Markov property of solutions of SDEs.

For the SDE (5.1) or

$$dX_t = f(X_t,t)dt + g(X_t,t)dW_t, t \in [t_0, T], T > 0, \qquad (5.1)$$

where $f$ and $g$ satisfy the assumptions posited in Section 5.1, the solution $X_t$, with $X(t_0) = X_0$, is **stochastically bounded** if for each $\varepsilon > 0$ there exists a $\gamma_\varepsilon = \gamma_\varepsilon(t_0, X_0) > 0$ such that

$$\inf_{t \in [t_0, T]} P(|X_t| \leq \gamma_\varepsilon) > 1 - \varepsilon. \qquad (5.41)$$

If in (5.41) $\gamma_\varepsilon$ depends only on $X_0$, then $X_t$ is said to be **uniformly stochastically bounded**. Inequality (5.4.1) implies that $\{X_t\}_{t \in [t_0, T]}$ does not exhibit any tendency for large fluctuations in this process to beget even larger ones, that is, the average frequency of large fluctuations does not increase over time or large fluctuations in $X$ are observed infrequently (Gard, 1988). A **sufficient condition for a process $\{X_t\}_{t \in [t_0, T]}$ to be stochastically bounded** is that for some $p > 0$, the **$p$th-order moment** $E|X_t|^P$ is bounded on the interval $[t_0, T]$ (more on this condition shortly).

Next, suppose $f$ and $g$ satisfy conditions (5.3) and (5.4) of Theorem 5.1 and $\{X_t\}_{t\in[t_0,T]}$ is a solution of (5.1). Then $X_t$ is **continuous** on $[t_0, T]$ if there is a constant $c > 0$ such that

$$|X_t - X_r|^2 \le c|t - r|, t_0 \le r, t \le T,$$ (5.42)

where we can write

$$X_t - X_r = \int_r^t f(X_s, s)\, ds + \int_r^t g(X_s, s)\, dW_s.$$

Now, in the preceding sufficient condition for stochastic boundedness set $p = 2$. Then there is an $M > 0$ such that $E|X_s|^2 \le M, 0 \le s \le T$, and thus it can be shown that (5.42) holds if we take $c = 2k(T+1)(1+M)$.

We now turn to the topic of *moment inequalities* and the *calculation of pth-order moments* $E|X_t|^p$ of the solution of the SDE (5.1), with $[t_0, T], T < +\infty$, and $X(t_0) = X_0$. We begin with

**Theorem 5.3**  Existence of Moments (Arnold, 1974)

Suppose that the assumptions of the existence and uniqueness theorem (Theorem 5.1) are satisfied and that $E|X_0|^{2p} < +\infty$, where $p$ a positive integer. Then for $X_t$ a solution to the SDE (5.1) on $[t_0, T], T < +\infty$,

i. $E|X_t|^{2p} \le \left(1 + E(X_0)^{2p}\right)e^{c(t-t_0)}$; and

ii. $E|X_t - X_0|^{2p} \le D\left(1 + E(X_0)^{2p}\right)(t - t_0)^p e^{c(t-t_0)},$ (5.43)

where $c = 2p(2p+1)k^2$ and $D$ are constants (dependent upon $p$, $k$, and $T - t_0$).

Furthermore,

a. $E|X_t - X_s|^{2p} \le c_1|t - s|^p, t_0 \le s, t \le T, c_1$ constant.
   For $E|X_0|^2 < +\infty$ and $p = 1$, $\lim_{t\to s} |X_t - X_s|^2 = 0$.
b. Let $p \ge 2, X(t_0) = X_0 \in L^p(\Omega; R)$, and assume that *the linear growth condition* (5.4) is satisfied. Then the $p$th-order moment

$$E|X_t|^p \le 2^{\frac{p-2}{2}}(1 + E|X_0|^p)e^{p\alpha(t-t_0)}, t \in [t_0, T],$$ (5.44)

   where $\alpha = k + k^2(p-1)/2$.
c. The solution $X_t$ of the linear SDE (5.8) has, for all $t \in [t_0, T]$, a $p$th-order moment $E|X_t|^p$ if and only if $E|X_0|^p < +\infty$.
d. For the autonomous linear homogeneous SDE,

$$dX_t = bX_t\, dt + eX_t\, dW_t, X(t_0) = X_0,$$

$$E|X_t|^p = E|X_0|^p \exp\left\{p\left(b - e^2/2\right)(t - t_0) + \frac{p^2}{2}e^2(t - t_0)\right\}.$$ (5.45)

e. If $X \sim N(E(X), V(X)) = N(\mu, \sigma^2)$, then, for every $p > 0$,

$$E\left(e^X\right)^p = e^{p\mu + p^2\sigma^2/2}. \tag{5.46}$$

This is because

$$\frac{1}{\sqrt{2\pi}\sigma} \int_{-\infty}^{+\infty} e^{\left[px - (x-\mu)\frac{2}{2\sigma^2}\right]} dx = e^{p\mu + p^2\sigma^2/2}.$$

(For $p = 1$, obviously $E(e^x) = e^{\mu + e^2/2}$.)

Continuing with our moment inequalities, we have

**Theorem 5.4**   Suppose $p \geq 2, g(t) \in \mathcal{m}^2([0, T])$, and $E\int_0^T |g(s)|^p \, ds < +\infty$. Then

$$E\left|\int_0^t g(s)dW(s)\right|^p \leq \left(\frac{p(p-1)}{2}\right)^{\frac{p}{2}} T^{\frac{p-2}{2}} E\int_0^T |g(s)|^p \, ds.$$

For $p = 2$,

$$E\left|\int_0^t g(s)dW(s)\right|^2 = E\int_0^T |g(s)|^2 ds$$

(Itô's isometry).

**Example 5.8**   For the Brownian bridge process (5.38) or

$$X_t = (1-t)\int_0^t \frac{1}{1-s} dW_s, t \in [0,1],$$

find $E(X_t)$ and $V_t(X)$.[3]

---

3 As an aid (review) for solving the current and the next few example problems, we offer some useful expectation/stochastic integral results as follows:

1. $E(W_t) = 0, V(W_t) = t, \int_0^t (dW_s)^2 = \int_0^t ds.$

2. $E\left(\int_0^t g(s)dW_s\right) = 0, E\left|\int_0^t g(s)dW_s\right|^2 = E\int_0^t |g(s)|^2 ds.$

3. $E\left(\int_0^t X_s \, dW_s\right) = 0, E\left|\int_0^t X_s \, dW_s\right|^2 = \int_0^t X_s^2 ds.$

4. $E\left(\int_0^t W_s dW_s\right) = 0, E\left|\int_0^t W_s dW_s\right|^2 = \int_0^t E|W_s|^2 ds = \int_0^t sds.$

5. $E\left(\int_0^t s dW_s\right) = 0, E\left|\int_0^t s dW_s\right|^2 = \int_0^t s^2 ds.$

6. $E(e^{aW_t}) = e^{\frac{a^2}{2}t}, E(e^{-W_t}) = e^{t/2}.$ In general, $E(e^{g(W_t)}) = e^{E[g(W_t)^2]/2}.$

7. For $t \in [0, T], E\left(\int_0^T e^{-W_t} dt\right) = \int_0^T e^{t/2} dt = 2(e^{T/2} - 1).$

$$E(X_t) = (1-t)E\left(\int_0^t \frac{1}{1-s} dW_s\right) = 0.$$

$$V(X_t) = E|X_t - E(X_t)|^2 = E|X_t|^2$$

$$= E\left|(1-t)\int_0^t \frac{1}{1-s} dW_s\right|^2$$

$$= (1-t)^2 \int_0^t \frac{1}{1-s} ds$$

$$= (1-t)^2 \left(\frac{1}{1-t} - 1\right) = t(1-t). \blacksquare$$

**Example 5.9**  For the geometric BM process (5.18) or

$$X_t = X_0 \exp\left\{\left(\mu - \frac{\sigma^2}{2}\right)t + \sigma W_t\right\}, X_0 \text{ arbitrary,}$$

find $E(X_t)$ and $V_t(X)$. For $p > 0$ and $E|X_0|^p \neq 0$, let us start by finding

$$E|X_t|^p = E\left\{|X_0|^p \exp\left[p\left(\mu - \frac{\sigma^2}{2}\right)t + p\sigma W_t\right]\right\}$$

$$= \exp\left[p\left(\mu - \frac{(1-p)\sigma^2}{2}\right)t\right] E\left\{|X_0|^p \exp\left[-\frac{p^2\sigma^2}{2}t + p\sigma W_t\right]\right\}$$

$$= \exp\left[p\left(\mu - \frac{(1-p)\sigma^2}{2}\right)t\right] E|X_0|^p.$$

Then, for $p = 1$,

$$E|X_t| = e^{\mu t} E|X_0| = X_0 e^{\mu t};$$

and for $p = 2$,

$$E|X_t|^2 = e^{2\left(\mu + \frac{\sigma^2}{2}\right)t} E|X_0|^2$$

$$= X_0^2 e^{2\mu t} e^{\sigma^2 t}.$$

Then

$$V(X_t) = E|X_t|^2 - (E|X_t|)^2$$

$$= X_0^2 e^{2\mu t}\left(e^{\sigma^2 t} - 1\right). \blacksquare$$

**Example 5.10** For the OU process (5.34) or

$$X_t = X_0 e^{-at} + \sigma \int_0^t e^{-a(t-s)} dW_s,$$

determine $E(X_t)$ and $V_t(X)$.

$$E(X_t) = E\left(X_0 e^{-at} + \sigma e^{-at} \int_0^t e^{-as} dW_s\right)$$

$$= E(X_0)e^{-at} + \sigma e^{-at} E\left(\int_0^t e^{as} dW_s\right) = E(X_0)e^{-at}.$$

$$V(X_t) = E|X_t - E(X_t)|^2 = E\left|X_0 e^{-at} + \sigma e^{-at} \int_0^t e^{as} dW_s - e^{-at}E(X_0)\right|^2$$

$$= E\left|e^{-at}(X_0 - E(X_0)) + \sigma e^{-at} \int_0^t e^{as} dW_s\right|^2$$

$$= e^{-2at} E|X_0 - E(X_0)|^2 + \sigma^2 e^{-2at} E\left|\int_0^t e^{as} dW_s\right|^2$$

$$= e^{-2at} V(X_0) + \sigma^2 e^{-2at} \int_0^t e^{2as} ds$$

$$= e^{-2at} V(X_0) + \frac{\sigma^2}{2a}\left(1 - e^{-2at}\right).$$

Looking at the structure of the equation for $X_t$, we can see that, for arbitrary $X_0$,

$$\lim_{t \to +\infty} e^{-at} X_0 = 0 \text{ a.s.}$$

and $\sigma \int_0^t e^{-a(t-s)} dW_s$ is $N(0,(\sigma^2/2a)(1-e^{-2at}))$. Hence, the distribution of $X_t \to N(0,\sigma^2/2a)$ as $t \to +\infty$ with $X_0$ arbitrary. If $X_0$ is normally distributed or constant, then $X_t$ is also normally distributed, that is, if $X_0 \sim N(0,\sigma^2/2a)$, then $X_t \sim N(0,\sigma^2/2a)$. ■

**Example 5.11** Consider item no. 6 of footnote 3 $\left(E(e^{aW_t}) = e^{(a^2/2)t}\right)$. It is instructive to verify this equality using Itô's formula (4.18).

Given the SDE $dX_t = f(X_t,t)dt + g(X_t,t)dW_t$, set $f(X_t,t) = 0, g(X_t,t) = 1, X_t = W_t$, so that $dX_t = dW_t$, and choose $V(X_t,t) = e^{aW_t}$. From Itô's formula,

$$dV = \left(0 + 0 + \frac{1}{2}(1)a^2 e^{aW_t}\right) dt + a e^{aW_t} dW_t,$$

$$V_t = V_0 + \frac{a^2}{2} \int_0^t e^{aW_t} ds + a \int_0^t e^{aW_t} dW_t,$$

and thus

$$E(V_t) = E(V_0) + \frac{\alpha^2}{2} \int_0^t E(V_s)\,ds$$

$\left( E\left( \int_0^t e^{\alpha W_t} dW_t \right) = 0 \right.$ since, for any nonanticipatory random variable $V_t$,

$E\left( \int_0^t V_s dW_s \right) = 0$). Then

$$\frac{dE(V_t)}{dt} = \frac{\alpha^2}{2} E(V_t)$$

(via the fundamental theorem). Hence,

$$\frac{dE(V_t)}{E(V_t)} = \frac{\alpha^2}{2} dt,$$

$$\ln E(V_t) = \frac{\alpha^2}{2} t,$$

and thus $E(V_t) = e^{(\alpha^2/2)t} = E(e^{\alpha W_t})$. ∎

### 5.3.3  Solutions to SDEs as Markov Processes

We now turn to the examination of solutions to SDEs as Markov processes. In particular, the solution $X_t$ of the SDE (5.1) $(dX_t = f(X_t,t)dt + g(X_t,t)dW_t, t \geq 0)$ is a Markov process, with the probability law for any such process completely determined by the initial distribution

$$P_0(B) = P(X_0 \in B), B \in \mathcal{B} \tag{5.47}$$

(where the initial value $X(0) = X_0$ is an arbitrary random variable), and the transition probability

$$P(x,s;B,t) = P(X_t \in B | X_s = x), \tag{5.48}$$

where $0 \leq s \leq t \leq T$, and $B \in \mathcal{B}$. For $x$ and $s$ fixed, the transition probability

$$P(x,s;B,t) = P(X_t(x,s) \in B) \tag{5.48.1}$$

is exactly the distribution of the solution $X_t = X_t(x,s)$ of the integral equation

$$X_t = x + \int_s^t f(X_r,r)\,dr + \int_s^t g(X_r,r)\,dW_r, s \leq t. \tag{5.49}$$

Stated more formally, we have Theorem 5.5.

**Theorem 5.5** Let $X_t$ be a solution of the SDE (5.1) or (Arnold, 1974; Gard, 1988; Mao, 2007)

$$dX_t = f(X_t, t)dt + g(X_t, t)dW_t, t \geq 0,$$

with the coefficient functions $f$ and $g$ satisfying the conditions of the existence and uniqueness theorem (Theorem 5.1). Then $X_t$ is a Markov process whose initial and transition probabilities are given by (5.47) and (5.48.1), respectively, and where $X_t = X_t(x,s)$ is a solution to (5.49).

(It is important to note that while the distribution of $X_0$ determines the initial probability of the process $X_t$, the transition probabilities do not depend on $X_0$ but are completely specified by the coefficient function $f$ and $g$.) In order for the solution to exhibit the strong Markov property, the conditions/assumptions of Theorem 5.5 must be strengthened somewhat to yield Theorem 5.5a:

**Theorem 5.5a** Let $X_t$ be a solution of SDE (5.1) and assume that the coefficient functions $f$, $g$ are uniformly Lipschitz continuous and satisfy the linear growth condition (conditions (5.3) and (5.4), respectively, of Theorem 5.1) for all $x, y \in R$ and $t \geq 0$ (Mao, 2007). Then $X_t$ is a strong Markov process.

A Markov process is said to be **homogeneous** if the transition probabilities are stationary, that is,

$$P(x, s + u; B, t + u) = P(x, s; B, t), 0 \leq u \leq T - t,$$

is identically satisfied. Moreover, for fixed $x$ and $B$, the transition probability is a function of $t - s$ only or $P(x, s; B, t) = P(t - s, x, B)$.

Suppose a process in differential form is characterized by an autonomous SDE, that is, the coefficient functions $f(X_t, t) \equiv f(X_t)$ and $g(X_t, t) \equiv g(X_t)$ are independent of $t$ on $[0, T]$ so that

$$dX_t = f(X_t)dt + g(X_t)dW_t, t \geq 0. \tag{5.50}$$

Also, let $f(X_t), g(X_t)$ satisfy the conditions of the existence and uniqueness theorem (Theorem 5.1). Then the solution $X_t$ of (5.50) is a homogeneous Markov process with stationary transition probabilities. In addition, if $f(X_t), g(X_t)$ are uniformly Lipschitz continuous, with the linear growth condition satisfied, then the solution $X_t$ to (5.50) is a homogeneous strong Markov process.

To fully appreciate the *stochastic* nature of SDEs, we now assert that a solution of an SDE can be viewed as a specialized type of stochastic process called an **Itô diffusion**. More formally, a one-dimensional Markov process $\{X_t\}_{t \in [t_0, T]}$ having continuous sample paths with probability 1 is termed a **diffusion process** if its transition probability $P(x, s; B, t)$ is smooth in the sense that it satisfies the following conditions. For every $s \in [t_0, T], x \in R$, and $\varepsilon > 0$,

a. $\lim_{t\downarrow s}\dfrac{1}{t-s}\displaystyle\int_{|y-x|>\varepsilon} p(x,s;t,y)\,dy = 0;$

b. $\lim_{t\downarrow s}\dfrac{1}{t-s}\displaystyle\int_{|y-x|<\varepsilon} (y-x)p(x,s;t,y)\,dy = a(x,s);$ and (5.51)

c. $\lim_{t\downarrow s}\dfrac{1}{t-s}\displaystyle\int_{|y-x|<\varepsilon} (y-x)^2 p(x,s;t,y)\,dy = b(x,s)^2,$

where $p(x, s; t, y)$ is the **transition density**; $a(x, s)$ is the **drift coefficient** function; and $b(x, s)$ is the **diffusion coefficient function** of the diffusion process. Here, $a(x, s)$ may be viewed as the instantaneous rate of change in the mean of the process given that $X_s = x$; and $b(x, s)^2$ can be thought of as the instantaneous rate of change of the squared fluctuations of the process given that $X_s = x$. The importance of (5.51a) is that it ensures that the diffusion process will not exhibit instantaneous jumps over time. To summarize, a diffusion process is a Markov process that has its probability law determined by the drift and diffusion coefficients. Such coefficients represent the conditional infinitesimal mean and variance of the process, respectively.

We noted earlier that the solutions $X_t$ of SDEs such as (5.1) are Markov processes; such solutions can also be thought of as representing arbitrary diffusion processes that, essentially, are conversions of standard Wiener processes. In fact, the standard Wiener process is a one-dimensional diffusion process with drift coefficient $a(x,s) \equiv 0$ and diffusion coefficient $b(x,s) \equiv 1$. We close with Theorem 5.6.

**Theorem 5.6** Given the SDE (5.1) or $dX_t = f(X_t,t)dt + g(X_t,t)dW_t$, if the functions $f$ and $g$ satisfy the conditions of the existence and uniqueness theorem (Theorem 5.1), then any solution $X_t$ is a diffusion process on the interval $[0,T]$ with drift coefficient $f(X_t, t)$ and diffusion coefficient $g(X_t, t)^2$.[4]

## 5.4 SDEs and Stability

One of the difficulties associated with SDEs is that it may be impossible to find a solution that appears in a convenient and/or explicit form. Hence, it is

___

4 Conversely, suppose $Y_t$ is a diffusion process defined on some interval with $a(y, t)$ and $b(y, t)$ serving as drift and diffusion coefficients, respectively. Under what conditions will $Y_t$ satisfy an SDE? If the functions $a$ and $\sqrt{b}$ satisfy the conditions of Theorem 5.1 and if $\{W_t\}$ is *any* standard Wiener process, then the solution $X_t$ of the SDE $dX_t = a(X_t,t)dt + \sqrt{b(X_t,t)}dW_t, X(t_0) = Y(t_0)$ shares the same probability law as $Y_t$ so that $\{X_t\}$ and $\{Y_t\}$ are equivalent processes (although their sample paths may not coincide with probability 1). (See Gard (1988, pp. 89–92), for a stronger version of this converse question.)

important to focus on what *qualitative* information can be elicited about the solutions of SDEs without actually solving them. In this regard, in what follows we shall consider the notion of the *stability* of a solution. Simply stated, we shall consider the question of whether small changes in the initial conditions (state) or parameters of a dynamic system lead to small changes (*stability*) or to large changes (*instability*) in the solution. That is, the sample paths of a stable dynamic system that are "close" to each other at a particular point in time should remain so at all subsequent time points.

Our point of reference for studying the stability of solutions to SDEs is the deterministic stability theory developed by A. M. Lyapunov (1892) for ODEs (see also Bucy (1965) and Hahn (1967)). This will be our "jumping off point" for the assessment of stability for solutions to SDEs.

Lyapunov's *direct (second) method* for determining stability will be applied to an ODE of the form

$$dX_t = f(X_t, t)dt, X(t_0) = X_0, t \geq t_0, \tag{5.52}$$

where $X_t \in R$. To investigate this technique, let us first develop some basic definitions. Suppose that for every initial value $X_0$, there exists a unique global solution $X_t(t_0, X_0)$ defined on $[t_0, +\infty)$ and that $f(X_t, \cdot)$ is continuous. In addition, assume that $f(0, t) = 0$ for all $t \geq t_0$. Then (5.52) is said to have a **trivial** or **equilibrium solution** $X_t \equiv 0$ corresponding to the initial value $X_0 = 0$. This equilibrium is **stable** if for every $\varepsilon > 0$ there exists a $\delta_{\varepsilon, t_0} > 0$ such that $|X_t(t_0, X_0)| < \varepsilon$ for all $t \geq t_0$ whenever $|X_0| < \delta_{\varepsilon, t_0}$. Thus, small changes in $X_0$ do not produce large changes in the solution for $t \geq t_0$. Otherwise, the solution $X_t \equiv 0$ is said to be **unstable**. The equilibrium solution is termed **asymptotically stable** if it is stable and there exists a $\delta_{t_0}$ such that $lim_{t \to +\infty} X_t(t_0, X_0) = 0$ whenever $|X_0| < \delta_{t_0}$. If $\delta$ does not depend on $t_0$, then stability is characterized as **uniform**. In addition, if the preceding limit holds for all $X_0$, then $X_t \equiv 0$ is termed **globally asymptotically stable** (or **asymptotically stable in the large**).

To cover the basics of Lyapunov stability for ODEs, we note first that a continuous scalar-valued nondecreasing function $v(x)$ defined on a suitably restricted neighborhood of the zero point $N_h = \{x | |x| < h, h > 0\} \subset R$ is (Lyapunov) **positive definite** if $v(0) = 0$ and $v(x) > 0, x \in N_h$, for all $x \neq 0$. More broadly, a continuous function $v(x, t)$ defined on $N_h \times [t_0, +\infty)$ is positive definite if $v(0, t) \equiv 0$ and there exists a positive definite function $w(x)$ such that $v(x, t) \geq w(x)$ for $t \geq t_0$. A function $v$ is **negative definite** if $-v$ is positive definite. A continuous non-negative function $v(x, t)$ is termed **decrescent** (i.e., it has an arbitrarily small upper bound) if there exists a positive definite function $u(x)$ such that $v(x, t) \leq u(x)$ for $t \geq t_0$. (If the positive definite function $v(x)$ is independent of $t$, then it is decrescent.) The function $v(x, t)$ is **radially unbounded** if $lim_{|x| \to +\infty} inf_{t \geq t_0} v(x, t) \to +\infty$.

Now, suppose $X_t$ is a solution to (5.51) and $v(X_t, t)$ is a positive definite function having continuous first partial derivatives with respect to $t$ and $X_t$. In addition, let $v_t = v(X_t, t)$ depict a function of $t$ whose derivative, via (5.52), is

$$\frac{dV_t}{dt} = \frac{\partial v}{\partial X_t}\frac{dX_t}{dt} + \frac{\partial v}{\partial t} = \frac{\partial v}{\partial X_t}f(X_t, t) + \frac{\partial v}{\partial t}. \tag{5.53}$$

If $dV_t/dt \le 0$, then $X_t$ varies in a fashion that would not cause $V_t$ to increase, that is, the distance between $X_t$ and the trivial or equilibrium solution, as measured by $v(X_t, t)$, does not increase. If $dV_t/dt < 0$, then $V_t$ will decrease to zero and thus the distance from $X_t$ to the trivial solution concomitantly diminishes or $X_t \to 0$.

The preceding discussion serves as the foundation of the Lyapunov method and leads to the following (sufficient) condition for the stability of an equilibrium solution to an ODE. Specifically, we have Theorem 5.7.

**Theorem 5.7**

i. If there exists a positive definite function $V_t = v(X_t, t)$ with continuous first partial derivatives such that the derivative of $V_t$ along the solutions to (Lyapunov, 1892)

$$dX_t = f(X_t, t)dt, t \ge 0, f(0, t) \equiv 0, \tag{5.54}$$

satisfies

$$\frac{dV_t}{dt} = \frac{\partial v}{\partial X_t}f(X_t, t) + \frac{\partial v}{\partial t} \le 0 \tag{5.55}$$

for all $(X_t, t) \in N_h \times [t_0, +\infty)$, then the trivial or equilibrium solution to (5.52) is stable.

ii. If there exists a positive definite decrescent function $V_t = v(X_t, t)$ with continuous first partial derivatives such that $dV_t/dt$ taken along the solutions of (5.54) is negative definite for $(X_t, t) \in N_h \times [t_0, +\infty)$, then the trivial or equilibrium solution to (5.52) is asymptotically stable.

A function $v(X_t, t)$ that satisfies the stability conditions (i) and (ii) of Theorem 5.5a is termed a **Lyapunov function** corresponding to the ODE (5.52). For some ODEs, a Lyapunov function may be difficult to find. However, if an appropriate Lyapunov function can be readily determined, then the Lyapunov method can be an extremely valuable tool for determining if an equilibrium point is stable.

**Example 5.12** The ODE $dX_t/dt = f(X_t, t) = -X_t - X_t^2, X_t(0) = X_0 = 0, t \in [t_0, +\infty), X_t \in R$, has an equilibrium point at $X_t = 0$. Demonstrate that $v(X_t) = X_t$ serves as a Lyapunov function for this ODE. Here,

$$\frac{\partial v}{\partial X_t}f(X_t, t) = (1)\left(-X_t - X_t^2\right) = -X_t - X_t^2 \le -v(X_t).$$

Then along the trajectories of (5.54),

$$\frac{d}{dt} v(X_t(t_0,X_0),t) \le -v(X_t(t_0,X_0),t);$$

and thus

$$dv(X_t(t_0,X_0),t) \le -v(X_t(t_0,X_0),t)dt$$

or

$$v(X_t(t_0,X_0),t) \le v(X_0)e^{-(t-t_0)}$$

for *all* $X_0$ and $t \ge t_0$.[5] Hence, the trivial point $X_t = 0$ is uniformly globally asymptotic stable. ■

When we attempt to transfer the principles of Lyapunov stability for deterministic ODEs to their stochastic counterparts, certain compelling questions present themselves.

1. How should stochastic stability be defined?
2. How should Lyapunov functions be defined, and what properties should a stochastic Lyapunov function possess?
3. What should the condition $dV_t/dt \le 0$ be replaced by in order for us to make stability assessments?

To address these questions, let us first introduce a set of operational assumptions. Our investigation of the stability of solutions to SDEs will focus on Equation (5.1) or

$$dX_t = f(X_t,t)dt + g(X_t,t)dW_t, t \ge t_0, \tag{5.1}$$

where (a) the assumptions of the existence and uniqueness theorem (Theorem 5.1) are satisfied; (b) the initial value $X(t_0) = X_0 \in R$ is taken to be, with probability 1, a constant; and (c) for *any* $X_0$ independent of $W_t, t \ge t_0$, Equation (5.1) has a unique global solution $X_t(t_0, X_0)$ with continuous sample paths and finite moments; and $(d) f(0,t) = 0$ and $g(0,t) = 0$ for $t \ge t_0$ so that the unique solution $X_t \equiv 0$ corresponds to the initial value $X(t_0) = X_0 = 0$. Mirroring the deterministic case treated earlier, this solution will also be termed the **trivial** or **equilibrium solution**, but now to SDE (5.1).

For $X_t$ the solution to (5.1), let $V_t = v(X_t,t)$ be a positive definite function defined on $N_h \times R_+, N_h = \{X_t \| X_t | < h, h > 0\}$, and which is continuously twice differentiable in $X_t$ and once in $t$. Then the operator $L$ associated with (5.1) is

---

5 Since $dv/v = -dt, \ln v = -t + \ln C$, where $C$ is a constant of integration. Then $v = Ce^{-t}$ and, at $(t_0,X_0), C = v(X_0)e^{t_0}$ so that the solution readily follows.

$$L = \frac{\partial}{\partial t} + f(X_t, t)\frac{\partial}{\partial X_t} + \frac{1}{2}g(X_t, t)^2\frac{\partial^2}{\partial X_t^2};$$ (5.56)

and if $L$ acts on the function $V_t = v(X_t, t)$, then

$$Lv(X_t, t) = \frac{\partial v}{\partial t} + f(X_t, t)\frac{\partial v}{\partial X_t} + \frac{1}{2}g(X_t, t)^2\frac{\partial^2 v}{\partial X_t^2}.$$ (5.57)

Note that, via Itô's formula (4.18), if $X_t \in N_h$, then

$$dv(X_t, t) = Lv(X_t, t)dt + g(X_t, t)\frac{\partial v}{\partial X_t}dW_t.$$ (5.58)

In the presence of the random term $dW_t$ in (5.58) (since $g(X_t, t) \not\equiv 0$) a stable solution $X_t$ to (5.1) requires that, on the average, $E(dV_t) \leq 0$. But since $E(dV_t) = E(Lv(X_t, t)dt)$, this condition is satisfied if $Lv(X_t, t) \leq 0$ for all $t \geq 0$. Clearly, the inequality $dV_t/dt \leq 0$ required for stability in the deterministic case (with $g(X_t, t) \equiv 0$) is replaced by $Lv(X_t, t) \leq 0$ in order to make stochastic stability statements. Thus, the function $V_t = v(X_t, t)$ serves as **the stochastic Lyapunov function** corresponding to SDE (5.1).

Armed with these considerations, our discussion of stability of solutions of SDEs must now be framed in terms of probability, that is, suppose assumptions (a)–(d) are satisfied. Then the trivial or equilibrium solution $X_t \equiv 0$ is termed **stochastically stable** or **stable in probability** if for every $\varepsilon > 0$,

$$\lim_{X_0 \to 0} P\left(\sup_{t \in [t_0, +\infty)} |X_t(t_0, X_0)| \geq \varepsilon\right) = 0;$$ (5.59)

otherwise, the said solution is **stochastically unstable**. The equilibrium solution is **stochastically asymptotically stable** if it is stochastically stable and

$$\lim_{X_0 \to 0} P\left(\lim_{t \to +\infty} X_t(t_0, X_0) = 0\right) = 1.$$ (5.60)

The equilibrium solution is **globally stochastically asymptotically stable** or **stochastically asymptotic stable in the large** if it is stochastically stable and, for all $X_0 \in R$,

$$P\left(\lim_{t \to +\infty} X_t(t_0, X_0) = 0\right) = 1.$$ (5.61)

To extend the Lyapunov approach for determining the stability of ODEs to that of SDEs, let us consider the general stochastic stability theorem (Theorem 5.8).

**Theorem 5.8**

i. Suppose assumptions (a)–(d) are in effect and that there exists a positive definite stochastic Lyapunov function $v(X_t, t)$ defined on $N_h \times [t_0, +\infty)$ that is everywhere continuously twice differentiable in $X_t$ and once in $t$. In addition, $Lv(X_t, t) \leq 0, t \geq t_0, 0 \leq X_t \in N_h$, where $L$ is given by (5.56) ( or $Lv(X_t, t)$ is determined by (5.57)). Then the trivial or equilibrium solution of (5.1) is stochastically stable.

ii. If, in addition, $v(X_t, t)$ is decrescent and $Lv(X_t, t)$ is negative definite, then the equilibrium solution is stochastically asymptotically stable.

iii. If the assumptions of (ii) hold for a radially unbounded function $v(X_t, t)$ defined everywhere on $[t_0, +\infty) \times R$, then the equilibrium solution is globally stochastically asymptotically stable (Has'minskiy, 1980).

**Example 5.13**   Suppose an SDE has the form $dX_t = \mu X_t dt + \sigma X_t dW_t, t \geq 0$. If we select $V_t = v(X_t, t) = X_t^2$ as the stochastic Lyapunov function, then

$$Lv = \left(0 + \mu X_t (2X_t) + \frac{1}{2}\sigma^2 X_t^2 (2)\right)$$

$$= 2\left(\mu + \frac{1}{2}\sigma^2\right)v_t \leq 0$$

if $\mu + (1/2)\sigma^2 < 0$. If this inequality holds, then the trivial solution $X_t = 0$ is stochastically asymptotically stable in the large. ∎

# Appendix 5.A   Solutions of Linear SDEs in Product Form (Evans, 2013; Gard, 1988)

### 5.A.1   Linear Homogeneous Variety

Suppose we have a linear SDE of the form

$$dX_t = b_t X_t\, dt + e_t X_t dW_t, X(0) = X_0. \tag{5.A.1}$$

Our objective is to find a solution to (5.A.1) in **product form** or

$$X(t) = X_1(t) \cdot X_2(t), \tag{5.A.2}$$

where

  a. $dX_1(t) = e_t X_1(t)dW_t, X_1(0) = X_0;$

  b. $dX_2(t) = A_t\, dt + B_t\, dW_t, X_2(0) = 1,$ (5.A.3)

where the process coefficients $A_t$, $B_t$ are, as yet, undetermined.

From the *product rule* (4.A.6) and Itô's multiplication table (Table 4.A.1),

$$dX = d(X_1 \cdot X_2)$$
$$= X_1 \, dX_2 + X_2 \, dX_1 + dX_1 dX_2$$
$$= X_1 \, dX_2 + X_2 \, dX_1 + (e_t X_1 \, dW_t)(A_t dt + B_t \, dW_t)$$
$$= X_1 \, dX_2 + X_2 \, dX_1 + e_t X_1 B_t \, dt \qquad (5.A.4)$$
$$= X_1 \, dX_2 + X_2 \, (e_t X_1 \, dW_t) + e_t X_1 B_t \, dt$$
$$= e_t X \, dW_t + X_1 (dX_2 + e_t B_t dt).$$

Next, select $A_t$ and $B_t$ in (5.A.3b) so that

$$dX_2 + e_t B_t \, dt = b_t X_2 \, dt. \qquad (5.A.5)$$

(Under this equality, (5.A.4) becomes $dX = e_t X dW_t + b_t X dt$, which is consistent with (5.A.1).) Given (5.A.5), for (5.A.3b) to hold, set $B_t \equiv 0$ and take $A_t = b_t X_2$ so that (5.A.3b) becomes

$$dX_2 = b_t X_2 \, dt, X_2(0) = 1. \qquad (5.A.6)$$

With this expression nonrandom, upon integrating we obtain $ln X_2 = \int_0^t b_s \, ds$ or

$$X_2 = e^{\int_0^t b_s ds}. \qquad (5.A.7)$$

To solve (5.A.3a) or $dX_1(t) = e_t X_1(t) dW_t, X_1(0) = X_0$, let us employ Itô's formula (4.18). To this end, we have, for $V(X_1(t), t) = ln X_1(t)$ and $g = e_t X_1(t)$,

$$dV = \left( 0 + 0 - \frac{1}{2} e_t^2 X_1(t)^2 \left( \frac{1}{X_1(t)^2} \right) \right) dt + e_t X_1(t) \left( \frac{1}{X_1(t)} \right) dW_t$$

or

$$d \, ln X_1(t) = -\frac{1}{2} e_t^2 \, dt + e_t \, dW_t,$$
$$\int_0^t d \, ln X_1(t) = -\frac{1}{2} \int_0^t e_s^2 \, ds + \int_0^t e_s \, dW_s,$$
$$ln X_1(t) - ln X_1(0) = -\frac{1}{2} \int_0^t e_s^2 \, ds + \int_0^t e_s \, dW_s, \qquad (5.A.8)$$
$$X_1(t) = X_0 \, exp \left\{ \int_0^t e_s \, dW_s - \frac{1}{2} \int_0^t e_s^2 \, ds \right\}.$$

Upon combining (5.A.7) and (5.A.8), our *product form solution* to (5.A.1) is

$$X(t) = X_1(t) \cdot X_2(t)$$

$$= X_0 \, exp \left\{ \int_0^t b_s ds + \int_0^t e_s \, dW_s - \frac{1}{2} \int_0^t e_s^2 \, ds \right\}. \tag{5.A.9}$$

### 5.A.2  Linear Variety

More generally, suppose our linear SDE now has the form

$$dX_t = (a_t + b_t X_t)dt + (c_t + e_t X_t)dW_t, X(0) = X_0. \tag{5.A.10}$$

Again a solution of the product form

$$X(t) = X_1(t) \cdot X_2(t) \tag{5.A.11}$$

is sought, where

$$
\begin{aligned}
&\text{a. } dX_1(t) = b_t X_1(t)dt + e_t X_1(t)dW_t, X(0) = 1; \\
&\text{b. } dX_2(t) = A_t \, dt + B_t \, dW_t, X_2(0) = X_0,
\end{aligned}
\tag{5.A.12}
$$

and again the process coefficients $A_t, B_t$ are, at this point, unknown.
Again invoking the product rule (4.A.6) and Table 4.A.1,

$$dX = d(X_1 \cdot X_2)$$

$$= X_1 \, dX_2 + X_2 \, dX_1 + dX_1 \, dX_2$$

$$= X_1 \, dX_2 + X_2 \, dX_1 + (b_t X_1 \, dt + e_t X_1 \, dW_t)(A_t \, dt + B_t \, dW_t)$$

$$= X_1 \, dX_2 + X_2 \, dX_1 + e_t X_1 \, B_t \, dt$$

$$= X_1(A_t \, dt + B_t \, dW_t) + X_2 \, (b_t X_1 \, dt + e_t X_1 \, dW_t) + e_t X_1 B_t \, dt$$

$$= X_1(A_t \, dt + B_t \, dW_t) + b_t Xdt + e_t XdW_t + e_t X_1 B_t dt. \tag{5.A.13}$$

Now, select $A_t, B_t$ in (5.A.12b) so that

$$X_1(A_t \, dt + B_t \, dW_t) + e_t X_1 B_t \, dt = a_t \, dt + c_t \, dW. \tag{5.A.14}$$

(If this equality holds, then (5.A.13) becomes $dX = a_t \, dt + e_t \, dW_t + b_t Xdt + e_t XdW_t = (a_t + b_t X)dt + (c_t + e_t X)dW_t$, which is consistent with (5.A.10).) So given (5.A.14), for this expression to hold, set

$$
\begin{aligned}
&A_t = (a_t - e_t c_t)X_1^{-1}; \\
&B_t = c_t X_1^{-1}.
\end{aligned}
\tag{5.A.15}
$$

We determined earlier that an expression such as (5.A.12a) integrates to

$$X_1(t) = exp\left\{\int_0^t e_s dW_s + \int_0^t \left(b_s - \frac{1}{2}e_s^2\right) ds\right\}$$

(5.A.16)

(e.g., use Example 5.1 as a guide). Given (5.A.15), (5.A.12b) integrates to

$$X_2(t) = X_0 + \int_0^t (a_s - e_s c_s) X_1(s)^{-1} ds + \int_0^t c_s X_1(s)^{-1} dW_s.$$

(5.A.17)

Combining (5.A.16) and (5.A.17) renders the *product form solution* to (5.A.10)

$$X(t) = X_1(t) \cdot X_2(t)$$

$$= exp\left\{\int_0^t e_s dW_s + \int_0^t \left(b_s - \frac{1}{2}e_s^2\right) ds\right\}$$

$$\times \left\{X_0 + \int_0^t (a_s - e_s c_s) X_1(s)^{-1} ds + \int_0^t c_s X_1(s)^{-1} dW_s\right\},$$

(5.A.18)

where

$$X_1(s)^{-1} = exp\left\{-\int_0^s e_u dW_u - \int_0^s \left(b_u - \frac{1}{2}e_u^2\right) du\right\}.$$

(5.A.19)

A more conventional representation of this solution appears as

$$X(t) = \Phi(t)\left\{X_0 + \int_0^t (a_s - e_s c_s)\Phi_s^{-1} ds + \int_0^t c_s \Phi_s^{-1} dW_s\right\},$$

(5.A.20)

where

$$\Phi(t) = exp\left\{\int_0^t e_s dW_s + \int_0^t \left(b_s - \frac{1}{2}e_s^2\right) ds\right\};$$

(5.A.21)

$$\Phi_s^{-1} = exp\left\{-\int_0^s e_u dW_u - \int_0^s \left(b_u - \frac{1}{2}e_u^2\right) du\right\}.$$

(5.A.22)

## Appendix 5.B  Integrating Factors and Variation of Parameters

In this section, we shall consider two additional techniques that can be utilized to solve SDEs, namely the use of integrating factors and the method of variation of parameters.

### 5.B.1 Integrating Factors

Consider a class of SDEs of the form

$$dX_t = f(X_t, t) \, dt + c(t) X_t \, dW_t, X(0) = X_0, \tag{5.B.1}$$

where $f : R \times R \to T$ and $C : R \to R$ are given continuous deterministic functions. Given (5.B.1), let the **integrating factor** be expressed as

$$\rho_t = exp \left[ -\int_0^t c(s) dW_s + \frac{1}{2} \int_0^t c(s)^2 ds \right].$$

Then

$$d(\rho_t X_t) = \rho_t f(X_t, t) dt \tag{5.B.2}$$

represents an exact SDE. To see this, suppose $Y_t = \rho_t X_t$. Then $X_t = \rho_t^{-1} Y_t$, and thus

$$
\begin{aligned}
dY_t = d(\rho_t X_t) &= \rho_t f(X_t, t) dt \\
&= \rho_t f\left(\rho_t^{-1} Y_t, t\right) dt.
\end{aligned}
\tag{5.B.3}
$$

Here, (5.B.3) represents a deterministic differential equation in $Y_t$ as a function of $t$, that is, as (5.B.2) reveals, multiplying both sides by the integrating factor, we obtain a deterministic differential equation.

For example, let us solve

$$dX_t = \frac{1}{X_t} dt + \alpha X_t dW_t, X(0) = X_0, \alpha \text{ a constant.}$$

Clearly, this is an SDE of the form (5.1) with $f(X_t, t) = (1/X_t), g(X_t, t) = \alpha X_t$, and, $c(t) = \alpha$. Hence,

$$
\begin{aligned}
\rho(t) &= exp \left[ -\int_0^t c(s) dW_s + \frac{1}{2} \int_0^t c(s)^2 ds \right] \\
&= exp \left( -dW_t + \frac{1}{2}\alpha^2 t \right),
\end{aligned}
$$

$$X_t = \rho_t^{-1} Y_t = exp \left( \alpha W_t - \frac{1}{2}\alpha^2 t \right) Y_t,$$

$$\frac{dY_t}{dt} = \rho_t f\left(\rho_t^{-1} Y_t, t\right)$$

$$= exp \left( -\alpha W_t + \frac{1}{2}\alpha^2 t \right) \left[ \frac{1}{exp(\alpha W_t - (1/2)\alpha^2 t) Y_t} \right], \text{and}$$

$$Y_t dY_t = exp \left( 2 \left( -\alpha W_t + \frac{1}{2}\alpha^2 t \right) \right) dt.$$

Then

$$\int_0^t Y_t dY_t = \int_0^t \exp\left(-2\alpha W_t + \alpha^2 t\right) dt,$$

$$Y_t^2 = Y_0^2 + 2\int_0^t \exp\left(-2\alpha W_t + \alpha^2 t\right) dt,$$

$$Y_t = \left[Y_0^2 + 2\int_0^t \exp\left(2\alpha W_t + \alpha^2 t\right) dt\right]^{\frac{1}{2}},$$

$$\rho_t X_t = \left[Y_0^2 + 2\int_0^t \exp\left(-2\alpha W_t + \alpha^2 t\right) dt\right]^{\frac{1}{2}},$$

and thus

$$X_t = \exp\left(\alpha W_t - \frac{1}{2}\alpha^2 t\right)\left[X_0^2 + 2\int_0^t \exp\left(-2\alpha W_t + \alpha^2 t\right) dt\right]^{\frac{1}{2}},$$

where $Y_0 = \rho_0 X_0 = X_0$.

### 5.B.2 Variation of Parameters

Suppose our objective is to obtain a solution to the geometric BM SDE (5.15) or to

$$dX_t = \mu X_t\, dt + \sigma X_t\, dW_t,$$

with $\mu$ and $\sigma$ constant. The general procedure that we shall follow in order to accomplish this task has two basic steps.

1. Transform the SDE into an equivalent integral form and integrate *directly* and arrive at an *artificial solution* with c representing an arbitrary constant of integration. (The resulting solution is termed artificial since $X_t$ does not satisfy the original SDE.)

   For instance, from the preceding equation let us write

$$\frac{dX_t}{X_t} = \mu dt + \sigma dW_t,$$

$$\int \frac{dX_t}{X_t} = \int dt + \sigma \int dW_t,$$

$$\ln X_t = \mu_t + \sigma W_t + c,$$

   so that the artificial solution is

$$X_t = e^{\mu_t + \sigma W_t + c}.$$

2. Replace $c$ by the function $c(t)$ (the variation of the parameter $c$ with $t$),

$$X_t = e^{\mu_t + \sigma W_t + c(t)},$$

and apply Itô's formula (4.18) to this expression so as to obtain

$$dX_t = X_t \left( \mu + c'(t) + \frac{\sigma^2}{2} \right) dt + \sigma X_t dW_t.$$

Substituting the last term from the original SDE gives

$$\left( c'(t) + \frac{\sigma^2}{2} \right) dt = 0$$

or

$$c(t) = -\frac{\sigma^2}{2} t + c_1, c_1 \text{ a constant.}$$

Then

$$X_t = e^{\mu t + \sigma W_t + c(t)}$$

$$= e^{\mu t + \sigma W_t - \frac{\sigma^2}{2} t + c_1}.$$

For $t = 0, X_0 = e^{c_1}$ so that our final solution is

$$X_t = X_0 e^{\left( \mu - \frac{\sigma^2}{2} \right) t + \sigma W_t}.$$

# 6

# Stochastic Population Growth Models

## 6.1 Introduction

It is a well-known fact that all biological populations exhibit some form of stochastic behavior and that environmental noise should thus be an integral component of any dynamic population model. Such *noise* is generally introduced as a stochastic (Wiener) process and, in particular, as a Brownian motion (BM) process.

In general, a dynamic model of a population will have *deterministic* and *stochastic* components that operate simultaneously. Deterministic elements serve to make the level of the response variable predictable from the initial conditions, while stochastic elements are typically attributable to the following sources:

1. **Demographic stochasticity**—refers to random factors that might impact personal reproduction, mortality, etc., and that are thus taken to be individual specific. Since this variety of stochasticity operates independently among population elements, it tends to "average itself out" in larger populations; its impact is mostly an issue for small populations. Under demographic stochasticity, random factors affect the individual birth and death processes.
2. **Environmental stochasticity**—refers to the effect on local and global populations (both large and small) of factors such as the weather, major accidents/catastrophes, epidemics, natural disasters, crop failures, and international (political, financial, and societal) dislocations. It is assumed that these factors operate independently over the long run. For environmental stochasticity, random fluctuations in the environment affect the entire population.
3. **Mensuration stochasticity**—refers to the notion that measurement error resulting from determining, say, the size of a population at various points in time is a legitimate source of uncertainty that may result in biased assessments of some phenomenon.
4. **Informational stochasticity**—refers to the possibility that critical errors of omission or specifications may be made when trying to model reality, either

*Stochastic Differential Equations: An Introduction with Applications in Population Dynamics Modeling*, First Edition. Michael J. Panik.
© 2017 John Wiley & Sons, Inc. Published 2017 by John Wiley & Sons, Inc.

through structural deficiencies or through inappropriate/unrealistic modeling assumptions.

The preceding discussion has indicated that it is imperative to explicitly take into account random phenomena in our modeling activities. This, ostensibly, will enable us to enhance the accuracy of our estimates/predictions of population growth (rates) and sustainability. This need for greater accuracy has prompted the use of stochastic process models, and these types of models can be solved via stochastic differential equations (SDEs), that is, the sample paths of the process can be described by an SDE.

The fundamental deterministic population growth model offered in Section 6.2 will serve as our starting point for developing population growth models that, for the most part, can be solved (in closed form) using SDEs: models that incorporate random effects attributed to, say, environmental fluctuations.

## 6.2 A Deterministic Population Growth Model

Let us express the basic continuous-time population growth model as

$$\frac{dN(t)}{dt} = rN(t), r = \text{constant}, \tag{6.1}$$

that is, the **instantaneous absolute growth rate** is proportional to the existing population level at time $t$, or the **instantaneous relative growth rate** of the population is

$$\frac{dN(t)/dt}{N(t)} = r = \text{constant}. \tag{6.2}$$

For a single-species biological population with no migration, we may view $r$ in the following fashion. Given that

$$\frac{dN(t)}{dt} = \text{births} - \text{deaths}, \tag{6.3}$$

if the births and deaths are each taken to be proportional to $N(t)$, then births = $bN(t)$ and deaths = $dN(t)$, where $b$ is the constant per capita birth rate and $d$ is the constant per capita death rate. Then (6.3) becomes

$$\frac{dN(t)}{dt} = bN(t) - dN(t) = (b-d)N(t)$$
$$= rN(t), \tag{6.4}$$

where $r = b - d$ is the average per capita number of offspring less the average per capita number of deaths per unit time. Hence, $r$ will be termed the constant **net birth rate** or the constant **per capita growth rate** of the population. The solution of the ordinary differential equation or ODE (6.4) is

$$N(t) = N_0 e^{rt}, \ N(0) = N_0. \tag{6.5}$$

So if $r > 0$, then $b > d$, and thus the population grows exponentially. If $r < 0$, then $b < d$, and thus the population declines exponentially.

Our introductory discussion (Section 6.1) has indicated that, for instance, environmental variability will affect a biological population in a random fashion. To model environmental effects, one possibility is to explicitly include additional variables (e.g., food supply, rainfall, and average temperature) into, say, Equation (6.4). In a nonstatic environment, the per capita birth and death rates ($b$ and $d$, respectively) could be expressed as functions of the aforementioned additional environmental variables. So as these supplemental variables change, then so do the per capita birth and death rates. However, the inclusion of additional explanatory variables into the expressions for $b$ and $d$ may unduly complicate the model. As an alternative, let us approximate the effect of environmental variability on population growth by varying the per capita birth and death rates in a random fashion. To this end, let us hypothesize that changes in the environment precipitate random changes in the population's per capita birth and death rates, or changes in the net birth rate $r$. Moreover, changes in $r$ are deemed independent of changes due to demographic, mensuration, and information stochasticity.

Section 6.3 introduces, in a very rudimentary fashion, how random effects are readily incorporated into an elementary deterministic population growth model. More complicated stochastic models then follow.

## 6.3 A Stochastic Population Growth Model

Let us return to the population growth model offered in Section 4.1. Given that $N(t)$ depicts population at time $t$, we posited that

$$\frac{dN(t)}{dt} = a(t)N(t), \ N(0) = N_0, \ t \geq 0, \tag{6.6}$$

where $a(t)$ is the relative population growth rate. Then, based upon the arguments underlying Equations (4.3)–(4.5.1), we can write (6.6) as

$$\frac{dN(t)}{dt} = (r(t) - \alpha(t) \cdot \text{noise})N(t), \ N(0) = N_0, \ t \geq 0. \tag{6.7}$$

If $r(t) = r = \text{constant}$, $\alpha(t) = \alpha = \text{constant}$, and $\text{noise} = dW(t)/dt$, with $W(0) = W_0 = 0$, then (6.7) becomes

$$\frac{dN(t)}{dt} = \left(r + \alpha \frac{dW(t)}{dt}\right)N(t), \ N(0) = N_0, \ t \geq 0, \tag{6.8}$$

or

$$dN(t) = rN(t)dt + \alpha N(t)dW(t), \ N(0) = N_0, \ t \geq 0. \tag{6.8.1}$$

Clearly, this SDE (which is autonomous, linear, and homogeneous) is of the form (5.15) and thus has a particular solution mirroring Equation (5.18) or

$$N(t) = N_0 e^{\left(r - \frac{1}{2}\alpha^2\right)t + \alpha W_t}, \tag{6.9}$$

a *geometric BM process*. (Note that the solution to (6.8.1) is also provided by Equation (5.A.9).)

As far as the long-term dynamic behavior of (6.9) is concerned,

i.   if $r > \frac{1}{2}\alpha^2$, then $N(t) \to +\infty$ as $t \to +\infty$;

ii.  if $r < \frac{1}{2}\alpha^2$, then $N(t) \to 0$ as $t \to +\infty$; and   (6.10)

iii. if $r = \frac{1}{2}\alpha^2$, then $N(t)$ fluctuates between arbitrarily large

and arbitrarily small values as $t \to +\infty$.

Suppose $N_0$ is independent of $W_t$, then via Example 5.9,

a.   $E(N(t)) = N_0 e^{rt}$;

b.   $V(N(t)) = N_0 e^{2rt}\left(e^{\alpha^2 t} - 1\right).$   (6.11)

## 6.4  Deterministic and Stochastic Logistic Growth Models

If the per capita rate of growth of a population depends upon population density, then the population cannot expand without limit. If a population is far from its growth limit, then it can possibly grow exponentially. However, when nearing its growth limit, the population growth rate declines significantly and reaches zero when that limit has been attained (Verhulst, 1838). (If the limit is exceeded, the population growth rate becomes negative so that the size of the population declines.) Verhulst proposed a long-run adjustment to exponential growth via a self-regulating mechanism (it compensates for overcrowding) that becomes operational when population becomes "too large" (in the sense that the population size approaches a "limiting population" that is determined by the carrying capacity of the environment).

To model this phenomenon, let us modify $r$ by multiplying it by an "overcrowding term" so that we obtain a density-dependent population growth equation, that is, using (6.2), set

$$\frac{dN(t)/dt}{N(t)} = r\left(1 - \frac{N(t)}{K}\right) = r(n(t)), \ N(0) = N_0, \tag{6.12}$$

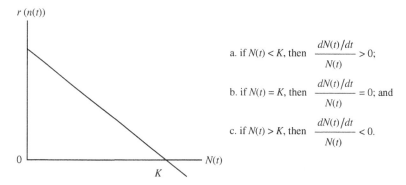

$r(n(t))$

a. if $N(t) < K$, then $\dfrac{dN(t)/dt}{N(t)} > 0$;

b. if $N(t) = K$, then $\dfrac{dN(t)/dt}{N(t)} = 0$; and

c. if $N(t) > K$, then $\dfrac{dN(t)/dt}{N(t)} < 0$.

**Figure 6.1** Per capita rate of growth is density dependent.

so that the instantaneous relative growth rate is a decreasing function of $N(t)$ or $r(n(t)) = r(1-(N(t)/K))$. As Figure 6.1 reveals, (6.12) implies a constant linear decrease in $r$ as population size increases, where $r$ is the constant maximum population growth rate when $N(t) = 0$, and $K$ represents the limiting size or **carrying capacity** of the population.[1]

The solution to the ODE (6.12) or to

$$dN(t) = r\left(1-\frac{N(t)}{K}\right)N(t)dt, \tag{6.12.1}$$

is termed the (deterministic) **logistic population growth function** and appears as

$$N(t) = \frac{KN_0}{N_0 + (K-N_0)e^{-rt}}, \quad N(0) = N_0, \, t \geq 0. \tag{6.13}$$

Given this expression, it is readily seen that

i. for $0 < N_0 < K$, populations increases asymptotically to $K$, that is, population follows a logistic curve;
ii. for $N_0 > K$, population decreases asymptotically to $K$;
iii. for $N_0 = K$, the population level $N(t) = K$ for all $t$; and
iv. for $N_0 = 0$, $N(t) = 0$ for all $t$.

Looking to the specification of the stochastic version of the logistic population growth model, let us augment (6.12.1) with the **multiplicative noise term**

---

1 How might $K$ be determined? One way to rationalize the choice for $K$ is to assume that *births* = $b_1 - b_2N(t)$ and *deaths* = $d_1 + d_2N(t)$, where $b_1$, $b_2$, $d_1$, and $d_2$ are positive scalars. Then,

$$\frac{dN(t)}{dt} = (births - deaths)N(t) = [b_1 - d_1 - (b_2 + d_2)N(t)]N(t)$$

$$= (b_1 - d_1)\left(1 - \frac{N(t)}{(b_1-d_1)/(b_2+d_2)}\right)N(t) = r\left(1 - \frac{N(t)}{K}\right)N(t),$$

where $r = b_1 - d_1 (> 0)$ and $K = (b_1 - d_1)/(b_2 + d_2)$.

$\alpha N(t)dW(t)$. (Note: **additive noise** appears as $\alpha dW(t)$.) For convenience, we shall work with the resulting equation in the form

$$dN_t = rN_t\left(1 - \frac{N_t}{K}\right)dt + \alpha N_t dW_t, \, N(0) = N_0, \, t \geq 0.^2 \tag{6.14}$$

Turning to Itô's equation (4.18) for solving the SDE $dX_t = f(X_t, t)$ $dt + g(X_t, t)dW_t, \, t \geq 0$, or

$$dV(X(t), t) = \left(V_t(X(t), t) + f(X(t), t)V_X + \frac{1}{2}g(X(t), t)^2 V_{XX}\right)dt$$
$$+ g(X(t), t)V_X dW(t),$$

let $V(N_t, t) = N_t^{-1}, V_N = -N_t^{-2}, V_{NN} = 2N_t^{-3}, f = rN_t\left(1 - \frac{N_t}{K}\right)$

$$= r\left(N_t - \frac{N_t^2}{K}\right), \text{ and } g = \alpha N_t.$$

$$dV = \left\{0 + r\left(N_t - \frac{N_t^2}{K}\right)(-N_t^{-2}) + \frac{1}{2}\alpha^2 N_t^2(2N_t^{-3})\right\}dt$$

$$+ \alpha N_t(-N_t^{-2})dW_t,$$

$$d(N_t^{-1}) = \left\{r\left(-N_t^{-1} + \frac{1}{K}\right) + \alpha^2 N_t^{-1}\right\}dt - \alpha N_t^{-1}dW_t. \tag{6.18}$$

To linearize (6.18), set $N_t^{-1} = y_t$ so that

$$dy_t = \left(\frac{r}{K} + (-r + \alpha^2)y_t\right)dt - \alpha y_t\, dW_t. \tag{6.19}$$

From Equations (5.A.20)–(5.A.22), (6.19) becomes, for $a = r/K, b = -r + \alpha^2$, $c = 0$, and $e = -\alpha$,

$$y_t = \Phi(t)\left\{y_0 + \int_0^t \left(\frac{r}{K}\right)\Phi_s^{-1}ds\right\}, \tag{6.20}$$

where

$$\Phi(t) = exp\left\{\int_0^t (-\alpha)dW_s + \int_0^t \left(-r + \frac{1}{2}\alpha^2\right)ds\right\}, \tag{6.21}$$

---

2 An alternative way of stochasticizing the deterministic logistic model is to write
$\frac{dN_t/dt}{N_t} = a(t)\left(1 - \frac{N_t}{K}\right), N(0) = N_0, t \geq 0$, 

(6.15)

and replace $a(t)$ in this expression by $r + \alpha(dW_t/dt)$ (see (6.6)–(6.8)) so that we obtain
$\frac{dN_t/dt}{N_t} = \left(r + \alpha\frac{dW_t}{dt}\right)\left(1 - \frac{N_t}{K}\right), N(0) = N_0, t \geq 0$, 

(6.16)

or
$dN_t = N_t\left(1 - \frac{N_t}{K}\right)(rdt + \alpha dW_t), N(0) = N_0, t \geq 0$. 

(6.17)

$$\Phi_s^{-1} = exp\left\{ -\int_0^s (-\alpha)dW_u - \int_0^s \left( -r + \frac{1}{2}\alpha^2 \right)du \right\}. \tag{6.22}$$

Evaluating (6.21) yields

$$\Phi(t) = exp\left\{ -\alpha W_t + \left( -r + \frac{1}{2}\alpha^2 \right)t \right\}, \tag{6.21.1}$$

and (6.22) gives

$$\Phi_s^{-1} = exp\left\{ \alpha W_s + \left( r - \frac{1}{2}\alpha^2 \right)s \right\}. \tag{6.22.1}$$

Then a substitution of (6.21.1) and (6.22.1) into (6.20) renders

$$y_t = exp\left\{ -\alpha W_t - \left( r - \frac{1}{2}\alpha^2 \right)t \right\}$$

$$\times \left\{ y_0 + \frac{r}{K}\int_0^t exp\left[ \left( r - \frac{1}{2}\alpha^2 \right)s + \alpha W_s \right]ds \right\}$$

or

$$N_t = \frac{exp\{(r - (1/2)\alpha^2)t + \alpha W_t\}}{\left\{ N_0^{-1} + (r/K)\int_0^t exp\left[ (r - (1/2)\alpha^2)s + \alpha W_s \right]ds \right\}}, \tag{6.23}$$

the **stochastic logistic population growth equation**, where $y_0 = N_0^{-1}$. (See Appendix 6.A for a discussion of both the deterministic and stochastic logistic population growth equations with an **Allee effect**.)

Since the stochastic logistic model is widely used in areas such as biology, engineering, and so on, a useful exercise is to determine the mean and variance of the process (6.23). To this end, we first rewrite this expression as

$$N_t = \frac{K exp[(r - (1/2)\alpha^2)t + \alpha W_t]}{\left\{ (K/N_0) + r\int_0^t exp\left[ (r - (1/2)\alpha^2)s + \alpha W_s \right]ds \right\}}. \tag{6.23.1}$$

Then employing the expectation rules presented in footnote 3 of Chapter 5, we can write

a.
$$E\left( exp\int_0^t \alpha W_s ds \right) = exp\int_0^t E\left[ \frac{(\alpha W_s)^2}{2} \right] ds$$

$$= exp\int_0^t \left( \frac{\alpha^2 s}{2} \right) ds;$$

b. $E\left[ exp\{ (r - \frac{1}{2}\alpha^2)t + \alpha W_t \} \right] = e^{rt}$; and

c. from (6.22.1),

$$E\left\{ exp\left( r\int_0^t \Phi_s^{-1} ds \right) \right\} =$$

$$E\left\{ exp\left( r\int_0^t \left[ \left(r - \frac{1}{2}\alpha^2\right)s + \alpha W_s \right] ds \right) \right\} =$$

$$exp\left\{ r\int_0^t E\left[ \left(r - \frac{1}{2}\alpha^2\right)s + \alpha W_s \right] ds \right\} =$$

$$exp\left( r\int_0^t (rs)ds \right) = e^{rt} - 1.$$

Given these results, it follows, from (6.23.1), that

$$E(N_t) = \frac{K}{e^{-rt}\{(K/N_0) + (e^{rt} - 1)\}} = \frac{K}{1 + [(K/N_0) - 1]e^{-rt}}.$$

Next, since $V(N_t) = E(N_t^2) - E(N_t)^2$, we first obtain

$E(N_t^2)$ as

$$E(N_t^2) = \frac{K^2 e^{(2r + \alpha^2)t}}{\{(K/N_0) + (e^{rt} - 1)\}^2}.$$

Then

$$V(N_t) = K^2 e^{2rt}\left\{ \frac{K}{N_0} + (e^{rt} - 1) \right\}^{-2} \left( e^{\alpha^2 t} - 1 \right).$$

## 6.5 Deterministic and Stochastic Generalized Logistic Growth Models

We may generalize the deterministic logistic population growth law (Richards, 1959) by considering the modified growth equation

$$\frac{dN_t}{dt} = \beta N_t \left[ 1 - \left( \frac{N_t}{N_\infty} \right)^r \right] \tag{6.24}$$

or

$$\frac{dN_t}{dt} = \beta N_t + \eta N_t^{r+1}, \quad N(0) = N_0, \quad t \geq 0, \tag{6.24.1}$$

where $N_\infty$ represents an upper horizontal asymptote (the **saturation level**) when $t \to +\infty$. As (6.24) indicates, the instantaneous absolute population growth rate is proportional to $N_t$ times a **feedback term**. A glance back at Equation (6.24.1) reveals an ODE (Bernoulli's equation) that integrates to the **Richards growth function** or the **deterministic generalized logistic equation**

$$N_t = N_\infty \left[ 1 + \left( \left( \frac{N_0}{N_\infty} \right)^{-r} - 1 \right) e^{-\beta r t} \right]^{-\frac{1}{r}}, \; N(0) = N_0, \; t \geq 0. \tag{6.25}$$

To obtain a stochastic generalized logistic growth equation, let us insert the **multiplicative noise term** $\alpha N_t dW_t$ into (6.24) so as to determine the SDE as follows:

$$dN_t = \beta N_t \left( 1 - \left( \frac{N_t}{N_\infty} \right)^r \right) dt + \alpha N_t dW_t. \tag{6.26}$$

Next, if we divide both sides of this equation by $N_\infty$ and set $Y_t = N_t / N_\infty$, then (6.26) becomes

$$d \left( \frac{N_t}{N_\infty} \right) = \beta \left( \frac{N_t}{N_\infty} \right) \left( 1 - \left( \frac{N_t}{N_\infty} \right)^r \right) dt + \alpha \left( \frac{N_t}{N_\infty} \right) dW_t$$

or

$$dY_t = \beta Y_t (1 - Y_t^r) dt + \alpha Y_t dW_t. \tag{6.27}$$

To solve this autonomous, nonlinear SDE, we shall employ the **reduction method** (see Appendix 6.B for details). To determine an appropriate transformation function $Z_t = U(Y_t)$, let us find, from (6.B.11),

$$A(Y_t) = \frac{f(Y_t)}{g(Y_t)} - \frac{1}{2} \frac{dg(Y_t)}{dY_t},$$

where $f(Y_t) = \beta Y_t (1 - Y_t^r)$ and $g(Y_t) = \alpha Y_t$. Then

$$A(Y_t) = \frac{\beta}{\alpha} (1 - Y_t^r) - \frac{1}{2} \alpha \neq 0$$

so that

$$\frac{dA(Y_t)}{dY_t} = -r \frac{\beta}{\alpha} Y_t^{r-1}.$$

In addition, with

$$g(Y_t) \frac{dA(Y_t)}{dY_t} = -r\beta Y_t^r$$

and

$$\frac{d}{dY_t} \left( g(Y_t) \frac{dA(Y_t)}{dY_t} \right) = -r^2 \beta Y_t^{r-1},$$

it follows from (6.B.14) that $b_2 = -r\alpha$ so that (6.B.13) is satisfied. Then from (6.B.15a) and (6.B.9),

$$U(Y_t) = Ce^{-r\alpha B(Y_t)} = Ce^{-r \ln \left( \frac{Y_t}{Y_0} \right)} = Y_t^{-r}$$

(given that $C$ and $Y_0$ are arbitrary) is our transformation function for the reduction process.

To determine the coefficients of (6.B.6), we look to (6.B.7). From (6.B.7a),

$$fU_Y + \frac{1}{2}g^2 U_{YY} = \beta Y_t \left(1 - Y_t^r\right)\left(-rY_t^{-r-1}\right) + \frac{1}{2}\alpha^2 Y_t^2 r(r+1)Y_t^{-r-2}$$

$$= \left[\frac{1}{2}\alpha^2 r(r+1) - \beta r\right]Z_t + \beta r,$$

where $Z_t = Y_t^{-r}$. Similarly, from (6.B.7b),

$$gU_y = -\alpha r Y_t^{-r} = -\alpha r Z_t.$$

Hence, the reduced form of Equation (6.27) is the autonomous linear SDE

$$dZ_t = \left[\left(\frac{1}{2}\alpha^2 r(r+1) - \beta r\right)Z_t + \beta r\right]dt - \alpha r Z_t dW_t. \tag{6.28}$$

To obtain the solution to (6.28), let us look to the general form offered for the solution to a linear SDE of the type

$$(AL)\ dZ_t = (a + bZ_t)dt + (c + eZ_t)dW_t.$$

(See Table 5.1.) Specifically, from Equation (5.A.20),

$$Z_t = \Phi_t \left\{ Z_0 + (a - ec)\int_0^t \Phi_s^{-1} ds + c \int_0^t \Phi_s^{-1} dW_s \right\}, \tag{6.29}$$

where

$$\Phi_t = exp\left\{ e\int_0^t dW_s + \left(b - \frac{1}{2}e^2\right)\int_0^t ds \right\}; \tag{6.30}$$

$$\Phi_s^{-1} = exp\left\{ -e\int_0^s dW_u - \left(b - \frac{1}{2}e^2\right)\int_0^s du \right\}. \tag{6.31}$$

In terms of (6.28),

$$a = \beta r,\ b = \frac{1}{2}\alpha^2 r(r+1) - \beta r,\ c = 0,\ \text{and}\ e = -\alpha r.$$

Evaluating (6.30) and (6.31) yields, respectively,

$$\Phi_t = exp\left\{ -\alpha r W_t + \left(\frac{1}{2}\alpha^2 r - \beta r\right)t \right\}\text{and} \tag{6.30.1}$$

$$\Phi_t^{-1} = exp\left\{ \alpha r W_s - \left(\frac{1}{2}\alpha^2 r - \beta r\right)s \right\}. \tag{6.31.1}$$

Hence, (6.29) becomes

$$Z_t = \Phi_t \left\{ Z_0 + \beta r \int_0^t \Phi_s^{-1} ds \right\}. \tag{6.29.1}$$

Then substituting (6.30.1) and (6.31.1) into (6.29.1) gives

$$Z_t = exp\left\{ r\left[ \left(\frac{1}{2}\alpha^2 - \beta\right)t - \alpha W_t \right] \right\}$$
$$\times \left\{ Z_0 + \beta r \int_0^t exp\left\{ r\left[ \left(-\frac{1}{2}\alpha^2 + \beta\right)s + dW_s \right] ds \right\} \right\}. \tag{6.29.2}$$

Since $Z_t = Y_t^{-r}$, an application of the reverse transformation $Y_t = Z_t^{-1/r}$ to (6.29.2) yields

$$Y_t = \Phi_t^{-1/r} \left\{ Y_0^{-r} + \beta r \int_0^t \Phi_s^{-1} ds \right\}^{-\frac{1}{r}}. \tag{6.29.3}$$

Since $N_t = N_\infty Y_t$, (6.29.3) becomes

$$N_t = N_\infty \Phi_t^{-1/r} \left\{ \left(\frac{N_0}{N_\infty}\right)^{-r} + \beta r \int_0^t \Phi_s^{-1} ds \right\}^{-\frac{1}{r}}, \tag{6.32}$$

the **stochastic generalized logistic population growth equation**.

## 6.6 Deterministic and Stochastic Gompertz Growth Models

Suppose we posit that the instantaneous rate of growth of the population $N_t$ at time $t$ is proportional to $ln N_\infty - ln N_t$ or

$$\frac{dN_t/dt}{N_t} = \beta(ln N_\infty - ln N_t) \tag{6.33}$$

(here we have a linear relationship between the instantaneous growth rate and $ln N_t$). The solution to this ODE is the **deterministic Gompertz (1825) population growth model**

$$N_t = N_\infty \, exp\{-\alpha \, exp(-\beta t)\}$$
$$= N_\infty e^{-\alpha e^{-\beta t}}, N(0) = N_0, t \geq 0, \tag{6.34}$$

where $\alpha = ln(N_\infty/N_0)$.[3]

---

3 Note that from (6.34), we can obtain
$$\frac{dN_t/dt}{N_t} = \alpha \beta e^{-\beta t}$$
so that
$$ln\left(\frac{dN_t/dt}{N_t}\right) = ln(\alpha\beta) - \beta t.$$
We thus have a linear relationship between the logarithm of the instantaneous growth rate and $t$.

To obtain the stochasticized version of the deterministic Gompertz model, let us rewrite (6.33). Set $X_t = N_t/N_\infty$. Then

$$\frac{dX_t}{dt} = \frac{1}{N_\infty}\frac{dN_t}{dt},$$

and thus (6.33) becomes

$$\frac{dN_t/dt}{N_\infty} = \beta\frac{N_t}{N_\infty}ln\left(\frac{N_\infty}{N_t}\right)$$

or

$$\frac{dX_t}{dt} = \beta X_t \, ln X_t^{-1} = -\beta X_t \, ln X_t. \tag{6.35}$$

Next, if we insert the multiplicative noise term $kX_t dW_t$ into (6.35), we obtain the autonomous nonlinear SDE.

$$dX_t = -\beta X_t \, ln X_t dt + kX_t dW_t, \tag{6.36}$$

where $f(X_t) = -\beta X_t \, ln X_t$ and $g(X_t) = kX_t$.

To solve (6.36), let us apply the reduction method developed in Appendix 6.A. To this end, it can be shown that the requisite transformation function has the form $Z_t = U(X_t) = ln X_t.$[4]

Turning to (6.A.7) we have,

$$fU_X + \frac{1}{2}g^2 U_{XX} = -\beta \, ln X_t - \frac{1}{2}k^2 = -\beta Z_t - \frac{1}{2}k^2, \, gU_X = k,$$

and thus the reduced form of Equation (6.36) is

$$dZ_t = \left(-\frac{1}{2}k^2 - \beta Z_t\right)dt + kdW_t, \tag{6.37}$$

an autonomous linear SDE.

Looking to the general form for the solution to a linear SDE (see Table 5.1), we have

$$(AL) \, dZ_t = (a + bZ_t)dt + (c + eZ_t)dW_t.$$

In this regard, from Equation (5.A.20),

$$Z_t = \Phi_t\left\{Z_0 + (a - ec)\int_0^t \Phi_s^{-1}ds + c\int_0^t \Phi_s^{-1}dW_s\right\}, \tag{6.38}$$

where

$$\Phi_t = exp\left\{e\int_0^t dW_s + \left(b - \frac{1}{2}e^2\right)\int_0^t ds\right\}; \tag{6.39}$$

---

4 From (6.A.11), $A(X_t) = -\frac{\beta}{k}ln X_t - \frac{1}{2k}$. Then $g(dA/dX_t) = -\beta$ and, thus, from (6.A.14), $b_2 = 0$. Then (6.A.15b) yields $U(X_t) = b_1 B(X_t) + C = (b_1/k)ln X_t$, where $b_1 = k$ and $C = 0$.

$$\Phi_s^{-1} = exp\left\{ -e\int_0^s dW_u - \left(b - \frac{1}{2}e^2\right)\int_0^s du \right\}.$$ (6.40)

In terms of Equation (6.37),

$$a = -\frac{1}{2}k^2, b = -\beta, c = k, \text{ and } e = 0.$$

Evaluating (6.39) and (6.40) yields, respectively,

$$\Phi_t = exp\{-\beta t\} \text{ and}$$
$$\Phi_s^{-1} = exp\{\beta s\}.$$

Hence, (6.38) becomes

$$Z_t = \Phi_t \left\{ Z_0 + a\int_0^t \Phi_s^{-1} ds + c\int_0^t \Phi_s^{-1} dW_s \right\}$$

$$= e^{-\beta t}\left\{ Z_0 - \frac{1}{2\beta}k^2\left(e^{\beta t} - 1\right) + k\int_0^t e^{\beta s} dW_s \right\}$$

or

$$\ln X_t = e^{\beta t}\left\{ Z_0 - \frac{1}{2\beta}k^2\left(e^{\beta t} - 1\right) + k\int_0^t e^{\beta s} dW_s \right\}.$$

Hence,

$$X_t = exp\left\{ e^{-\beta t}\left[ \ln X_0 - \frac{1}{2\beta}k^2\left(e^{\beta t} - 1\right) + k\int_0^t e^{\beta s} dW_s \right] \right\},$$ (6.41)

the **stochastic Gompertz growth function**.

## 6.7 Deterministic and Stochastic Negative Exponential Growth Models

Let the time profile of the values of a variable $N$ reflect the operation of the **deterministic negative exponential model**

$$N_t = N_\infty\left(1 - e^{-\beta t}\right), t \geq 0,$$ (6.42)

where the parameters $N_\infty$ and $\beta$ are both positive. Here, $N_\infty$ is the **limit to growth parameter** ($lim_{t \to +\infty} N_t = N_\infty$), while $\beta$ is the growth rate parameter. For $t = 0$, the initial value of $N$ is $N_0 = 0$. It is readily shown that

$$\frac{dN_t/dt}{N_t} = \beta\left(\frac{N_\infty}{N_t} - 1\right),$$ (6.43)

that is, the instantaneous rate of growth in $N_t$ at time $t$ approaches zero as $t \to +\infty$ and that this growth rate is proportional to the **feedback term** $(N_\infty / N_t) - 1$.

To determine the stochastic version of the deterministic negative exponential model, let us write (6.43) as

$$dN_t = \beta(N_\infty - N_t)dt. \tag{6.43.1}$$

Then upon inserting a multiplicative noise term of the form $\alpha N_t dW_t$ into (6.43.1), we obtain the autonomous linear SDE

$$dN_t = (\beta N_\infty - \beta N_t)dt + \alpha N_t dW_t. \tag{6.44}$$

A glance at Table 5.1 reveals that for a general linear SDE of the form

$$(AL) \ dN_t = (a + bN_t)\,dt + (c + eN_t)dW_t,$$

the requisite solution (see Equation (5.A.20)) appears as

$$N_t = \Phi_t \left\{ N_0 + (a - ec) \int_0^t \Phi_s^{-1} ds + c \int_0^t \Phi_s^{-1} dW_s \right\}, \tag{6.45}$$

where

$$\Phi_t = \exp\left\{ e \int_0^t dW_s + \left( b - \frac{1}{2}e^2 \right) \int_0^t ds \right\}; \tag{6.46}$$

$$\Phi_s^{-1} = \exp\left\{ -e \int_0^s dW_u - \left( b - \frac{1}{2}e^2 \right) \int_0^s du \right\}. \tag{6.47}$$

In terms of Equation (6.44),

$$a = \beta N_\infty, \ b = -\beta, \ c = 0, \text{ and } e = \alpha.$$

Evaluating (6.46) and (6.47) yields, respectively,

$$\Phi_t = \exp\left\{ \alpha W_t - \left( \beta + \frac{1}{2}\alpha^2 \right)t \right\} \text{ and}$$

$$\Phi_s^{-1} = \exp\left\{ -\alpha W_s + \left( \beta + \frac{1}{2}\alpha^2 \right)s \right\}.$$

Hence, (6.45) becomes

$$N_t = \exp\left\{ \alpha W_t - \left( \beta + \frac{1}{2}\alpha^2 \right)t \right\}$$
$$\times \left\{ N_0 + \beta N_\infty \int_0^t \exp\left\{ \left( \beta + \frac{1}{2}\alpha^2 \right)s - \alpha W_s \right\} ds \right\}, \tag{6.48}$$

the **stochastic negative exponential growth equation**.

## 6.8 Deterministic and Stochastic Linear Growth Models

Suppose the growth process is described by the **deterministic linear growth function**

$$N_t = N_0 + \beta t, \ t \geq 0, \tag{6.49}$$

where

$$\frac{dN_t}{dt} = \beta = \text{constant}$$

or

$$dN_t = \beta dt. \tag{6.50}$$

If we insert the multiplicative noise term $\alpha N_t dW_t$ into (6.50), then we obtain the autonomous linear SDE

$$dN_t = \beta dt + \alpha N_t dW_t. \tag{6.51}$$

Looking to Table 5.1, we see that the general SDE

$$(\text{AL}) \ dN_t = (a + bN_t)dt + (c + eN_t)dW_t$$

has the solution (using Equation (5.A.20))

$$N_t = \Phi_t \left\{ N_0 + (a - ec) \int_0^t \Phi_s^{-1} ds + c \int_0^t \Phi_s^{-1} dW_s \right\}, \tag{6.52}$$

where

$$\Phi_t = exp \left\{ e \int_0^t dW_s + \left( b - \frac{1}{2}e^2 \right) \int_0^t ds \right\}; \tag{6.53}$$

$$\Phi_s^{-1} = exp \left\{ -e \int_0^s dW_u - \left( b - \frac{1}{2}e^2 \right) \int_0^s du \right\}. \tag{6.54}$$

In terms of Equation (6.51),

$$a = \beta, \ b = c = 0, \text{ and } e = \alpha.$$

Evaluating (6.53) and (6.54) yields, respectively,

$$\Phi_t = exp \left\{ \alpha W_t - \frac{1}{2}\alpha^2 t \right\} \text{ and }$$

$$\Phi_s^{-1} = exp \left\{ -\alpha W_s + \frac{1}{2}\alpha^2 s \right\}.$$

Hence, (6.52) becomes

$$N_t = exp\left\{ \alpha W_t - \frac{1}{2}\alpha^2 t \right\}$$

$$\times \left\{ N_0 + \beta \int_0^t exp\left( \frac{1}{2}\alpha^2 s - \alpha W_s \right) ds \right\}, \tag{6.55}$$

the **stochastic linear growth equation with multiplicative noise.**

Note that if the additive noise term $\alpha dW_t$ is inserted into (6.50), we obtain the autonomous linear SDE in the narrow sense.

$$dN_t = \beta dt + \alpha dW_t. \tag{6.56}$$

That is, from Table 5.1, we obtain the (ALNS) equation

$$dN_t = adt + cdW_t$$

with solution

$$N_t = X_0 + a \int_0^t ds + c \int_0^t dW_s. \tag{6.57}$$

In terms of Equation (6.56),

$$a = \beta, b = e = 0, \text{ and } c = \alpha.$$

Then clearly, (6.57) renders

$$N_t = X_0 + \beta \int_0^t ds + \alpha \int_0^t dW_s$$

$$= X_0 + \beta t + \alpha W_t, \tag{6.58}$$

the **stochastic linear growth equation with additive noise.**

## 6.9 Stochastic Square-Root Growth Model with Mean Reversion

Although the **mean-reverting square-root SDE** (Cox et al., 1985; Feller, 1951a, 1951b)

$$dX_t = (\alpha - \beta X_t)dt + \sigma\sqrt{X_t}dW_t, X(0) = X_0 > 0, \tag{6.59}$$

was introduced by Feller (1951a, 1951b) in the context of population growth,[5] it has, of late, come to be known as the Cox, Ingersol, and Ross (**CIR) SDE model** for describing short-run interest rate dynamics. Here, $\alpha$ (the **speed of**

---

5 Suppose the random variable $X_t$ depicts population size (or gene frequency). Then, as noted by Feller, for the growth of large populations, a continuous approximation is appropriate, and this leads to diffusion-type processes, that is, for a growing population, the process in its later stages converges to a diffusion process mirroring simple growth. Hence, we should expect, in the limit, diffusion equations with drift, and thus normal distributions for the deviation of $X_t$ from some equilibrium point.

**adjustment**), $\beta$, and $\sigma$ are all taken to be strictly positive. Under the restriction $2\alpha > \sigma^2$ for all $t$ a.s., the **CIR process** (the solution to (6.59)) is well defined and strictly positive; otherwise, it is nonnegative. Moreover, the drift coefficient $\alpha - \beta X_t$ ensures mean reversion of $X_t$ toward its long-term level $\alpha/\beta$, and $\sigma\sqrt{X_t}$ is the diffusion coefficient describing volatility. Since the diffusion coefficient is singular at the origin, it follows that the initial nonnegative $X_t$ value can never turn negative for increasing $t$. (This is appealing when the $X_t$ process is used to describe interest rate behavior, or since interest rates must always remain and fluctuate around a long-term trend.)

To solve the SDE given by (6.59), let us first rewrite it as

$$dX_t + \beta X_t\, dt = \alpha\, dt + \sigma\sqrt{X_t}dW_t. \tag{6.59.1}$$

For $e^{\beta t}$, an integrating factor, the preceding expression becomes

$$d\left(e^{\beta t}X_t\right) = \alpha e^{\beta t}dt + \sigma e^{\beta t}\sqrt{X_t}dW_t,$$

and thus

$$\int_0^t d\left(e^{\beta t}X_t\right) = \alpha\int_0^t e^{\beta t}dt + \sigma\int_0^t e^{\beta u}\sqrt{X_u}dW_u.$$

Then it is readily verified that

$$X_t = \frac{\alpha}{\beta} + e^{-\beta t}\left(X_0 - \frac{\alpha}{\beta}\right) + \sigma e^{-\beta t}\int_0^t e^{\beta u}\sqrt{X_u}dW_u, \tag{6.60}$$

the **stochastic mean-reverting square-root growth equation** or, as it is now commonly called, the **CIR growth equation**.

Since the CIR model has been widely used to explore the behavior of short-term interest rates, it is important to consider some of its salient features. In particular, we shall examine its mean and variance. To this end, since $dW_u \sim N(0, du)$, it follows from (6.60) that

$$E(X_t) = \frac{\alpha}{\beta} + e^{-\beta t}\left(X_0 - \frac{\alpha}{\beta}\right);$$

$$V(X_t) = E\left[(X_t - E(X_t))^2\right]$$

$$= E\left[\left(\sigma e^{-\beta t}\int_0^t e^{\beta u}\sqrt{X_u}dW_u\right)^2\right]$$

$$= \sigma^2 e^{-2\beta t}E\left[\left(\int_0^t e^{\beta u}\sqrt{X_u}dW_u\right)^2\right] \tag{6.61}$$

$$= \sigma^2 e^{-\beta t}\int_0^t e^{2\beta u}E(X_u)du$$

via Itô's isometry. Then, upon substituting

$$E(X_u) = \frac{\alpha}{\beta} + e^{-\beta u}\left(X_0 - \frac{\alpha}{\beta}\right)$$

into (6.61) and integrating, we obtain

$$V(X_t) = X_0\left[\frac{\sigma^2\left(e^{-\beta t} - e^{2\beta t}\right)}{\beta}\right] + \frac{\alpha\sigma^2\left(1 - e^{-2\beta t}\right)}{2\beta^2}. \tag{6.61.1}$$

An extension of (6.59) has been offered by Chan, Karolyi, Longstaff, and Sanders or CKLS (1992) and appears as

$$dX_t = (\alpha - \beta X_t)dt + \sigma X_t^\gamma dW_t, X(0) = X_0 > 0, \tag{6.62}$$

the **CKLS mean-reverting gamma SDE**, where the parameter $\gamma$ measures the degree of nonlinearity of the relationship between $X_t$ and its volatility. Given that $\sigma$ must always be positive, $X_t \in (0.5, 1)$ if $\alpha, \beta > 0$ and $\gamma > 0.5$. In fact, when $\gamma \in (0.5, 1)$, $X_t > 0$ for all $t \geq 0$ a.s.

## Appendix 6.A   Deterministic and Stochastic Logistic Growth Models with an Allee Effect

Let us examine an important modification of the logistic function (Jiang et al., 2005; Krstić and Jovanović, 2010). Specifically, it has been observed (Allee, 1931) that in high-density populations, mutually positive cooperation effects were significant (i.e., average well-being was high). Similarly, as a population decreases in volume, there could be insufficient membership to achieve these benefits. So while cooperation could enhance population performance, it was subject to the risk that, at low population densities, its loss could lead to an increased chance of extinction—a declining population may reach a density at which the per capita growth rate will be zero before population density itself reaches zero. Hence, any future declines will cause the growth rate to become negative, and this will likely be followed by extinction.

In a social context, the intrinsic growth rate might be negative at low population densities due to the possibility of diminished reproductive opportunity; it is again negative at higher densities due to intraspecific competition due to overcrowding. Thus an **Allee effect** or **underpopulation effect** implies a positive association between population density and the reproduction and survival of individuals. That is, the per capita growth rate has a positive correlation with population size at low population densities. It arises when the intrinsic growth rate increases at low densities, reaches a maximum, and then declines in the direction of high population densities (Figure 6.A.1). It is important to recognize two distinct variations of the Allee effect: the **strong Allee effect** introduces a

**Figure 6.A.1** Allee effect.

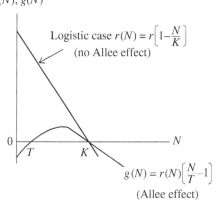

Per capita growth rates
$r(N), g(N)$

Logistic case $r(N) = r\left(1-\dfrac{N}{K}\right)$
(no Allee effect)

$g(N) = r(N)\left[\dfrac{N}{T}-1\right]$
(Allee effect)

population threshold (the minimal size of the population required to survive) that the population must exceed in order to grow; the **weak Allee effect** does not admit any such threshold.

To model the Allee effect, let us form what is called the **deterministic sparsity-impacted logistic per capita growth equation**

$$\frac{dN_t}{dt}\frac{1}{N_t} = g(N_t)$$

$$= r\left(1-\frac{N_t}{K}\right)\left(\frac{N_t}{T}-1\right), \tag{6.A.1}$$

where $T$ is the **sparsity parameter** or **Allee threshold** $(0 < T < K)$ and $((N_t/T)-1)$ is the **sparsity component** of (6.A.1). A population with an initial size $N_0$ less than $T$ tends to zero, while a population with $N_0 > T$ tends to $K$. Hence, $K$ may be viewed as a *stable* equilibrium point, while $T$ depicts an *unstable* equilibrium. In this regard, a *strong* Allee effect exists if the extinction equilibrium $N_t = 0$ is stable given that $K$ is a positive stable equilibrium point. (Note that Equation (6.A.1) displays a **multiplicative Allee effect**. An **additive Allee effect** can be written as

$$\frac{dN_t}{dt} = N_t\left[r\left(1-\frac{N_t}{K}\right)-\frac{m}{N_t+b}\right]$$

$$= rN_t\left(1-\frac{N_t}{K}\right)-\frac{mN_t}{N_t+b}, \quad 0 < m < 1, 0 < b < 1,$$

where $m$ and $b$ are termed "Allee constants." For $m > br$, we have a *strong* Allee effect; if $m < br$, the *weak* Allee effect is obtained.)

To randomize Equation (6.A.1), let us write

$$dN_t = \frac{r}{KT} N_t (K - N_t)(N_t - T)(rdt + \sigma dW_t), \quad t \geq 0, N(0) = N_0, \qquad (6.A.2)$$

where $\{W_t\}_{t \geq 0}$ is a standard BM process. Here, (6.A.2) is termed the **sparsity-impacted logistic SDE**. (See footnote 2 to this chapter for a justification of this stochastic specification.) For $N_t \neq 0$, $N_t \neq K$, and $N_t \neq T$ in (6.A.2), let us apply Itô's formula to the expression $V = ln|(K - N_t)/(N_t - T)| = ln|K - N_t| - ln|N_t - T| = V_1 - V_2$:

$$dV = dV_1 - dV_2$$

$$= V_1' \, dN_t + \frac{1}{2} V_1'' \, (dN_t)^2 - \left( V_2' \, dN_t + \frac{1}{2} V_2'' \, (dN_t)^2 \right)$$

$$= -\frac{dN_t}{K - N_t} - \frac{(dN_t)^2}{2(K - N_t)^2} - \frac{dN_t}{N_t - T} + \frac{(dN_t)^2}{2(N_t - T)^2}.$$

A substitution of (6.A.2) into this equation (remembering that $(dt)^2 = 0 = dt \, dW_t$ and $(dW_t)^2 = dt$) yields

$$d\ln\left|\frac{K - N_t}{N_t - T}\right| = -\frac{N_t}{KT}(K - T)(vdt + \sigma dW_t)$$

$$+ \frac{1}{2}\left(\frac{N_t}{KT}\right)^2 (K - T)(K + T - 2N_t)\sigma^2 dt \qquad (6.A.3)$$

$$= -\frac{K - T}{KT} N_t \left\{ \left[ r - \sigma^2 N_t \left( \frac{T + K - 2N_t}{KT} \right) \right] dt + \sigma dW_t \right\}.$$

Then

$$\frac{K - N_t}{N_t - T} = Ce^{At}, \quad C \text{ a constant}, \qquad (6.A.4)$$

where

$$A_t = -\frac{K - T}{KT} \int_0^t N_s \left\{ \left[ r - \sigma^2 N_s \left( \frac{T + K - 2N_s}{KT} \right) \right] ds + \sigma dW_s \right\}. \qquad (6.A.5)$$

When we take into consideration the dependence upon the initial population size $N_0$, the following two cases present themselves:

1. If $T < N_0 < K$, then $C = (K - N_0/N_0 - T) > 0$ a.s., and thus $T < N_t < K$, $t \geq 0$, so that (6.A.4) becomes

$$N_t = \frac{K + CT \, e^{A_t}}{1 + C \, e^{A_t}}. \qquad (6.A.6)$$

2. If $0 < N_0 < T$, then $C = (K - N_0/N_0 - T) < 0$ a.s., and thus $0 < N_t < T$ $(< K), t \geq 0$, so that (6.A.4) yields

$$N_t = \frac{K - CT \, e^{A_t}}{1 - C \, e^{A_t}}. \tag{6.A.7}$$

Clearly, (6.A.2) does not have a closed-form solution.

For any initial value $N_0$ such that $0 < N_0 < K$, it can be shown (Krstić and Jovanović, 2010) that there exists a unique, uniformly continuous positive solution to (6.A.2). For instance, let $T < N_0 < K$. For $T < N_t < K$, let $\{X_t\}_{t \geq 0}$ be a stochastic process defined by

$$X_t = ln\left(\frac{K - N_t}{N_t - T}\right). \tag{6.A.8}$$

Applying Itô's formula to (6.A.8) renders

$$dX_t = f(X_t)dt + g(X_t)dW_t, \ t \geq 0, \tag{6.A.9}$$

where

$$f(X_t) = -\frac{(K - T)(K + T \, e^{X_t})}{TK(1 + e^{X_t})}\left[r - \sigma^2 \frac{(K - T)(K + T \, e^{X_t})(e^{X_t} - 1)}{2TK(1 + e^{X_t})^2}\right],$$

$$g(X_t) = -\left[\frac{(K - T)(K + T \, e^{X_t})}{TK(1 + e^{X_t})}\right]\sigma$$

for all $t \geq 0$ and $X_0 = \bar{X}_0 = ln(K - N_0/N_0 - T)$. Under some mild restrictions (see Krstić and Jovanović (2010)), (6.A.9) has a unique continuous solution $X_t, t \geq 0$, satisfying $X_0 = \bar{X}_0$. Since (from (6.A.8))

$$N_t = \frac{K + T \, e^{X_t}}{1 + e^{X_t}},$$

we can readily demonstrate that $N_t$ is a positive and continuous solution to (6.A.2), that is,

$$dN_t = d\left(\frac{K + Te^{X_t}}{1 + e^{X_t}}\right)$$

$$= -(K - T)\frac{e^{X_t}}{(1 + e^{X_t})^2}\left[dX_t + \left(\frac{1 - e^{X_t}}{1 + e^{X_t}}\right)(dX_t)^2\right]$$

$$= \frac{(K - T)^2}{TK}\left(\frac{e^{X_t}}{(1 + e^{X_t})^2}\right)\left(\frac{K + Te^{X_t}}{1 + e^{X_t}}\right)(rdt + \sigma dW_t)$$

$$= N_t\left(1 - \frac{N_t}{K}\right)\left(\frac{N_t}{T} - 1\right)(rdt + \sigma dW_t), t \geq 0.$$

If $0 < N_0 < T$, then define the process $\{X_t\}_{t \geq 0}$ as

$$N_t = ln\left(\frac{K - N_t}{N_t}\right). \tag{6.A.10}$$

Then from an application of Itô's formula to (6.A.10), we obtain

$$dX_t = \tilde{f}(X_t)dt + \tilde{g}(X_t)dW_t, \ t \geq 0, \tag{6.A.11}$$

where

$$\tilde{f}(X_t) = -\frac{K - T(e^{X_t} + 1)}{T(e^{X_t} + 1)}\left[r - \sigma^2\frac{(e^{X_t} - 1)(K - T(e^{X_t} - 1))}{2T(e^{X_t} - 1)}\right],$$

$$\tilde{g}(X_t) = -\left(\frac{K - T(e^{X_t} - 1)}{T(e^{X_t} - 1)}\right)\sigma$$

for all $t \geq 0$ and $X_0 = \bar{X}_0$. With

$$N_t = \frac{K}{e^{X_t} + 1}$$

(from (6.A.10)), it follows that $N_t$ is a positive and continuous solution to (6.A.2) since

$$dN_t = d\left(\frac{K}{e^{X_t} + 1}\right)$$

$$= -\left(\frac{Ke^{X_t}}{(e^{X_t} + 1)^2}\right)dX_t + \left(\frac{Ke^{X_t}\left(e^{X_t} - 1\right)}{2(e^{X_t} + 1)^3}\right)(dX_t)^2$$

$$= \left(\frac{Ke^{X_t}(K - T(e^{X_t} + 1))}{T(e^{X_t} + 1)^3}\right)(rdt + \sigma dW_t)$$

$$= N_t\left(1 - \frac{N_t}{K}\right)\left(\frac{N_t}{T} - 1\right)(rdt + \sigma dW_t), \ t \geq 0.$$

Suppose $\sigma = 0$ in (6.A.5) so that

$$A_t = -\frac{r(K - T)}{KT}\int_0^t N_s \, ds = -\frac{r(K - T)}{KT}N_t t.$$

Then, from (6.A.4),

$$\frac{K - N_t}{N_t - T} = Ce^{-\left[\frac{r(K-T)}{KT}N_t t\right]}. \tag{6.A.12}$$

## Appendix 6.B   Reducible SDEs

Our objective herein is to ascertain whether a *nonlinear* SDE (Gard, 1988; Gihman and Skorokhod, 1972; Kloeden and Platen, 1999) of the form

$$dX_t = f(X_t, t)dt + g(X_t, t)dW_t \qquad (6.B.1)$$

can be reduced, under a suitable change of variables or transformation of the form $Y_t = U(X_t, t)$, to a linear SDE in $Y_t$ or

$$dY_t = (a_1(t) + a_2(t)Y_t)dt + (b_1(t) + b_2(t)Y_t)dW_t. \qquad (6.B.2)$$

If $\partial U/\partial X_t \neq 0$, then it can be shown that a solution $X_t$ of (6.B.1) has the form $X_t = V(Y_t, t)$, where $Y_t$ is provided by (6.B.2) for appropriately defined coefficients $a_1(t)$, $a_2(t)$, $b_1(t)$, and $b_2(t)$.

Applying Itô's formula (4.18) to $U(X_t, t)$ yields

$$dU(X_t, t) = \left( U_t + fU_X + \frac{1}{2}g^2 U_{XX} \right)dt + gU_X dW_t, \qquad (6.B.3)$$

with all coefficients and partial derivatives in this expression evaluated at $(X_t, t)$. This will render (6.B.1) **reducible** or yield a linear SDE of the form (6.B.2) if

$$
\begin{aligned}
&\text{a.}\quad U_t + fU_X + \frac{1}{2}g^2 U_{XX} = a_1(t) + a_2(t)U(X_t, t);\\
&\text{b.}\quad gU_X = b_1(t) + b_2(t)U(X_t, t).
\end{aligned}
\qquad (6.B.4)
$$

Looking to greater specificity, suppose (6.B.1) is an *autonomous* nonlinear SDE or

$$dX_t = f(X_t)dt + g(X_t)dW_t. \qquad (6.B.5)$$

Then the transformation function takes the form $Y_t = U(X_t)$ and can be used to reduce (6.B.5) to the autonomous linear SDE

$$dY_t = (a_1 + a_2 Y_t)dt + (b_1 + b_2 Y_t)dW_t. \qquad (6.B.6)$$

In this case, (6.B.4) becomes

$$
\begin{aligned}
&\text{a.}\quad fU_X + \frac{1}{2}g^2 U_{XX} = a_1 + a_2\, U(X_t);\\
&\text{b.}\quad gU_X = b_1 + b_2\, U(X_t).
\end{aligned}
\qquad (6.B.7)
$$

If $g(X_t) \neq 0$ and $b_2 \neq 0$, then, from (6.B.7b), we obtain

$$U(X_t) = Ce^{b_2 B(X_t)} - \frac{b_1}{b_2}, \qquad (6.B.8)$$

where

$$B(X_t) = \int_{X_0}^{X_t} \frac{ds}{g(s)} \qquad (6.B.9)$$

and $C$ and $X_0$ are arbitrary constants of integration. Substituting (6.B.8) into (6.B.7a) yields

$$C\left(b_2 A(X_t) + \frac{1}{2}b_2^2 - a_2\right)e^{b_2 B(X_t)} = a_1 - a_2\left(\frac{b_1}{b_2}\right), \qquad (6.B.10)$$

where

$$A(X_t) = \frac{f(X_t)}{g(X_t)} - \frac{1}{2}\frac{dg(X_t)}{dX_t}. \qquad (6.B.11)$$

Using (6.B.10), we can derive (with a multiplicity of steps omitted) the expression

$$b_2\frac{dA}{dX_t} + \frac{d}{dX_t}\left(g(X_t)\frac{dA}{dX_t}\right) = 0. \qquad (6.B.12)$$

Clearly, this equation holds if $dA/dX_t = 0$, or if

$$\frac{d}{dX_t}\left(\frac{(d/dX_t)(g(X_t)(dA/dX_t))}{dA/dX_t}\right) = 0, \qquad (6.B.13)$$

given that, from (6.B.12), $b_2$ is chosen so that

$$b_2 = -\frac{(d/dX_t)(g(X_t)(dA/dX_t))}{dA/dX_t}. \qquad (6.B.14)$$

Hence, a necessary and sufficient condition for the reducibility of an autonomous nonlinear SDE to an autonomous linear SDE is Equation (6.B.12), that is, (6.B.12) characterizes the conditions that $f$ and $g$ must satisfy for reducibility. Now

a.  if $b_2 \neq 0$, select $U(X_t) = Ce^{b_2 B(X_t)}$; and

b.  if $b_2 = 0$, select $U(X_t) = b_1 B(X_t) + C,$

$$(6.B.15)$$

where $b_1$ must satisfy (6.B.7b).

Once $U(X_t)$ has been determined, evaluating (6.B.7a) enables us to determine (6.B.2). Then applying the inverse transformation allows us to obtain the solution to (6.B.1).

**Example 6.B.1**  Given the SDE

$$dX_t = -\frac{1}{2}e^{-2X_t}dt + e^{-X_t}dW_t,$$

we have $f(X_t) = -(1/2)e^{-2X_t}$ and $g(X_t) = e^{-X_t}$. Then from (6.B.11),

$$A(X_t) = -\frac{1}{2}e^{-X_t} + \frac{1}{2}e^{-X_t} \equiv 0$$

so that (6.B.12) holds for *any* $b_2$. Set $b_2 = 0$ and $b_1 = 1$. Then a solution of (6.B.7b) is obtained as $gU_X = 1$, $U_X = e^{X_t}$, $dU(X_t) = e^{X_t}dX_t$, and thus $U(X_t) = e^{X_t}$ via (6.B.15b). Substituting $U(X_t)$ into (6.B.7a) yields

$$0 = a_1 + a_2 e^{X_t},$$

which holds for $a_1 = a_2 = 0$. Hence, $Y_t = U(X_t) = e^{X_t}$ and (6.B.2) reduces to the linear SDE $dY_t = dW_t$, which has the solution

$$Y_t = W_t + Y_0 = W_t + e^{X_0}.$$

Since $Y_t = e^{X_t}$, it follows that $\ln Y_t = X_t$ so that the original SDE (6.B.1) has the solution

$$X_t = \ln\left(W_t + e^{X_0}\right). \blacksquare$$

What if the transformation function $Y_t = U(X_t)$ is known in advance? (Remember that we are still dealing with the autonomous case.) If $U(X_t)$ is specified up front, then the second-order Taylor expansion of $U$ appears as

$$dY_t = U_X \, dX_t + \frac{1}{2} U_{XX} (dX_t)^2, \tag{6.B.16}$$

where $dX_t = f(X_t)dt + g(X_t)dW_t$. A substitution of this latter equality into (6.B.16) and simplifying (with the aid of Table 4.A.1) will then yield an autonomous linear SDE.

**Example 6.B.2** Our objective is to solve the autonomous nonlinear SDE

$$dX_t = f(X_t)dt + g(X_t)dW_t$$

$$= -\frac{1}{2} e^{-2X_t} dt + e^{-X_t} dW_t$$

given that we know the transformation function to be $Y_t = U(X_t) = e^{X_t}$ (see the preceding example problem). Then from (6.B.16),

$$dY_t = e^{X_t} dX_t + \frac{1}{2} e^{X_t} (dX_t)^2$$

$$= e^{X_t}\left(-\frac{1}{2} e^{-2X_t} dt + e^{-X_t} dW_t\right)$$

$$+ \frac{1}{2} e^{X_t}\left(-\frac{1}{2} e^{-2X_t} dt + e^{-X_t} dW_t\right)^2$$

$$= -\frac{1}{2} e^{-X_t} dt + dW_t + \frac{1}{2} e^{X_t}\left(e^{-2X_t} dt\right)$$

$$= dW_t,$$

which is the same linear SDE obtained in the preceding example. The solution to this SDE then follows as usual. $\blacksquare$

# 7

# Approximation and Estimation of Solutions to Stochastic Differential Equations

## 7.1   Introduction

It may not be possible to determine an explicit or closed-form solution to the **Itô's stochastic initial value problem** depicted by the stochastic differential equation (SDE)

$$dX_t = f(X_t, t)dt + g(X_t, t)dW_t, X(t_0) \equiv X_0, t \in [t_0 + \infty). \tag{7.1}$$

Hence, we must employ numerical methods for calculating approximations to problems such as (7.1). In this regard, we must use what is called a discrete-time approximation or **difference method** to iteratively approximate a solution to (7.1), that is, this method utilizes a recursive algorithm that outputs the values of a discrete-time approximation at the given discretization points of a finite subinterval $[t_0, T]$ of $[t_0, +\infty)$. While the approximation is made only at the discretization points, we will always view a discrete-time approximation as a "continuous-time process" defined on $[t_0, T]$.

Looking to specifics, two essential tasks present themselves. First, the continuous-time SDE (7.1) is replaced by a discrete-time difference equation that generates the values $Y_0(= X_0), Y_1, Y_2, ..., Y_N$ ($N$ a positive integer) used to approximate, at given discrete-time instants $t_0 < t_1 < t_2 < \cdots < t_N$, the solution values $X(t_i; X_0, t_0), i = 1, ..., N$. As one might have already surmised, to obtain a "good" approximate solution, the time increments $\Delta_i = t_{i+1} - t_i, i = 0, 1, 2, ...,$ should be "sufficiently small."

Second, given the stochastic nature of (7.1), the approximations $Y_i, i = 1, ..., N$ should mirror the sample paths or trajectories of the solution $X_t$ of this SDE; samples of random processes $\{Y_i\}$ are to be (iteratively) generated so as to approximate the solution process sample values $\{X_t\}$. So not only are the $\Delta_i$s required, the standard Brownian motion (BM) or Wiener increments $\Delta W_i$

*Stochastic Differential Equations: An Introduction with Applications in Population Dynamics Modeling*, First Edition. Michael J. Panik.

are needed as well. The Wiener increments are viewed as sample random variables that are normally distributed with a mean of zero and variance $\Delta_i$ (see Section 3.5.1).

How should the $Y_i$ sample values be obtained? An approximate sample path of SDE (7.1) is regarded as corresponding to a given sample path or trajectory of the underlying Wiener process. In addition, this correspondence can be realized by simulating the BM increments $\Delta W_i$ via a normal pseudorandom number generator. In fact, the iterative schemes to be implemented can be thought of as "sample path simulation processes"—we simulate the entire trajectory of the solution process $\{X_t\}$ of (7.1) over the entire discretization of $[t_0, T]$. For convenience, we can choose a set of equally spaced discretization instants, for example, for $N$, a positive integer, $t_{i+1} - t_i = (T - t_0)/N \equiv \Delta, i = 1, ..., N$, is the step size for the iteration scheme. Then we obtain the independent random increments $\Delta W_i = W(t_{i+1}) - W(t_i) \sim N(0, \Delta)$ (or, equivalently, $\Delta W_i \sim \sqrt{\Delta} N(0, 1)$), $i = 1, ..., N$, of the Wiener process $\{W_t\}_{t \in [t_0, T]}$. (For these increments, $E(\Delta W_i) = 0$ and $E[(\Delta W_i)^2] = \Delta$.) Hence, the sampling of normal variates to approximate this Wiener process in (7.1) is executed by generating a set of Gaussian or normal pseudorandom numbers. Once the random increments $\Delta W_i$ are generated, we can use a recursive method to calculate the sample values $\{Y_i, i = 0, 1, ..., N\}$ at the points $t_i, i = 0, 1, ..., N$, of the time discretization. The types of recursive schemes that have been proven successful in approximating (7.1) are the subject of the Sections 7.2, 7.4, and 7.5.

## 7.2 Iterative Schemes for Approximating SDEs

Once $\Delta, \Delta W_i, i = 1, ..., N$, and $Y_0$ have been specified, the approximate solutions $Y_1, Y_2, ..., Y_N$ can be calculated using an appropriate recursive formula. What sorts of recursive formulas typically work well in practice? What is our measure or yardstick for "working well?" In what follows, we shall first focus on the Euler–Maruyama (EM) and Milstein approximations to (7.1). The efficiency of these routines in rendering "good" approximations will be addressed by considering two criteria of optimality—strong and weak (orders of) convergence. We next explore "local linearization techniques" (those of Ozaki (1992, 1993) and Shoji and Ozaki (1997, 1998)) for approximating, via simulations, solutions of SDEs. Here, criteria for "working well," in terms of "goodness of approximation," are provided by an assessment of the rates of convergence (in the $L_p$ sense) of the one-step and multistep approximation errors.

### 7.2.1 The EM Approximation

The simplest discrete-time recursive routine used to approximate an Itô process of the form (7.1) is the **EM approximation scheme**. Given the time discretization $t_0 < t_1 < t_2 < \cdots < t_N = T$ of $[t_0, T]$, the EM approximation is a continuous-time stochastic process $Y = \{Y_t\}_{t\in[t_0, T]}$ satisfying the iterative scheme

$$Y_{i+1} = Y_i + f(Y_i, t_i)(t_{i+1} - t_i) + g(Y_i, t_i)(W(t_{i+1}) - W(t_i))$$
$$= Y_i + f(Y_i, t_i)\Delta_i + g(Y_i, t_i)\Delta W_i, \ i = 1, 2, ..., N-1, \tag{7.2}$$

where $Y(t_i) \equiv Y_i$, $Y_0 = X_0$, $\Delta_i = t_{i+1} - t_i$, and $\Delta W_i = W(t_{i+1}) - W(t_i)$. Under a regime of equidistant discretization times, we have $\Delta_i \equiv \Delta = (T - t_0)/N$.

What is the justification for the structure of (7.2)? Remember that the solution of (7.1) in integral form is

$$X_t = X_0 + \int_{t_0}^t f(X_s, s)\, ds + \int_{t_0}^t g(X_s, s)\, dW_s, \ t \in [t_0, T]. \tag{7.3}$$

Clearly, (7.3) suggests that the integral equation for the true solution increment $\Delta X_i$ should take the form

$$X_{i+1} - X_i = \int_{t_i}^{t_{i+1}} f(X_t, t)\, dt + \int_{t_i}^{t_{i+1}} g(X_t, t)\, dW_t, \tag{7.4}$$

and, in turn, (7.4) suggests that the difference equation scheme (7.2) should prove to be a satisfactory approximation technique since each term on the right-hand side of this expression approximates the corresponding term on the right-hand side of (7.4). So in view of (7.4), the length of the time discretization subinterval $[t_i, t_{i+1}]$ is

$$\Delta_i = \int_{t_i}^{t_{i+1}} dt;$$

and the $N(0, \Delta_i)$ distributed increment of the Wiener process $\{W_t\}$ on $[t_i, t_{i+1}]$ is

$$\Delta W_i = \int_{t_i}^{t_{i+1}} dW_t = W(t_{i+1}) - W(t_i).$$

If we (1) admit equidistant discretization times so that $\Delta_i \equiv \Delta$, $i = 1, ..., N-1$, (2) set $\Delta W_i \sim \sqrt{\Delta}Z$, where $N(0,1)$ is Gaussian distributed, and (3) assume that the coefficient functions $f$ and $g$ depend only on $Y_i$ and not on $t_i$, then (7.2) can be rewritten, for approximation purposes, as

$$Y_{i+1} = Y_i + f(Y_i)\Delta + g(Y_i)\sqrt{\Delta}Z, \ i = 1, 2, ..., N-1. \tag{7.2.1}$$

(For a more formal look at the construction of Equation (7.2.1), see Appendix 7.B.)

In sum, given an Itô process $\{X_t\}_{t\in[t_0, T]}$ solution of the SDE (7.1), the EM approximation of $X$ is a continuous-time stochastic process $\{Y_i\}$ satisfying the recursive scheme (7.2) (or (7.2.1)). To simulate $\{Y_i\}$, and thus the approximate sample path solution, one only needs to simulate the increments of the BM or

Wiener process $\{\Delta W_i\}$. In fact, the EM approximation uses the same BM realization as the true solution itself. What criterion should be used to assess the performance of the EM routine?

### 7.2.2 Strong and Weak Convergence of the EM Scheme

We noted earlier that the EM routine is a computational device for approximating an Itô process $X_t$ by the values $Y_i$ generated by (7.2) (or (7.2.1)). Since $X(T)$ and $Y(T)$ are both random variables, it is reasonable to use an expression such as $E|X(T) - Y(T)|$ to measure the degree of precision (error) of the approximation. More specifically, a time-discretized approximation $Y_\Delta$ of a continuous-time process $X$ **converges with strong order** $\gamma$ to the solution $X$ at time $T$ if there exists a constant $C$ (not depending on $\Delta$) such that

$$E|X(T) - Y_\Delta(T)| \le C\Delta^\gamma \tag{7.5}$$

for $N$ chosen large enough so that $\Delta = (T - t_0)/N \in (0, 1)$, where $X(T)$ is the true solution at time $T$ and $Y_\Delta(T)$ the approximation. Given this error bound, it can be demonstrated that the strong order of convergence for the EM scheme is $\gamma = 1/2$ (which holds under appropriate conditions of $f$ and $g$ in (7.1), that is, $f$ and $g$ satisfy uniform growth and Lipschitz conditions with respect to $X$ and are Hölder continuous of order $1/2$ in $t$).

The strong order of convergence criterion (7.5) indicates the rate at which the "mean endpoint error" decreases as $\Delta \to 0$. A less restrictive criterion—one that considers the rate of decrease of the "error of means"— involves the notion of "weak convergence." Specifically, a discrete-time approximation $Y_\Delta$ of a continuous time process $X$ **converges weakly of order** $\beta$ to the solution $X$ at time $T$ if there exists a continuously differentiable polynomial function $h$ and a constant $C_h$ (independent of $\Delta$) such that

$$|E(h(X_T)) - E(h(Y_\Delta(T)))| \le C_h \Delta^\beta, \Delta \in (0,1). \tag{7.6}$$

The EM routine converges with weak order $\beta = 1$.

In general, under some smoothness conditions on $f$ and $g$ in (7.1), iterative approximation schemes of a particular order that converge strongly typically have a higher order of weak convergence. That is, if the step size $\Delta$ decreases $k$-fold, then the approximation error decreases by a factor $k^\gamma$. So if, for instance, the order is 0.5 and we want to decrease the error by 100 times, we must make the step size $100^2 = 10,000$ times smaller.

### 7.2.3 The Milstein (Second-Order) Approximation

The EM approximation scheme has strong order of convergence $\gamma = 1/2$. Following Milstein (1978), it is possible to raise the strong order of convergence of the EM method to 1 by introducing a correction to the stochastic increment in (7.2). That is, we can augment the EM scheme (7.2) by introducing the term

$$\frac{1}{2}g\frac{\partial g}{\partial X}\left\{(\Delta W_i)^2 - \Delta_i\right\}\tag{7.7}$$

from the Itô–Taylor expansion (Appendix 7.A). Here, (7.7) is derived from the truncation of the stochastic Taylor expansion of the integral form of the solution $X_t$ of (7.1).[1] Hence, the **Milstein scheme** appears as

$$Y_{i+1} = Y_i + f(y_i, t_i)\Delta_i + g(y_i, t_i)\Delta W_i$$
$$+ \frac{1}{2}g(y_i, t_i)\frac{\partial g(y_i, t_i)}{\partial X}\left\{(\Delta W_i)^2 - \Delta_i\right\}, i = 0, 1, ..., N-1,\tag{7.8}$$

where $Y(t_i) \equiv Y_i$, $Y_0 = X_0$, and $t \in (t_0, T)$. If we again set $\Delta_i = \Delta$ (we have equal time discretizations), $i = 1, ..., N-1$, and take $\Delta W_i \sim \sqrt{\Delta}Z$, then, if the coefficient functions $f$ and $g$ depend only on $Y_i$ and not on $t_i$, (7.8) can be rewritten, for approximation purposes, as

$$Y_{i+1} = Y_i + f(Y_i)\Delta + g(Y_i)\sqrt{\Delta}Z$$
$$+ \frac{1}{2}g(Y_i)\frac{\partial g(Y_i)}{\partial X}\Delta(Z^2 - 1), i = 1, 2, ..., N-1.\tag{7.8.1}$$

(Appendix 7.B offers a much more straightforward derivation of (7.8.1).) Clearly, Milstein's approximation routine admits an additional term at each iteration relative to the EM technique. Milstein's approximation (7.8) (or (7.8.1)), which has strong order of convergence $\gamma = 1$, converges to the Itô solution of (7.1); it thus converges to the correct stochastic solution faster than the EM scheme as the step size $\Delta \to 0$. Milstein's method is identical to the EM method if there is no "$X$" term in the diffusion portion of (7.1). Both the EM and Milstein schemes render successively improved approximations as the step size decreases.

---

1 From Equation (7.A.13), let us isolate the term

$$\int_{t_0}^{t}\int_{t_0}^{s}L^1g(X_Z)dW_Z\,dW_s \approx g(X_t)\frac{\partial g(X_t)}{\partial X}\int_{t_0}^{t}(W_s - W_{t_0})\,dW_s$$
$$= g\frac{\partial g}{\partial X}\left[\int_{t_0}^{t}W_s\,dW_s - W_{t_0}(W_t - W_{t_0})\right] = g\frac{\partial g}{\partial X}\left[\frac{1}{2}W_t^2 - \frac{1}{2}W_{t_0}^2 - \frac{1}{2}(t-t_0) - W_{t_0}W_t + W_{t_0}^2\right]$$
$$= \frac{1}{2}g\frac{\partial g}{\partial X}\left[(W_t - W_{t_0})^2 - \Delta\right] = \frac{1}{2}g\frac{\partial g}{\partial X}\left\{(\Delta W_t)^2 - \Delta\right\}.$$

**Example 7.1**  The SDE for the geometric BM model was previously given as

$$dX_t = \mu X_t \, dt + \sigma X_t \, dW_t, \, X(0) \equiv X_0.$$

From (7.2.1) (or from (7.B.5) since $Y_{t+1} = Y_{t+\Delta}$), if we set $f(X_i) = \mu X_i$ and $g(X_i) = \sigma X_i$, then it is readily verified that the EM iteration scheme for approximating this SDE is

$$Y_{i+1} = Y_i + \mu Y_i \Delta + \sigma Y_i \sqrt{\Delta} Z, \, i = 1, ..., N - 1. \tag{7.9}$$

Also, from Example 5.2, we have

$$d \ln X_t = \left( \mu - \frac{1}{2}\sigma^2 \right) dt + \sigma dW_t. \tag{7.10}$$

Under the EM approximation of this expression,

$$\ln Y_{i+1} = \ln Y_i + \left( \mu - \frac{1}{2}\sigma^2 \right) \Delta + \sigma \sqrt{\Delta} Z \tag{7.11}$$

so that

$$Y_{i+1} = Y_i \exp \left[ \left( \mu - \frac{1}{2}\sigma^2 \right) \Delta + \sigma \sqrt{\Delta} Z \right], \, i = 1, ..., N - 1. \tag{7.12}$$

In a similar vein, using (7.8.1) (or using (7.B.12)), we can demonstrate that the Milstein approximation to the given SDE for the geometric BM process is

$$Y_{i+1} = Y_i + \mu Y_i \Delta + \sigma Y_i \sqrt{\Delta} Z$$
$$+ \frac{1}{2}\sigma^2 \Delta \left( Z^2 - 1 \right), \, i = 1, ..., N - 1. \tag{7.13}$$

Moreover, the Milstein approximation of (7.10) is

$$\ln Y_{i+1} = \ln Y_i + \left( \mu - \frac{1}{2}\sigma^2 \right) \Delta + \sigma \sqrt{\Delta} Z, \, i = 1, ..., N - 1, \tag{7.14}$$

which is identical to (7.11), the EM approximation of (7.10). Hence, the Milstein scheme improves the approximation of the geometric BM process $X_t$, but does not improve the approximation of $\ln X_t$ for this process. ∎

**Example 7.2**  Approximate the Cox, Ingersol, and Ross (CIR) SDE

$$dX_t = (\alpha - \beta X_t)dt + \sigma \sqrt{X_t} dW_t, \, X(t_0) \equiv X_0 > 0,$$

using the EM and Milstein iteration schemes. For this SDE,

$f(X_t) = \alpha - \beta\, X_t$ and $g(X_t) = \sigma\sqrt{X_t}$. For the EM routine we have,

$$Y_{i+1} = Y_i + (\alpha - \beta)Y_i\Delta + \sigma\sqrt{Y_i}\sqrt{\Delta}Z, \; i = 1, \ldots, N-1; \tag{7.15}$$

and for the Milstein process, we have

$$Y_{i+1} = Y_i + (\alpha - \beta)Y_i\Delta + \sigma\sqrt{Y_i}\sqrt{\Delta}Z$$

$$+ \frac{1}{4}\sigma^2\Delta\left(Z^2 - 1\right), \; i = 1, \ldots, N-1. \; \blacksquare \tag{7.16}$$

## 7.3 The Lamperti Transformation

A desirable feature of the sample path trajectories that result from the approximation of SDEs is their *stability*. To achieve this characteristic, we now turn to an examination of a technique used to remove any state or level-dependent noise from these trajectories. In addition, this will be accomplished by transferring nonlinearities from the diffusion coefficient function $g$ to the drift coefficient function $f$. To this end, suppose we have an SDE of the form

$$dX_t = f(X_t, t)dt + g(X_t)dW_t, \; X(t_0) \equiv X_0, \tag{7.17}$$

associated with the Wiener or BM process $\{X_t\}_{t \in [t_0,\, +\infty)}$, where, as specified, the diffusion coefficient function $g(X_t)$ depends only on $X_t$ and not on $t$. The gist of the Lamperti transformation is that, under suitable conditions (see Appendix 7.C), an SDE such as (7.17) can always be transformed into one with a *unitary* diffusion coefficient.

Specifically, under the **Lamperti transformation**

$$Y_t = F(X_t) = \left.\int \frac{1}{g(u)}du\right|_{u = X_t}, \tag{7.18}$$

the process $\{Y_t\}_{t \in [t_0,\, +\infty)}$ solves the transformed SDE

$$dY_t = f_F(Y_t, t)dt + dW_t, \; Y(t_0) \equiv Y_0, \tag{7.19}$$

where the transformed drift coeffunction $f_F(Y_t, t)$ has the form

$$f_F = \frac{f(F^{-1}(Y_t), t)}{g(F^{-1}(Y_t))} - \frac{1}{2}g_X\left(F^{-1}(Y_t)\right)$$

$$= \frac{f(X_t, t)}{g(X_t)} - \frac{1}{2}g_X(X_t) \tag{7.20}$$

and the transformed drift coefficient is 1.

**Example 7.3** Let $\{X_t\}$ follow a geometric BM process with SDE

$$dX_t = \mu X_t dt + \sigma X_t dW_t,$$

where $\mu$ and $\sigma$ are constants. What is the SDE associated with the Lamperti-transformed process $\{Y_t\}$? From (7.19), the transformed SDE is

$$dY_t = \left(\frac{\mu X_t}{\sigma X_t} - \frac{1}{2}\sigma\right)dt + dW_t$$

$$= \left(\frac{\mu}{\sigma} - \frac{1}{2}\sigma\right)dt + dW_t.$$

(Note: from (7.18) we have,

$$Y_t = F(X_t) = \int \frac{1}{\sigma u}du \bigg|_{u = X_t} = \frac{1}{\sigma}\ln X_t,$$

and thus $X_t = F^{-1}(Y_t) = e^{\sigma Y_t}$ so that $X_t = e^{\sigma(\ln X_t/\sigma)} = X_t$ as required. Clearly,

$$f_F = \frac{f(F^{-1}(Y_t), t)}{g(F^{-1}(Y_t))} - \frac{1}{2}g_x\left(F^{-1}(Y_t)\right)$$

$$= \frac{\mu e^{\sigma Y_t}}{\sigma e^{\sigma Y_t}} - \frac{1}{2}\sigma = \frac{\mu}{\sigma} - \frac{1}{2}\sigma. \blacksquare$$

**Example 7.4** Let the SDE

$$dX_t = (\mu_1 + \mu_2 X_t)dt + \sigma X_t dW_t$$

be associated with the Wiener process $\{X_t\}$, where $\mu_1, \mu_2$, and $\sigma$ are constants. What is the SDE for the Lamperti process $\{Y_t\}$? Using (7.19),

$$dY_t = \left(\frac{\mu_1 + \mu_2 X_t}{\sigma X_t} - \frac{1}{2}\sigma\right)dt + dW_t$$

$$= \left(\frac{\mu_1}{\sigma X_t} + \frac{\mu_2}{\sigma} - \frac{1}{2}\sigma\right)dt + dW_t$$

or, from the preceding example problem (wherein $X_t = e^{\sigma Y_t}$),

$$dY_t = \left(\frac{\mu_1}{\sigma}e^{-\sigma Y_t} + \frac{\mu_2}{\sigma} - \frac{1}{2}\sigma\right)dt + dW_t. \blacksquare$$

**Example 7.5**  Given the SDE (7.17) for the process $\{X_t\}_{t\in[t_0,\ +\infty)}$, determine the EM and Milstein iteration schemes for the transformed process $\{Y_t\}_{t\in[t_0,\ +\infty)}$ with SDE (7.19). For the EM routine (using (7.2.1)),

$$Y_{i+1} = Y_i + \left(\frac{f(Y_i, t_i)}{g(Y_i)} - \frac{1}{2}g_X(Y_i)\right)\Delta + \sqrt{\Delta}Z; \tag{7.21}$$

and for the Milstein scheme (from (7.8.1)),

$$Y_{i+1} = Y_i + \left(\frac{f(Y_i, t_i)}{g(Y_i)} - \frac{1}{2}g_X(Y_i)\right)\Delta + \sqrt{\Delta}Z$$
$$+ \frac{1}{2}g(Y_i)g_X(Y_i)\Delta\left(Z^2 - 1\right). \tag{7.22}$$

Let us look at these results from a slightly different perspective. In particular, given $Y_t = F(X_t)$ from (7.18), let us write the inverse transformation $X_t = F^{-1}(Y_t) = G(Y_t)$. From the Lamperti transformation,

$$F'(X_t) = \frac{1}{g(X_t)} \text{ and } F''(X_t) = -\frac{g_X(X_t)}{g(X_t)^2}; \tag{7.23}$$

and from the inverse transformation,

$$G'(Y_t) = \frac{1}{F'(G(Y_t))} = g(G(Y_t)) \text{ and}$$

$$G''(Y_t) = -\frac{1}{F'(G(Y_t))^2}F''(G(Y_t))G'(Y_t)$$
$$= -g(Y_t)^2\left(-\frac{g_X(G(Y_t))}{g(G(Y_t))^2}\right)g(G(Y_t)) \tag{7.24}$$
$$= g(G(Y_t))g_X(G(Y_t)).$$

Next,

1. Let $Y_t = F(X_t)$ be determined by the Lamperti transformation. Then under $dX_t = f(X_t, t)dt + g(X_t)dW_t$, $F$ is also an Itô process with differential

$$dF = \left(F'(X_t)f(X_t, t) + \frac{1}{2}F''(X_t)g(X_t)^2\right)dt + F'(X_t)g(X_t)dW_t$$

(see Itô's formula (4.18)). Then substituting from (7.23), the preceding expression becomes

$$dF = \left( \frac{f(X_t, t)}{g(X_t)} - \frac{1}{2}g_x(X_t) \right) dt + dW_t$$

or

$$\Delta Y_t = \left( \frac{f(X_t, t)}{g(X_t)} - \frac{1}{2}g_x(X_t) \right) \Delta + \sqrt{\Delta}Z,$$

that is, Itô's formula applied to transformation function $\Gamma$ yields the EM scheme for the transformed process (7.21).

2. Let $X_t = G(Y_t)$. The Taylor expansion of the inverse transformation can be written as

$$dG = G'(Y_t)dY_t + G''(Y_t)(dY_t)^2 + \text{higher order } \Delta Y_t \text{ terms.}$$

Then from (7.24), the preceding expression can be written (using Table A.4.1) as

$$dG = g(G(Y_t))dY_t + \frac{1}{2}g(G(Y_t))g_x(G(Y_t))(dY_t)^2$$

$$+ \text{ higher order } \Delta Y_t \text{ terms}$$

$$= g(G(Y_t))\left[ \left( \frac{f(X_t, t)}{g(X_t)} - \frac{1}{2}g_x(X_t) \right) dt + dW_t \right]$$

$$\frac{1}{2}g(G(Y_t))g_x(G(Y_t))dt + \text{higher order } \Delta Y_t \text{ terms}$$

$$= f(X_t, t)\Delta + g(X_t)\sqrt{\Delta}Z + \text{higher order } \Delta Y_t \text{ terms.}$$

If we ignore the higher order $\Delta Y_t$ terms in the preceding expression, then the Taylor expansion of the inverse transformation $G$ under (7.19) yields the EM scheme for the untransformed process. ∎

**Example 7.6**  Given the CIR SDE

$$dX_t = (\alpha - \beta X_t)dt + \sigma\sqrt{X_t}dW_t$$

associated with the Wiener process $\{X_t\}$, determine the SDE for the Lamperti process $\{Y_t\}$. As a first step, let us utilize (7.18) to find

$$Y_t = F(X_t) = \int \frac{1}{g(u)}du \bigg|_{u = X_t}$$

$$= \frac{2}{\sigma}X_T^{1/2}.$$

Then

$$X_t = \frac{\sigma^2}{4} Y_t^2 = F^{-1}(Y_t)$$

and, from (7.19),

$$dY_t = f_F(Y_t, t)dt + dW_t$$

$$= \left[ \frac{\alpha - \beta X_t}{\sigma \sqrt{X_t}} - \frac{1}{2} \left( \frac{1}{2} \sigma X_t^{-1/2} \right) \right] dt + dW_t.$$

A substitution of $F^{-1}(Y_t)$ into this expression yields

$$dY_t = \frac{1}{2Y_t} \left( \frac{4\alpha}{\sigma^2} - \beta Y_t - 1 \right) dt + dW_t. \blacksquare$$

## 7.4 Variations on the EM and Milstein Schemes

Section 7.2 has focused on the EM and Milstein iteration schemes. However, these routines are not the "only games in town." Other iteration techniques abound. For instance (in what follows, let $Y(t_0) \equiv Y_0, f(Y_i) \equiv f$, and $g(Y_i) \equiv g$),

1. Strong Order 1.5 Taylor Scheme (Kloeden and Platen, 1999)

$$Y_{i+1} = Y_i + f\Delta + g\Delta W_i + \frac{1}{2} g g_X \left( (\Delta W_i)^2 - \Delta \right)$$

$$+ f_X g \Delta Z_i + \frac{1}{2} \left( f f_X + \frac{1}{2} g^2 f_{XX} \right) (\Delta)^2$$

$$+ \left( f g_X + \frac{1}{2} g^2 g_{XX} \right) \left( (\Delta W_i)\Delta - \Delta Z_i \right)$$

$$+ \frac{1}{2} g \left( g g_{XX} + g_X^2 \right) \left( \frac{1}{3} (\Delta W_i)^2 - \Delta \right) \Delta W_i,$$

(7.25)

where $\Delta W_i = \sqrt{\Delta} Z$, $\Delta Z_i$ is $N(0, (1/3)\Delta^3)$, and $COV(\Delta W_i, \Delta Z_i) = (1/2)\Delta^2$. The random variable $\Delta Z_i$ is specified as

$$\Delta Z_i = \frac{1}{2} \Delta \left( \Delta W_i + \frac{\Delta V_i}{\sqrt{3}} \right),$$

where $\Delta V_i$ is chosen independently from $\sqrt{\Delta} Z$, that is, all of the pairs $(\Delta V_i, \Delta W_i)$ are independent for all $i$.

2. Weak Order 2.0 Taylor Scheme (Kloeden and Platen, 1999)

$$Y_{i+1} = Y_i + f\Delta + g\Delta W_i + \frac{1}{2}gg_x\left((\Delta W_i)^2 - \Delta\right)$$

$$+ f_x g\Delta Z_i + \frac{1}{2}\left(ff_x + \frac{1}{2}f_{xx}g^2\right)(\Delta)^2 \tag{7.26}$$

$$+ \left(fg_x + \frac{1}{2}g_{xx}g^2\right)\left((\Delta W_i)\Delta - \Delta Z_i\right),$$

where $\Delta W_i = \sqrt{\Delta}Z$ and $\Delta Z_i$ is distributed as in the preceding strong order 1.5 Taylor routine.

3. Weak Order 2.0 Predictor–Corrector Scheme (Gard, 1988; Kloeden and Platen, 1999; Platen, 1980, 1981).

For a predictor–corrector method, the **predictor** $\bar{Y}_{i+1}$ is inserted into the **corrector equation** to give the next iteration $Y_{i+1}$.

Step 1. Consider the predictor

$$\bar{Y}_{i+1} = Y_i + f\Delta + \lambda_i + \frac{1}{2}f_x g\left(\Delta\hat{W}\right)\Delta + \frac{1}{2}\left(ff_x + \frac{1}{2}f_{xx}g^2\right)\Delta^2, \tag{7.27}$$

where

$$\lambda_i = g\Delta\hat{W} + \frac{1}{2}gg_x\left[\left(\Delta\hat{W}\right)^2 - \Delta\right] + \frac{1}{2}\left(fg_x + \frac{1}{2}g^2g_{xx}\right)\left(\Delta\hat{W}\right)\Delta$$

and $\Delta\hat{W} \sim N(0, \Delta)$.

Step 2. Choose the corrector

$$Y_{i+t} = Y_i + \frac{1}{2}[(\bar{Y}_{i+1}) + f]\Delta + \lambda_i. \tag{7.28}$$

4. Derivative-Free Weak Order 2.0 Predictor–Corrector Scheme (Kloeden and Platen, 1999)

Step 1. Determine the predictor

$$Y_{i+t} = Y_i + \frac{1}{2}[(f(u)) + f]\Delta + \emptyset_i, \tag{7.29}$$

where the **supporting value**

$$u = Y_i + f\Delta + g\Delta\hat{W},$$

$$\emptyset_i = \frac{1}{4}[g(u^+) + g(u^-) + 2g]\Delta\hat{W}$$

$$+ \frac{1}{4}[g(u^+) - g(u^-)]\left[\left(\Delta\hat{W}\right)^2 - \Delta\right]\Delta^{-\frac{1}{2}},$$

with **supporting values**

$$u^+ = Y_i + f\Delta + g\sqrt{\Delta},$$
$$u^- = Y_i + f\Delta - g\sqrt{\Delta},$$

and $\Delta\hat{W} \sim N(0, \Delta)$.

Step 2. Calculate the corrector

$$Y_{i+1} = Y_i + \frac{1}{2}[f(\bar{Y}_{i+1}) + f]\Delta + \varnothing_i. \tag{7.30}$$

5. Weak Order 2.0 Milstein Scheme (Milstein, 1978)

Milstein's "second scheme" involves the approximation

$$Y_{i+1} = Y_i + \left(f - \frac{1}{2}gg_X\right)\Delta + g\sqrt{\Delta}Z + \frac{1}{2}gg_X\Delta t\, Z^2$$

$$+ (\Delta)^{3/2}\left(\frac{1}{2}fg_X + \frac{1}{2}f_Xg + \frac{1}{2}g^2g_{XX}\right)Z$$

$$+ (\Delta)^2\left(\frac{1}{2}ff_X + \frac{1}{4}f_{XX}g^2\right).$$

## 7.5 Local Linearization Techniques

### 7.5.1 The Ozaki Method

Given that the evolution of some dynamic phenomenon can be described by a continuous-time stochastic process, with the latter modeled as a nonlinear SDE, the **Ozaki local linearization method** approximates the original SDE by a linear one (Ozaki, 1992, 1993). This method enables us to obtain a sample path or trajectory of the approximated process. By discretizing the sample path, we can obtain the discretized version of the process.

Looking to particulars, Ozaki's technique is designed for time-homogeneous SDEs; it involves the approximation of the drift function of a nonlinear SDE by a linear function. Under this approximation, we should obtain a marked improvement over a simple constant approximation employed by, say, the EM scheme. Suppose we have a one-dimensional continuous-time process characterized by the one-dimensional SDE

$$dX_t = b(X_t)dt + g(X_t)dW_t, \, t \geq 0, \tag{7.31}$$

where $b$ and $g$ are twice continuously differentiable functions of $X$ and $W_t$ is standard BM. Since this SDE can be transformed (via Itô's formula (4.18)) into one with a constant diffusion term,[2] we need only consider the SDE

$$dX_t = b(X_t)dt + \sigma dW_t. \tag{7.31.1}$$

Assume that the SDE (7.31.1) is deterministic (the stochastic nature of $X_t$ is initially ignored). Our objective is to locally approximate $b(X_t)$ by a linear function of $X_t$. So from $dX_t/dt = b(X_t)$, we can write

$$\frac{d^2 X_t}{dt^2} = b_X(X_t)\frac{dX_t}{dt}, \tag{7.32}$$

where $b_X(X_t)$ is taken to be constant in the suitably restricted interval $[t, t + \Delta t)$. Upon integrating both sides of (7.32), first from $t$ to $u \in [t, t + \Delta t)$, and then from $t$ to $t + \Delta t$, we obtain the differential equation

$$X_{t+\Delta t} = X_t + \frac{b(X_t)}{b_X(X_t)}\left(e^{b_X(X_t)\Delta t} - 1\right).^3 \tag{7.33}$$

To translate (7.33) back to the SDE (7.31.1), we again use local linearization by setting $b(X_t) = K_t X_t$, where $K_t$ is constant over the interval $[t, t + \Delta t)$. Thus,

$$dX_t = K_t X_t \, dt + \sigma dW_t. \tag{7.34}$$

---

2 Let $Y_t \equiv \phi(X_t)$, where $\phi$ is twice continuously differentiable in $X_t$ and $(\phi')(X_t)g(X_t) = \sigma, \sigma$ a constant. Then by Itô's formula, the new process $Y_t$ satisfies the SDE

$$dY_t = \left(b\phi' + \frac{g^2}{2}\phi''\right)dt + g\phi'dW_t$$

$$= h(X_t)dt + \sigma dW_t.$$

3 From (7.32) or, for convenience, $f''(X_t) = b_X(X_t)f'(X_t)$ we have

$$\int_t^u \frac{f''(X_t)}{f'(X_t)}dt = b_X(X_t)\int_t^u dt,$$

$$\ln\left(\frac{f'(X_u)}{f'(X_t)}\right) = b_X(X_t)(u-t)$$

so that

$$f'(X_u) = f'(X_t)e^{b_X(X_t)(u-t)}.$$

Set $f'(X_u) = dX_u/du$. Then

$$\int_t^{t+\Delta t} dX_u = \int_t^{t+\Delta t} f'(X_t)\,e^{b_X(X_t)(u-t)}du$$

and thus

$$X_u\big]_t^{t+\Delta t} = \frac{f'(X_t)}{b_X(X_t)}\left(e^{b_X(X_t)(u-t)}\right)\Big]_t^{t+t\Delta t} \quad \text{or (7.33).}$$

Clearly, this SDE is of the autonomous linear narrow sense (ALNS) variety with $a = 0$ (see Table 5.1). Its solution is

$$X_{t+\Delta t} = X_t e^{K_t \Delta t} + \sigma \int_t^{t+\Delta t} e^{K_t(t+\Delta t - s)} dW_s. \tag{7.35}$$

It now remains to determine $K_t$. From (7.35), the conditional expectation of $X_{t+\Delta t}$ with respect to $t$, or $E(X_{t+\Delta t}|X_t) = X_t e^{K_t \Delta t}$, coincides with the state of (7.33) at time $t + \Delta t$ so that

$$X_t e^{K_t \Delta t} = X_t + \frac{b(X_t)}{b_X(X_t)} \left( e^{b_X(X_t)\Delta t} - 1 \right)$$

or

$$K_t = \frac{1}{\Delta t} \ln \left[ 1 + \frac{b(X_t)}{X_t b_X(X_t)} \left( e^{b_X(X_t)\Delta t} - 1 \right) \right]. \tag{7.36}$$

Given that the stochastic integral

$$\int_t^{t+\Delta t} e^{K_t(t+\Delta t - s)} dW_s$$

is distributed as $N(0, V_t)$, where

$$V_t = \sigma^2 \left( \frac{e^{2K_t \Delta t} - 1}{2K_t} \right), \qquad ^4$$

the discrete time model of (7.34), which approximates (7.31.1) at $X_t (\neq 0)$, with $b_X(X_t) \neq 0$, is

$$X_{t+\Delta t} = A_t X_t + B_t W_{t+\Delta t},$$

$$A_t = e^{K_t \Delta t},$$

$$B_t = \sigma \left( \frac{e^{2K_t \Delta t} - 1}{2K_t} \right)^{\frac{1}{2}}, \text{ and} \tag{7.37}$$

$$K_t = \frac{1}{\Delta t} \ln \left[ 1 + \frac{b(X_t)}{X_t b_X(X_t)} \left( e^{b_X(X_t)\Delta t} - 1 \right) \right].$$

### 7.5.2 The Shoji–Ozaki Method

The **Shoji–Ozaki (SO) local linearization technique** is an extension of the basic local linearization method of Ozaki to the case where the drift coefficient

---

4 Utilize footnote 3 of Chapter 5 to obtain this result.

depends upon both $X_t$ and $t$. In addition, the diffusion function is to have $X_t$ as its argument (Shoji, 1998; Shoji and Ozaki, 1997, 1998). In this regard, consider the nonhomogeneous SDE

$$dX_t = f(X_t, t)\, dt + g(X_t)\, dW_t, \; t \ge 0, \tag{7.38}$$

where $f$ is taken to be twice continuously differentiable in $X$ and continuously differentiable in $t$, and $g$ is continuously differentiable in $X$. In addition, by virtue of the argument offered in footnote 2, we need only consider the nonhomogeneous SDE

$$dX_t = f(X_t, t)\, dt + \sigma\, dW_t, \tag{7.39}$$

where $\sigma$ is constant.

The development of this "new" (SO) local linearization method is motivated by that of the basic Ozaki approach. For the latter, the drift function $f(X_t)$ was locally approximated by a linear function of $X_t$. In the SO local linearization process, we now focus on the local behavior of the drift function $f(X_t, t)$, which can be expressed, via Itô's formula, by the differential of $f(X_t, t)$ or

$$df = \left( \frac{\partial f}{\partial t} + \frac{\sigma^2}{2} \frac{\partial^2 f}{\partial X_t^2} \right) dt + \frac{\partial f}{\partial X_t} dX_t. \tag{7.40}$$

To linearize $f$ with respect to $X_t$ and $t$, suppose that $\partial^2 f / \partial X_t^2$, $\partial f / \partial X_t$, and $\partial f / \partial t$, each evaluated at the point $(X_t, t)$, are all constant on the suitably restricted time interval $[s, s + \Delta s)$. Hence, (7.40) can be rewritten as

$$f(X_t, t) - f(X_s, s) = \left( \frac{\partial f}{\partial t} + \frac{\partial^2}{2} \frac{\partial^2 f}{\partial X_t^2} \right)(t - s) + \frac{\partial f}{\partial X_t}(X_t - X_s) \tag{7.41}$$

or

$$f(X_t, t) = L_s X_t + M_s t + N_s, \tag{7.42}$$

where

$$L_s = \frac{\partial f}{\partial X_t}(X_s, s),$$

$$M_s = \frac{\partial f}{\partial t}(X_s, s) + \frac{\sigma^2}{2} \frac{\partial^2 f}{\partial X_t^2}(X_s, s), \text{ and}$$

$$N_s = f(X_s, s) - \frac{\partial f}{\partial X_t}(X_s, s)X_s - \left( \frac{\partial f}{\partial t}(X_s, s) + \frac{\sigma^2}{2} \frac{\partial^2 f}{\partial X_t^2}(X_s, s) \right)s$$

$$= f(X_s, s) - \frac{\partial f}{\partial X_t}(X_s, s)X_s - M_s s.$$

So instead of working with (7.39), we can focus on the linear SDE

$$dX_t = (L_s X_t + M_s t + N_s)dt + \sigma dW_t, \ t \geq s. \tag{7.43}$$

Next, consider the transformed process $Y_t = e^{-L_s t} X_t$ or, from Itô's formula (4.18),

$$dY_t = \left[ -L_s e^{-L_s t} X_t + (L_s X_t + M_s t + N_s) e^{-L_s t} \right] dt + \sigma e^{-L_x t} dW_t$$

$$= (M_s t + N_s) e^{-L_s t} dt + \sigma e^{-L_s t} dW_t.$$

Then

$$\int_s^t dY_t = \int_s^t (M_s u + N_s) e^{-L_s u} du + \int_s^t e^{-L_s u} dW_u$$

or

$$Y_t = Y_s + \int_s^t (M_s u + N_s) e^{-L_s u} du + \int_s^t e^{-L_s u} dW_u$$

$$= Y_s + \frac{M_s}{L_s^2} \left[ -e^{-L_s t} (L_s t + 1) + e^{-L_s s} (L_s s + 1) \right] \tag{7.44}$$

$$- \frac{N_s}{L_s} \left( e^{-L_s t} - e^{-L_s s} \right) + \int_s^t e^{-L_s u} dW_u.$$

Substituting $Y_t = e^{-L_s t}$ and $Y_s = e^{-L_s s}$ into (7.44) and simplifying yields the discretized process of $X_t$ or

$$X_t = X_s + \frac{f(X_s, s)}{L_s} \left( e^{L_s(t-s)} - 1 \right) + \frac{M_s}{L_s^2} \left[ \left( e^{L_s(t-s)} - 1 \right) - L_s(t-s) \right]$$

$$+ \sigma \int_s^t e^{L_s(t-u)} dW_u. \tag{7.45}$$

Since the preceding approximation holds locally or on $[s, s + \Delta s)$, (7.45) can be rewritten as

$$X_{s+\Delta s} = X_s + \frac{f(X_s, s)}{L_s} \left( e^{L_s \Delta s} - 1 \right) + \frac{M_s}{L_s^2} \left[ \left( e^{L_s \Delta s} - 1 \right) - L_s \Delta s \right]$$

$$+ \sigma \int_s^{s+\Delta s} e^{L_s(s + \Delta s - u)} dW_u. \tag{7.45.1}$$

We know that the last term on the right-hand side of (7.45.1) is distributed as $N(0, V_s)$, where

$$V_s = \sigma^2 \left( \frac{e^{2L_s \Delta s} - 1}{2L_s} \right).$$

Hence, we may express the discrete time model of (7.43), which approximates (7.39) at $X_s (\neq 0)$, with $L_s \neq 0$, as

$$X_{s+\Delta s} = A(X_s)X_s + B(X_s)W_{s+\Delta s},$$  (7.46)

where

$$A(X_s) = 1 + \frac{f(X_s, s)}{X_s L_s}\left(e^{L_s \Delta s} - 1\right) + \frac{M_s}{X_s L_s^2}\left[\left(e^{L_s \Delta s} - 1\right) - L_s \Delta s\right]$$

and

$$B(X_s) = \sigma\left(\frac{e^{2L_x \Delta s} - 1}{2L_s}\right)^{\frac{1}{2}}.$$

**Example 7.7**   Given the homogeneous SDE

$$dX_t = \theta X_t^2 dt + \sigma dW_t, \ t \geq 0,$$

with $\theta$ and $\sigma$ constant, determine the discretized process associated with (1) the Ozaki local linearization method and (2) the SO local linearization method.

1. Given that $b(X_t) = 2\theta X_t$ and $b_X(X_t) = 2\theta$, we may utilize (7.37) to obtain

$$K_t = \frac{1}{\Delta t}\ln\left[1 + X_t\left(e^{2\theta \Delta t} - 1\right)\right],$$

and thus we can readily determine that

$$A_t = e^{K_t \Delta t}$$

and

$$B_t = \sigma\left(\frac{e^{2K_t \Delta t}}{2K_t}\right)^{\frac{1}{2}}.$$

Hence, the Ozaki local linearization method renders the process discretization

$$X_{t+\Delta t} = A_t X_t + B_t W_{t+\Delta t}.$$

2. From (7.46), $f(X_s, s) = \theta X_s^2$; $L_s = 2\theta X_s$; and $M_s = \theta \sigma^2$. Then

$$A(X_s) = 1 + \frac{1}{2}\left(e^{2\theta X_s \Delta s} - 1\right) + \frac{\sigma^2}{4\theta X_s^3}\left[\left(e^{2\theta X_s \Delta s} - 1\right) - 2\theta X_s \Delta s\right];$$

and

$$B(X_s) = \sigma\left(\frac{e^{4\theta X_s \Delta s} - 1}{4\theta X_s}\right)^{\frac{1}{2}}.$$

Hence, the SO local linearization method yields the process discretization

$$X_{s+\Delta s} = A(X_s)X_s + B(X_s)W_{s+\Delta s}. \blacksquare$$

### 7.5.3 The Rate of Convergence of the Local Linearization Method

We noted in Section 7.5.2 that Shoji and Ozaki (1997, 1998) have offered an upgrade of the Ozaki (1992, 1993) local linearization method used to facilitate the discrete approximation of continuous-time processes (and, as we shall see later on, estimate the parameters of the same) (Shoji, 1998). The following question arises: How close does the process approximated by the SO method come to the true process? To answer this question, we obviously need to assess the goodness of the approximation of the local linearization technique by considering the rate of convergence of the approximation error.

In this regard, two types of approximation errors are defined, the errors resulting from one-step and multistep approximations, respectively. (While the one-step approximation error must be considered when estimating parameters, the rate of convergence of the multistep approximation error is of paramount importance when constructing a sample path of a stochastic process. This is because the multistep approximation error reflects the cumulative error of the local linearization.) Looking to specifics, let us define the one-step and multistep approximation errors as the difference between the state of the true process and that of the approximating process given that the current state of the true stochastic process, $X_s$, is known and bounded. Moreover, the evaluation of the rate of convergence will be cast in $L_p$ terms. To this end, Shoji (1998) offers two theorems. The first theorem involves the rate of convergence of the one-step approximation in the $L_p$ sense.

**Theorem 7.1**  Shoji's Theorem 1
Suppose $X_t$ and $\tilde{X}_t$ are the true stochastic process and the approximate process derived from the SO local linearization method, respectively. Let a $p$th order error of one-step-ahead prediction be defined as $E_s|X_t - \tilde{X}_t|^p$, with $s \leq t \leq T$. Then the rate of convergence of $\left(E_s|X_t - \tilde{X}_t|^p\right)^{1/p}$ is 2, or

$$E_s|X_t - \tilde{X}_t|^p = O\left((t-s)^{2p}\right), {}^5$$

---

5 The **large $O$ notation** is utilized to represent the limiting behavior of a function when its argument tends to some value (or to $+\infty$). That is, let $f(x)$ and $g(x)$ be real-valued functions defined on the same real domain, and let $x$ tend to a limit. Then $f(x) = O(g(x))$ means that

$|f(x)/g(x)|$ remains bounded, or

$|f(x)| \leq M|g(x)|$ for all $x \geq x_0$ and all $M > 0$.

Thus $f(x) = O(1)$ implies that $f(x)$ is itself a bounded function; $u_n = O(1)$ implies that $u_n$ is the $n$th term of a bounded sequence or $|u_n| \leq M$ for all $M > 0$. Similarly, $f(h) = O(h^p)$ means that $|h|^{-p} f(h)$ is bounded as $h \to 0$. In sum, $O(g(x))$ is the quantity whose ratio to $g(x)$ remains bounded as $x$ tends to a limit.

where $E_s$ represents conditional expectation at time $s$.

We next state a theorem that pertains to the convergence of the multistep approximation error.

**Theorem 7.2**  Shoji's Theorem 2
Let $t$ be fixed for $s \le t \le T$ and $\{t_k\}_{0 \le k \le n}$ be a $n$-partition of the time interval $[s, t]$ with $\Delta t = t_k - t_{k-1}$. Consider a step-by-step approximation of the integration

$$X_t - X_s = \int_s^t f(X_u, u)\, du + \sigma \int_s^t dW_u$$

by using the SO local linearization method. Then the convergence of the step-by-step approximation is $O(\Delta t)$ in the $L_p$ sense.

Generally speaking, the results of numerical experiments lead to the conclusion that the errors induced by the local linearization method are much smaller than the errors generated by, say, the EM method. In fact, differences between these two methods for multistep approximations are very pronounced. The efficiency of the local linearization method in multistep approximations follows from the structure of the stochastic integration process, that is, in the local linearization method, the stochastic integration component uses the information on increments of a BM process with a discrete time interval; but in the EM method, the sum of the increments of a BM process is reduced to $W_{t+1} - W_t$ (with all intermediate information lost). Hence, the local linearization method is preferred over the EM method. A process discretized by the local linearization technique is normally distributed, and thus we can readily simulate a random variable following this distribution. In short, samples generated by the local linearization method can be treated as a very good approximation to the realization of the true process.

## Appendix 7.A   Stochastic Taylor Expansions

Our objective here is to determine the stochastic analogue of the classical or deterministic Taylor expansion of a real-valued function $h(X_t): R \to R$, where $h$ is taken to be continuously differentiable with a linear growth bound (there exists a constant $K > 0$ such that $|h(X_t)|^2 \le K^2(1 + |X_t|^2)$ for all $t \in [t_0, T]$).
We begin by deriving the deterministic Taylor formula. To this end, our starting point is the solution $X(t) \equiv X_t$ of the initial value problem

$$\frac{dX_t}{dt} = f(X_t), X(t_0) = X_0, \tag{7.A.1}$$

$t \in [t_0, T]$, with $t_0 \leq t \leq T$. In integral form, the solution appears as

$$X_t = X_0 + \int_{t_0}^{t} f(X_s) \, ds. \tag{7.A.2}$$

Given $h(X_t)$, we have, via the chain rule,

$$\frac{dh(X_t)}{dt} = \frac{\partial}{\partial X} h(X_t) \cdot \frac{dX_t}{dt} = f(X_t) \cdot \frac{\partial}{\partial X} h(X_t).$$

If we employ the operator

$$L = f \frac{\partial}{\partial X}, \tag{7.A.3}$$

then the integral equation (7.A.2) can be written as

$$h(X_t) = h(X_0) + \int_{t_0}^{t} Lh(X_s) \, ds, \ t \in [t_0, T]. \tag{7.A.4}$$

(Note that when $h(X) \equiv X$, $Lh = f$ so that (7.A.4) becomes (7.A.2).)

Next, if we now apply (7.A.4) to $h(X_s)$ in the preceding integrand term, we obtain

$$h(X_s) = h(X_0) + \int_{t_0}^{s} Lh(X_z) \, dz.$$

Then the integral in (7.A.4) becomes

$$\int_{t_0}^{t} Lh(X_s) \, ds = \int_{t_0}^{t} \left\{ Lh(X_0) + \int_{t_0}^{s} L(Lh)(X_z) \, dz \right\} ds$$

In addition, thus (7.A.4) can be rewritten as

$$h(X_t) = h(X_0) + \int_{t_0}^{t} \left\{ Lh(X_0) + \int_{t_0}^{s} L(Lh)(X_z) dz \right\} ds$$

$$= h(X_0) + \int_{t_0}^{t} Lh(X_0) \, ds + \int_{t_0}^{t} \int_{t_0}^{s} L^2 h(X_z) \, dz ds \tag{7.A.5}$$

$$= h(X_0) + (t - t_0) Lh(X_0) + \int_{t_0}^{t} \int_{t_0}^{s} L^2 h(X_z) \, dz ds.$$

A second application of (7.A.4), but this time to the term $h(X_z)$ in the integrand of the double integral in (7.A.5), yields

$$h(X_z) = h(X_0) + \int_{t_0}^{z} Lh(X_u) \, du.$$

Then the double integral term in (7.A.5) becomes

$$\int_{t_0}^{t} \int_{t_0}^{s} L^2 h(X_z) \, dz \, ds = \int_{t_0}^{t} \int_{t_0}^{s} \left\{ L^2 h(X_0) + \int_{t_0}^{z} L^2 (Lh)(X_u) \right\} dz \, ds,$$

and thus (7.A.5) appears as

$$h(X_t) = h(X_0) + (t - t_0) Lh(X_0)$$

$$+ \int_{t_0}^{t} \int_{t_0}^{s} \left\{ L^2 h(X_0) + \int_{t_0}^{z} L^3 h(X_u) \, du \right\} dz \, ds$$

$$= h(X_0) + (t - t_0) Lh(X_0) + \int_{t_0}^{t} \int_{t_0}^{s} L^2 h(X_0) dz \, ds + \int_{t_0}^{t} \int_{t_0}^{s} \int_{t_0}^{z} L^3 h(X_0) du \, dz \, ds$$

$$= h(X_0) + (t - t_0) Lh(X_0) + L^2 h(X_0) \int_{t_0}^{t} (s - t_0) ds + \int_{t_0}^{t} \int_{t_0}^{s} \int_{t_0}^{z} L^3 h(X_0) du \, dz \, ds$$

$$= h(X_0) + (t - t_0) Lh(X_0) + \frac{1}{2} (t - t_0)^2 L^2 h(X_0) + R_3,$$

$$(7.A.6)$$

where

$$R_3 = \int_{t_0}^{t} \int_{t_0}^{s} \int_{t_0}^{z} L^3 (X_0) du \, dz \, ds$$

is called the third-order remainder term. If we continue iterating in this fashion, we obtain the **deterministic Taylor formula with integral form of the remainder**

$$h(X_t) = h(X_0) + \sum_{l=1}^{r} \frac{(t - t_0)^l}{l!} L^l h(X_0) + R_{r+1}, \, r = 1, 2, \ldots, \qquad (7.A.7)$$

$t \in [t_0, T]$, where the **(r + 1)st-order remainder term** is

$$R_{r+1} = \int_{t_0}^{t} \int_{t_0}^{s_{r+1}} \cdots \int_{t_0}^{s_2} L^{r+1} h(X_{s_1}) ds_1 \ldots ds_{r+1}.$$

(Note that we are implicitly assuming that $h$ is $r + 1$ times continuously differentiable.) If $L = \partial / \partial X$, the more familiar version of (7.A.7) is

$$h(X_t) = h(X_0) + \sum_{l=1}^{r} \frac{(t - t_0)^l}{l!} \frac{\partial^l h(X_0)}{\partial X^l} + R_{r+1}, \, r = 1, 2, 3, \ldots, \quad t \in [t_0, T].$$

$$(7.A.7.1)$$

We next look to the development of the stochastic Taylor formula with properties analogous to (7.A.7.1). As we shall now see, the said development is based on the iterated application of Itô's formula.[6] To this end, suppose we have the SDE $d(X_t) = f(X_t)dt + g(X_t)dW_t$, $t \in [t_0, T]$, or, in integral form,

$$X_t = X_0 + \int_{t_0}^t f(X_s)\, ds + \int_{t_0}^t g(X_s)\, dW_s, \tag{7.A.9}$$

where it is assumed that $f$ and $g$ are sufficiently smooth real-valued functions satisfying a linear growth bound (i.e., $|f(X_t)|^2 \le K^2 (1 + |X_t|^2)$ and $|g(X_t)|^2 \le K^2 (1 + |X_t|^2)$, respectively, for constant $K > 0$ and $t \in [t_0, T]$). For $h(X_t): R \to R$ twice continuously differentiable,

$$h(X_t) = h(X_0) + \int_{t_0}^t \left[ f(X_s)\frac{\partial}{\partial X}h(X_s) + \frac{1}{2}g(X_s)^2 \frac{\partial^2}{\partial X^2}h(X_s) \right] ds$$

$$+ \int_{t_0}^t g(X_s)\frac{\partial}{\partial X}h(X_s)\, dW_s, t \in [t_0, T]. \tag{7.A.10}$$

If we introduce the operators

$$L^0 = f\frac{\partial}{\partial X} + \frac{1}{2}g^2 \frac{\partial^2}{\partial X^2}, L^1 = g\frac{\partial}{\partial X}, \tag{7.A.11}$$

then (7.A.10) can be expressed as

$$h(X_t) = h(X_0) + \int_{t_0}^t L^0 h(X_s)ds + \int_{t_0}^t L^1 h(X_s)dW_s, t \in [t_0, T]. \tag{7.A.12}$$

(Note that for $h(X) = X$, $L^0 h = f$ and $L^1 h = g$ and thus (7.A.12) reduces to (7.A.9).)

Now, if we apply (7.A.12) to the functions $h = f$ and $h = g$ in (7.A.9), we obtain the **stochastic Taylor formula** or the **Itô–Taylor expansion** of the solution $X_t$ of (7.A.9) as follows:

$$X_t = X_0 + \int_{t_0}^t \left( f(X_0) + \int_{t_0}^s L^0 f(X_z)dz + \int_{t_0}^s L^1 f(X_z)dW_z \right) ds$$

$$+ \int_{t_0}^t \left( g(X_0) + \int_{t_0}^s L^0 g(X_z)dz + \int_{t_0}^s L^1 g(X_z)dW_z \right) ds \tag{7.A.13}$$

$$= X_0 + f(X_0)\int_{t_0}^t ds + g(X_0)\int_{t_0}^t dW_s + R,$$

---

6 For $Y_t = V(X_t, t), t \in [t_0, T]$, we can express the **Itô's formula** for any $0 \le s \le t \le T$ as

$$Y_t - Y_s = \int_s^t \left( \frac{\partial V}{\partial t} + f\frac{\partial V}{\partial X} + \frac{1}{2}g^2\frac{\partial^2 V}{\partial X^2} \right) du + \int_s^t g\frac{\partial V}{\partial X} dW_u. \tag{7.A.8}$$

where the remainder term $R$ appears as

$$R = \int_{t_0}^{t}\int_{t_0}^{s} L^0 f(X_z)dzds + \int_{t_0}^{t}\int_{t_0}^{s} L^1 f(X_z)dW_z ds$$

$$+ \int_{t_0}^{t}\int_{t_0}^{s} L^0 g(X_z)dzdW_s + \int_{t_0}^{t}\int_{t_0}^{s} L^1 g(X_z)dW_z dW_s.$$

In the preceding specification of the Itô–Taylor expansion, the Itô formula was applied only once. By continuously expanding the integrand functions of the multiple integrals in $R$, higher order stochastic Taylor expansions are obtained.

**Example 7.A.1** For $r = 2$, find the deterministic (second-order) Taylor expansion of the real-valued function $h(X_t) = ae^{bX_t}$, $t \in [t_0, T]$, at the point $X(t_0) = X_0 = 1$. From (7.A.7.1),

$$h(X_t) = h(X_0) + (t - t_0)\frac{\partial h(X_0)}{\partial X} + \frac{1}{2}(t - t_0)^2 \frac{\partial^2 h(X_0)}{\partial X^2} + R_3$$

$$= ae^b + (t - t_0)abe^b + \frac{1}{2}(t - t_0)^2 ab^2 e^b + R_3$$

$$= ae^b\left[1 + (t - t_0)b + \frac{1}{2}(t - t_0)^2 b^2\right] + R_3. \blacksquare$$

**Example 7.A.2** For $f(X_t) = \mu X_t$ and $g(X_t) = \sigma X_t$, determine the Itô–Taylor expansion (7.A.13) at $X_0 = 1$. It is readily verified that

$$X_t = X_0 + \mu(t - t_0) + \sigma(W_t - W_{t_0}) + R$$

$$= X_0 + \mu(t - t_0) + \sigma(W_t - W_{t_0}) + \int_{t_0}^{t}\int_{t_0}^{s} f\frac{\partial f(X_z)}{\partial X}dzds$$

$$+ \int_{t_0}^{t}\int_{t_0}^{s} g\frac{\partial f(X_z)}{\partial X}dW_z ds + \int_{t_0}^{t}\int_{t_0}^{s} f\frac{\partial g(X_z)}{\partial X}dzdW_s$$

$$+ \int_{t_0}^{t}\int_{t_0}^{s} g\frac{\partial g(X_z)}{\partial X}dW_z dW_s$$

$$= X_0 + \mu(t - t_0) + \sigma(W_t - W_{t_0}) + \int_{t_0}^{t}\int_{t_0}^{s} \mu^2 X_t\, dzds$$

$$+ \int_{t_0}^{t}\int_{t_0}^{s} \mu\sigma X_t\, dW_z ds + \int_{t_0}^{t}\int_{t_0}^{s} \mu\sigma X_t\, dzdW_s$$

$$+ \int_{t_0}^{t}\int_{t_0}^{s} \sigma^2 X_t\, dW_z dW_s$$

$$= X_0 + \mu(t-t_0) + \sigma(W_t - W_{t_0}) + \int_{t_0}^t \mu^2 X_t(s-t_0)ds$$

$$+ \int_{t_0}^t \mu\sigma X_t(W_s - W_{t_0})ds + \int_{t_0}^t \mu\sigma X_t(s-t_0)dW_s$$

$$+ \int_{t_0}^t \sigma^2 X_t(W_s - W_{t_0})dW_s$$

$$= X_0 + \mu(t-t_0) + \sigma(W_t - W_{t_0}) + \frac{1}{2}\mu^2 X_t(t-t_0)^2$$

$$+ \mu\sigma X_t(W_s - W_{t_0})(t-t_0) + \mu\sigma X_t(s-t_0)(W_t - W_{t_0})$$

$$+ \frac{1}{2}\sigma^2 X_t(W_t - W_{t_0})^2. \blacksquare$$

## Appendix 7.B   The EM and Milstein Discretizations

### 7.B.1   The EM Scheme

Suppose a stochastic process $\{X_t\}_{t\in[t_0, +\infty)}$ follows an SDE of the form (Maruyama, 1955)

$$dX_t = f(X_t)dt + g(X_t)dW_t, \; X(t_0) \equiv X_0, \tag{7.B.1}$$

where the coefficient functions $f$ and $g$ depend only on the argument $X_t$ and not on $t$ itself. For uniform time steps $\Delta_t \equiv \Delta$, integrating (7.B.1) from $t$ to $t + \Delta$ yields

$$X_{t+\Delta} = X_t + \int_t^{t+\Delta} f(X_u)\,du + \int_t^{t+\Delta} g(X_u)\,dW_u. \tag{7.B.2}$$

Looking to the second term on the right-hand side of (7.B.2), we have

$$\int_t^{t+\Delta} f(X_u)\,du \approx f(X_t)\int_t^{t+\Delta} du = f(X_t)\Delta; \tag{7.B.3}$$

and from the third term on the right-hand side of the same,

$$\int_t^{t+\Delta} g(X_u)\,dW_u \approx g(X_t)\int_t^{t+\Delta} dW_u$$

$$= g(X_t)(W_{t+\Delta} - W_t) \tag{7.B.4}$$

$$= g(X_t)\Delta W_t.$$

Since we know that $\Delta W_t \sim \sqrt{\Delta} N(0, 1) = \sqrt{\Delta} Z$, we can write the EM approximation as

$$X_{t+\Delta} = X_t + f(X_t)\Delta + g(X_t)\sqrt{\Delta}Z. \tag{7.B.5}$$

### 7.B.2 The Milstein Scheme

The Milstein approximation augments the EM scheme by a second-order correction term. That is, the accuracy of the approximation is increased by introducing Itô expansions of the coefficient functions $f(X_t) = f_t$ and $g(X_t) = g_t$. By Itô's formula (4.18), the SDEs for these functions are, respectively,

a. $df_t = \left( f_t f_t' + \frac{1}{2} g_t^2 f_t'' \right) dt + g_t f_t' \, dW_t$ and

b. $dg_t = \left( f_t g_t' + \frac{1}{2} g_t^2 g_t'' \right) dt + g_t g_t' \, dW_t.$

$$\tag{7.B.6}$$

The integral form of the $f_t$ and $g_t$ coefficient functions at times $t < s < t + \Delta$ are thus

a. $f_s = f_t + \int_t^s \left( f_u f_u' + \frac{1}{2} g_u^2 f_u'' \right) du + \int_t^s g_u f_u' \, dW_u$ and

b. $g_s = g_t + \int_t^s \left( f_u g_u' + \frac{1}{2} g_u^2 g_u'' \right) du + \int_t^s g_u g_u' \, dW_u.$

$$\tag{7.B.7}$$

A substitution of these expressions into (7.B.2) yields

$$X_{t+\Delta} = X_t + \int_t^{t+\Delta} \left[ f_t + \int_t^s \left( f_u f_u' + \frac{1}{2} g_u^2 f_u'' \right) du + \int_t^s g_u f_u' \, dW_u \right] ds$$

$$+ \int_t^{t+\Delta} \left[ g_t + \int_t^s \left( f_u g_u' + \frac{1}{2} g_u^2 g_u'' \right) du + \int_t^s g_u g_u' dW_u \right] ds. \tag{7.B.8}$$

If we ignore terms of order higher than one and retain the term involving $dW_u dW_s$, then (7.B.8) becomes

$$X_{t+\Delta} = X_t + f_t \int_t^{t+\Delta} ds + g_t \int_t^{t+\Delta} dW_s + \int_t^{t+\Delta} \int_t^s g_u g_u' \, dW_u \, dW_s. \tag{7.B.9}$$

(Note that the first three terms on the right-hand side of this expression constitute the EM approximation (7.B.5).) Looking to the last term on the right-hand side of (7.B.9), we have

$$\int_t^{t+\Delta}\int_t^s g_u g_u' \, dW_u dW_s \approx g_t g_t' \int_t^{t+\Delta}\int_t^s dW_u dW_s$$

$$= g_t g_t' \int_t^{t+\Delta} (W_s - W_t) \, dW_s$$

$$= g_t g_t' \left[ \int_t^{t+\Delta} W_s dW_s - W_t (W_{t+\Delta} - W_t) \right] \tag{7.B.10}$$

$$= g_t g_t' \left( \int_t^{t+\Delta} W_s dW_s - W_t W_{t+\Delta} + W_t^2 \right).$$

If we harken back to Example 4.1, we see that

$$\int_t^{t+X} W_s dW_s = \frac{1}{2} W_{t+\Delta}^2 - \frac{1}{2} W_t^2 - \frac{1}{2}\Delta. \tag{7.B.11}$$

Then a substitution of this result into (7.B.10) renders

$$\int_t^{t+\Delta}\int_t^s g_u g_u' \, dW_u dW_s \approx \frac{1}{2} g_t g_t' \left[ (W_{t+\Delta} - W_t)^2 - \Delta \right] \tag{7.B.10.1}$$

$$= \frac{1}{2} g_t g_t' \left[ (\Delta W_t)^2 - \Delta \right].$$

In addition with $\Delta W_t \sim \sqrt{\Delta} N(0, 1)$, we finally obtain, from (7.B.9) and (7.B.11),

$$X_{t+\Delta} = X_t + f_t \Delta + g_t \sqrt{\Delta} Z + \frac{1}{2} g_t g_t' \Delta (Z^2 - 1). \tag{7.B.12}$$

## Appendix 7.C   The Lamperti Transformation

Let $\{X_t\}_{t \in [0, +\infty)}$ be a Weiner process governed by the SDE $dX_t = f(X_t, t)dt + g(X_t)dW_t$, $X(0) \equiv X_0$, where $t$ is not an explicit argument of the diffusion coefficient function $g$ (Lamperti, 1964). For this type of SDE, we have Theorem C.1.

**Theorem C.1**   (Lamperti Transformation)
Suppose

$$Y = F(X_t, t) = \int \frac{1}{g(u)} du \Big|_{u = X_t}, \tag{7.C.1}$$

where $g(X_t) > 0$ for all $(X_t, t)$ and $F : S \to R$ is one-to-one onto. Then $Y_t$ has unit diffusion and is governed by the SDE

$$dY_t = \left( \frac{f(F^{-1}(Y), t)}{g(F^{-1}(Y))} - \frac{1}{2} g_x \left( F^{-1}(Y) \right) \right) dt + dW_t$$

$$= \left( \frac{f(X_t, t)}{g(X_t)} - \frac{1}{2} g_x(X_t) \right) dt + dW_t, \ Y(t_0) \equiv Y_0.$$

(7.C.2)

To verify this result, we need only call upon the Itô formula (4.A.5) or

$$V(X_t, t) = \left( V_t + fV_X + \frac{1}{2} g^2 V_{XX} \right) dt + gV_X dW_t.$$

An application of this expression to (7.C.1) yields, via the fundamental theorem,

$$dY_t = \left[ 0 + f(X_t, t) \left( \frac{1}{g(X_t)} \right) + \frac{1}{2} g(X_t)^2 \left( -\frac{g_x(X_t)}{g(X_t)^2} \right) \right] dt$$

$$+ g(X_t) \left( \frac{1}{g(X_t)} \right) dW_t$$

$$= \left( \frac{f(X_t, t)}{g(X_t)} \right) - \frac{1}{2} g_x(X_t, t) dt + dW_t.$$

# 8

# Estimation of Parameters of Stochastic Differential Equations

## 8.1 Introduction

Suppose we have a time-homogeneous stochastic differential equation (SDE) of the form

$$dX_t = f(X_t; \theta)dt + g(X_t; \theta)dW_t, \ t \geq 0, \tag{8.1}$$

and we seek to obtain an estimate of the unknown parameter vector $\theta \in \boldsymbol{\theta}$, where $\boldsymbol{\theta}$ is an open-bounded parameter set in $R^k$. While this task might, at first blush, seem rather straightforward, it is no trivial matter to carry out the actual estimation process. This is because a continuous-time diffusion can be observed only at discrete time points. Hence, the transition probability density function, and thus the associated likelihood function,[1] is not explicitly computable, that is, it does not have a closed-form representation. To address this issue, a variety of statistical procedures have been developed that, to varying degrees, have been successful in providing "good" estimates of $\theta$ from a single sample of observations at discrete times. In fact, a representative assortment of surveys on the estimation of $\theta$ have been offered by Cysne (2004), Shoji and Ozaki (1998), Hurn et al. (2006), Jensen and Poulson (2002), Durham and Gallant (2002), Nicolau (2004), Iacus (2008), Shoji and Ozaki (1997), and Sørensen (2004), to name but a few.

While the presentation herein pertaining to the estimation of diffusion process parameters will not be, by any stretch of the imagination, exhaustive in nature, the primary focus will be on the application of the **pseudo-maximum likelihood (PML) method**. This technique is fairly straightforward, is efficient, and has yielded some very encouraging results.

---

1 For a review of the concept of the likelihood function and the particulars of the maximum likelihood (ML) procedure, see Appendix 8.A.

*Stochastic Differential Equations: An Introduction with Applications in Population Dynamics Modeling*, First Edition. Michael J. Panik.
© 2017 John Wiley & Sons, Inc. Published 2017 by John Wiley & Sons, Inc.

## 8.2 The Transition Probability Density Function Is Known

To put the PML method in perspective, let us first consider the case in which Equation (8.1) is readily solvable and the transition probability density function is known and thus of closed form. In this circumstance, **exact ML estimation** is feasible, and the construction of the pseudo-likelihood function is not necessary.

**Example 8.1**   Let us rewrite (8.1) as

$$dX_t = \mu X_t dt + \sigma X_t dW_t, \, t \geq 0. \tag{8.2}$$

Clearly, this is the case of geometric Brownian motion (BM) first introduced in Example 5.2. If we set $Y_t = ln X_t$, then, by Itô's formula (4.18), $dY_t = \left(\mu - \frac{1}{2}\sigma^2\right)dt + \sigma dW_t$. Hence, the transition densities of $Y_t$ follow a normal distribution. Upon integrating this SDE, we get

$$ln X_t = ln X_0 + \left(\mu - \frac{1}{2}\sigma^2\right)t + \sigma W_t \tag{8.3}$$

or

$$X_t = X_0 \, e^{\left(\mu - \frac{1}{2}\sigma^2\right)t + \sigma W_t}, \tag{8.4}$$

with

$$E(X_t) = X_0 \, e^{\mu t}, \, V(X_t) = X_0^2 \, e^{2\mu t}\left(e^{\sigma^2 t} - 1\right)$$

(see Example 5.9).
   For $i = 1, 2, \dots$,

$$ln X(t_i) - ln X(t_{i-1}) = \left(\mu - \frac{1}{2}\sigma^2\right)(t_i - t_{i-1}) + \sigma(W(t_i) - W(t_{i-1})), t_i \geq 0, \tag{8.5}$$

where $W(t_i) - W(t_{i-1}) \sim N(0, t_i - t_{i-1})$ (see Equation (3.6iii)), and thus

$$X(t_i) = X(t_{i-1}) \, e^{\left[\left(\mu - \frac{1}{2}\sigma^2\right)(t_i - t_{i-1}) + \sigma\sqrt{t_i - t_{i-1}}Z_i\right]}, \tag{8.6}$$

where $Z_i \sim N(0, 1)$, $i = 1, 2, \dots$. If we set $x(t_i) = ln X(t_i) - ln X(t_{i-1})$, then

$$x(t_i) \sim N\left(\left(\mu - \frac{1}{2}\sigma^2\right)(t_i - t_{i-1}), \sigma^2(t_i - t_{i-1})\right),$$

that is, $x(t_i)$ has a log-normal[2] transition density of the form

$$p(x(t_i), \Delta t_i; \mu, \sigma) = \frac{1}{\sigma X(t_i)\sqrt{2\pi\Delta t_i}} \exp\left\{ -\frac{[x(t_i) - (\mu - (1/2)\sigma^2)\Delta t_i]^2}{2\sigma^2 \Delta t_i} \right\},$$

(8.7)

where $\Delta t_i = t_i - t_{i-1}$. So if the sample random variables $X(t_i)$, $i = 1,\ldots,n$, are drawn from a log-normal population with probability density function (8.7), then the likelihood function of the sample is

$$\mathcal{L}(\mu, \sigma; x(t_1), \ldots, x(t_n), \Delta t_i, n) = \prod_{i=1}^{n} p(x(t_i), \Delta t_i; \mu, \sigma)$$

$$= \left( \prod_{i=1}^{n} \frac{1}{X(t_i)} \right) (2\pi\sigma^2 \Delta t_i)^{-\frac{n}{2}} e^{-\frac{1}{2\sigma^2 \Delta t_i} \sum_{i=1}^{n} [x(t_i) - (\mu - \frac{1}{2}\sigma^2)\Delta t_i]^2},$$

and thus the log-likelihood function appears as

$$\ln\mathcal{L} = -\sum_{i=1}^{n} \ln X(t_i) - \frac{n}{2}\ln(2\pi) - \frac{n}{2}\ln(\sigma^2 \Delta t_i)$$

$$- \frac{1}{2\sigma^2 \Delta t_i} \sum_{i=1}^{n} \left[ x(t_i) - \left( \mu - \frac{1}{2}\sigma^2 \right)\Delta t_i \right]^2.$$

(8.8)

As required, we seek to obtain the values of $\mu$ and $\sigma$ (denote them as $\hat{\mu}$ and $\hat{\sigma}$) such that

$$\mathcal{L} = \mathcal{L}(\hat{\mu}, \hat{\sigma}; x(t_1), \ldots, x(t_n), \Delta t_i, n)$$

$$= \arg\max \mathcal{L}(\mu, \sigma; x(t_1), \ldots, x(t_n), \Delta t_i, n).$$

To this end, let us look to the first-order conditions for a maximum. Setting $\partial \ln\mathcal{L}/\partial\mu = 0$ and $\partial \ln\mathcal{L}/\partial\sigma^2 = 0$ yields, respectively, estimates of the log-normal mean ($\hat{m}$) and variance ($\hat{v}$) or

a. $\hat{m} = \dfrac{\sum_{i=1}^{n} x(t_i)}{n} = \left( \hat{\mu} - \dfrac{1}{2}\hat{\sigma}^2 \right)\Delta t_i$

and

(8.9)

b. $\hat{v} = \dfrac{\sum_{i=1}^{n} (x(t_i) - \hat{m})^2}{n} = \hat{\sigma}^2 \Delta t_i.$

Then estimates of the geometric BM parameters $\mu$ and $\sigma$ are, respectively,

---

2 For a discussion of the log-normal distribution, see Appendix 8.B.

a. $\hat{\mu} = \dfrac{\hat{m} + (1/2)\hat{v}}{\Delta t_i}$ and

b. $\hat{\sigma} = \sqrt{\dfrac{\hat{v}}{\Delta t_i}}.$

(8.10)

(It is convenient to set $\Delta t_i = \Delta = \text{constant} = {}^1/_n.$) ∎

**Example 8.2**   We noted in Equation (5.35) that the mean-reverting Ornstein–Uhlenbeck process is given by the SDE

$$dX_t = \alpha(\mu - X_t)dt + \sigma dW_t, X(0) = X_0,$$

and that a particular solution to the same is

$$X_t = X_0 e^{-\alpha t} + \mu(1 - e^{-\alpha t}) + \sigma \int_0^t e^{-\alpha(t-s)}dW_s \qquad (8.11)$$

(see Equation (5.36)). Since $W_t$ is a Wiener process,

$$\sigma \int_0^t e^{-\alpha(t-s)}dW_s \sim N\left(0, \frac{\sigma^2}{2\alpha}(1 - e^{-2\alpha t})\right)$$

(see Example 5.10). Hence, $X_t$ is normally distributed with

$$E(X_t) = \mu + (X_0 - \mu)e^{-\alpha t}, V(X_t) = \frac{\sigma^2}{2\alpha}(1 - e^{-2\alpha t}). \qquad (8.12)$$

For $\Delta t_i = t_i - t_{i-1}, i = 1, 2, ..., X(t_i)$ has a normal transition probability density function of the form

$$p(X(t_i), \Delta t_i; \alpha, \mu, \sigma) = (2\pi)^{-\frac{1}{2}} \left[\frac{\sigma^2}{2\alpha}(1 - e^{-2\alpha \Delta t_i})\right]^{-\frac{1}{2}}$$

$$\times exp\left\{-\frac{(X(t_i) - \mu - (X(t_{i-1}) - \mu)e^{-\alpha \Delta t_i})^2}{(\sigma^2/\alpha)(1 - e^{-2\alpha \Delta t_i})}\right\}.$$

(8.13)

If the sample random variables $X(t_i), i = 1, ..., n$, are drawn from a normal population with probability density function (8.13), then the likelihood function of the sample is

$$\mathcal{L}(\alpha,\mu,\sigma;x(t_1),\ldots,x(t_n),\Delta t_i,n) = \prod_{i=1}^{n} p(x(t_i),\Delta t_i;\alpha,\mu,\sigma)$$

$$= (2\pi)^{-\frac{n}{2}} \left(\frac{\sigma^2}{2\alpha}\right)^{-\frac{n}{2}} \prod_{i=1}^{n} \left(1-e^{-2\alpha\Delta t_i}\right)^{-\frac{1}{2}}$$

$$\times \exp\left\{ -\frac{\alpha}{\sigma^2} \sum_{i=1}^{n} \frac{\left(X(t_i)-\mu-(X(t_{i-1})-\mu)e^{-\alpha\Delta t_i}\right)}{1-e^{-\alpha\Delta t_i}} \right\},$$

$$(8.14)$$

and thus the log-likelihood function has the form

$$\ln\mathcal{L} = -\frac{n}{2}\ln(2\pi) - \frac{n}{2}\ln\left(\frac{\sigma^2}{2\alpha}\right) - \frac{1}{2}\sum_{i=1}^{n}\ln\left(1-e^{-2\alpha\Delta t_i}\right)$$

$$-\frac{\alpha}{\sigma^2}\sum_{i=1}^{n}\frac{\left(X(t_i)-\mu-(X(t_{i-1})-\mu)e^{-\alpha\Delta t_i}\right)^2}{1-e^{-2\alpha\Delta t_i}}.$$

$$(8.15)$$

As expected, the first-order conditions for a maximum of (8.14) ($\partial\ln\mathcal{L}/\partial\alpha = \partial\ln\mathcal{L}/\partial\mu = \partial\ln\mathcal{L}/\partial\sigma = 0$) generate a highly nonlinear system of simultaneous equations that typically can be solved by numerical (iterative) methods. ∎

**Example 8.3** We previously specified the Cox, Ingersol, and Ross (CIR) square-root SDE (see Example 6.9) as

$$dX_t = (\alpha - \beta X_t)dt + \sigma\sqrt{X_t}dW_t, X(0) = X_0 > 0,$$

where the CIR process is well defined if $\alpha, \beta$, and $\sigma$ are all positive and $2\alpha > \sigma^2$. Since this diffusion process has a known transition probability density function that exists in a closed or explicit form, we may use the exact ML technique to determine the parameter vector $\theta = (\alpha, \beta, \sigma)$.

It has been demonstrated by Feller (1951a) that the transition density of the CIR process has the following form: given $X_t$ at time $t$, the density of $X_{t+\Delta t}$ at time $t + \Delta t$ is

$$p(X_{t+\Delta t}|X_t,\Delta t;\alpha,\beta,\sigma) = ce^{-u-v}\left(\frac{v}{u}\right)^{\frac{q}{2}}I_q\left(2\sqrt{uv}\right),$$

$$(8.16)$$

where

$$c = \frac{2\beta}{\sigma^2(1-e^{-\beta\Delta t})}, u = cX_t e^{-\beta\Delta t}, v = cX_{t+\Delta t}, q = \frac{2\alpha}{\sigma^2} - 1,$$

$I_q(\cdot)$ is the modified **Bessel function** of the first kind of order $q$,

$$I_q(x) = \sum_{k=0}^{\infty} \left(\frac{x}{2}\right)^{2k+q} \left(\frac{1}{k!\Gamma(kq+1)}\right), x \in R,$$

and $\Gamma(\cdot)$ is the **gamma function**

$$\Gamma(z) = \int_0^{\infty} x^{z-1}e^{-x}dx, z \in R^+.$$

(For the density (8.16), the distribution function is **noncentral chi-squared with** $2q +2$ degrees of freedom and noncentrality parameter $2u$.)

Given that $\{X_t\}_{t \geq 0}$ is a Markov process, the likelihood function of the sample, for $\Delta = \Delta t = t_i - t_{i-1} =$ constant, $i = 1, ..., n$, is

$$\mathcal{L}(\alpha, \beta, \sigma; X(t_0), X(t_1), ..., X(t_n), \Delta, n) =$$

$$p(X(t_0); \alpha, \beta, \sigma) \prod_{i=1}^{n} p(X(t_i) | X(t_{i-1}), \Delta; \alpha, \beta, \sigma),$$

and the log-likelihood function appears as

$$ln\,\mathcal{L}(\alpha, \beta, \sigma; X(t_0), X(t_1), ..., X(t_n), \Delta, n) =$$

$$\sum_{i=1}^{n} ln\,p(X(t_i) | X(t_{i-1}), \Delta; \alpha, \beta, \sigma) \tag{8.17}$$

under the assumption that the distribution of the initial condition $X(t_0)$ is independent of $\theta = (\alpha, \beta, \sigma)$. Upon combining (8.16) and (8.17), we obtain the log-likelihood function for the CIR process

$$ln\,\mathcal{L}(\cdot) = nlnc + \sum_{i=1}^{n} \left\{ -u_{t_{i-1}} - v_{t_i} + \frac{q}{2}ln\left(\frac{v_{t_i}}{u_{t_{i-1}}}\right) + ln\left[I_q\left(2\sqrt{u_{t_{i-1}}v_{t_i}}\right)\right] \right\}, \tag{8.18}$$

where

$$u_{t_{i-1}} = cX(t_{i-1})e^{-\beta\Delta t} \text{ and } v_{t_i} = cX(t_i).$$

Then the exact ML estimates of the parameters of the CIR process are obtained as

$$\hat{\theta} = (\hat{\alpha}, \hat{\beta}, \hat{\sigma}) = arg \max_{\theta} ln\,\mathcal{L}(\cdot). \blacksquare$$

## 8.3 The Transition Probability Density Function Is Unknown

We now turn to the particulars of the PML method. What exactly is a pseudo-likelihood function? Think of it as an approximation to the joint probability distribution of a set of random variables; it replaces the true or exact likelihood when conducting an ML operation. That is, if some conditional moments of a diffusion process are known but not the true transition density ($p$) itself, then, under certain restrictions, it might be feasible to estimate $\theta$ from a **pseudo-transition density function $h$**. So while $h$ does not properly belong to the family of the exact transition density function, it is consonant in terms of the moments of $p$. The $h$ density is thus a surrogate for $p$ and is termed the "pseudo-transition density."

To obtain the **pseudo-likelihood function**, let us first note that a diffusion process $\{X_t\}_{t \geq 0}$ is a Markov process. In this regard, suppose $X(t_0) = X_0$ and $X(t_i) \equiv X_i, i = 1, \ldots, n$, is a sequence of $n + 1$ historical observations on the random variable $X_t$ that are sampled at nonstochastic dates $t_0 < t_1 < \cdots < t_n$. To generate PML estimates of $\theta$, we need to first form the joint density of the sample or the pseudo-likelihood function.

$$\mathcal{L}(\theta, X_1, \ldots, X_n, \Delta_i, n) = h(X_0; \theta) \prod_{i=1}^{n} h(\Delta_i, X_{i-1}, X_i; \theta),^3 \qquad (8.19)$$

where $\Delta_i = t_i - t_{i-1}, i = 1, \ldots, n, h(\Delta_i, X_{i-1}, X_i; \theta)$ is the pseudo-transition density function, and $h(X_0; \theta)$ is the density of the initial state. As usual, we shall work with the logarithm of the pseudo-likelihood function or

$$\ln \mathcal{L} = \ln h(X_0; \theta) + \sum_{i=1}^{n} \ln h(\Delta_i, X_{i-1}, X_i; \theta). \qquad (8.20)$$

If the initial condition $X_0$ does not depend on $\theta$, then (8.19) becomes

$$\mathcal{L}(\theta, X_1, \ldots, X_n, \Delta_i, n) = \prod_{i=1}^{n} h(\Delta_i, X_{i-1}, X_i; \theta), \qquad (8.19.1)$$

and thus (8.20) is rewritten as

$$\ln \mathcal{L} = \sum_{i=1}^{n} \ln h(\Delta_i, X_{i-1}, X_i; \theta). \qquad (8.20.1)$$

---

3 See Appendix 8.C for the derivation of this expression via the Markov property.

The PML estimate of $\theta$ is then

$$\hat{\theta} = \arg\max_{\hat{\theta} \in \theta} \sum_{i=1}^{n} \ln h(\Delta_i, X_{i-1}, X_i; \theta).$$

### 8.3.1 Parameter Estimation via Approximation Methods

In what follows, we shall focus on *process-based solutions* to the SDE.

$$dX_t = f(X_t; \theta)dt + g(X_t; \theta)dW_t, X(0) = X_0. \tag{8.21}$$

That is, we shall consider three discrete-time iteration schemes that are often employed to obtain a practicable pseudo-likelihood function. These are the Euler–Maruyama (EM) scheme, the Ozaki (1985) local linearization method, and the Shoji–Ozaki or SO (1994) new local linearization method. (In what follows, we shall assume, for convenience, that $\Delta t_i = \Delta = $ constant for all $i$.)

#### 8.3.1.1 The EM Routine

Let the preceding SDE (Equation (8.21)) appear as

$$dX_t = f(X_t; \theta)dt + \sigma dW_t, \tag{8.22}$$

where $g(X_t; \theta) = \sigma = $ constant, $\sigma$ is not one of the components of $\theta$, and $f(X_t; \theta)$ is piecewise constant (e.g., $f(X_t; \theta)$ is constant on a small time interval $[t, t + \Delta t)$). Given the EM scheme (7.2) or

$$X_{t+\Delta t} = X_t + f(X_t; \theta)\Delta + \sigma(W_{t+\Delta} - W_t), \tag{8.23}$$

we know that the increments $X_{t+\Delta} - X_t$ are independent Gaussian variables with mean $f(X_t; \theta)\Delta$ and variance $\sigma^2 \Delta$. Then the pseudo-transition density function of the discretized process appears as

$$h(\Delta, X_{i-1}, X_i; \theta, \sigma) = \frac{1}{\sqrt{2\pi\sigma^2\Delta}} e^{-\frac{(X_i - X_{i-1} - f(X_{i-1}; \theta)\Delta)^2}{2\sigma^2\Delta}}, \tag{8.24}$$

and thus the logarithm of the pseudo-likelihood function is

$$\ln \mathcal{L}(\theta, \sigma; X_1, \ldots, X_n, \Delta, n) = \ln h(X_0; \theta, \sigma) - \frac{n}{2}\ln(2\pi\sigma^2\Delta)$$

$$- \frac{1}{2}\sum_{i=1}^{n} \frac{(X_i - X_{i-1} - f(X_{i-1}; \theta)\Delta)^2}{\sigma^2\Delta}. \tag{8.25}$$

Then the PML estimates of $\theta$ and $\sigma$, denoted, respectively, as $\hat{\theta}$ and $\hat{\sigma}$, are those that maximize (8.25).[4]

**Example 8.4**  We noted earlier (Example 8.2) that the Ornstein–Uhlenbeck (OU) process is given by the SED

$$dX_t = \alpha(\mu - X_t)dt + \sigma dW_t. \tag{8.26}$$

Then approximating the solution to this SDE via the EM routine gives

$$X_{t+\Delta} = X_t + \alpha(\mu - X_t)\Delta + \sigma(W_{t+\Delta} - W_t), \tag{8.27}$$

where the increments $X_{t+\Delta} - X_t$ are independent Gaussian random variables with mean $\alpha(\mu - X_t)\Delta$ and variance $\sigma^2\Delta$. Then the pseudo-transition density of the discretized process has the form

$$h(\Delta, X_{i-1}, X_i; \alpha, \mu, \sigma) = \frac{1}{\sqrt{2\pi\sigma^2\Delta}} e^{-\frac{\left(X_i - X_{i-1} - \alpha(\mu - X_{i-1})\Delta\right)^2}{2\sigma^2\Delta}}, \tag{8.28}$$

and the associated pseudo log-likelihood function appears as

$$\ln\mathcal{L}(\alpha, \mu, \sigma; X_1, \ldots, X_n, \Delta, n) = \ln h(X_0; \alpha, \mu, \sigma) - \frac{n}{2}\ln\left(2\pi\sigma^2\Delta\right)$$
$$- \frac{1}{2}\sum_{i=1}^{n}\frac{\left(X_i - X_{i-1} - \alpha(\mu - X_{i-1})\Delta\right)^2}{2\sigma^2\Delta}. \blacksquare$$

$$\tag{8.29}$$

It is important to note that the approximation afforded by the EM routine is "good" (1) if $\Delta t$ is very small (otherwise the parameter estimates are biased), (2) if the following holds

**Polynomial Growth Condition**: there exist positive $L$ and $m$ values (both independent of $\theta$) such that

$$|f(X; \theta)| \leq L(1 + |X|^m), \theta \in \boldsymbol{\theta},$$

and (3) if $n\Delta_n^3 \to 0$. If the third condition holds, the PML estimator obtained from (8.25) is consistent and asymptotically efficient. (For an elaboration of

---

4 If (8.23) is replaced by

$$X_{t+\Delta} + X_t + f(X_t; \theta)\Delta + g(X_t; \theta)(W_{t+\Delta} - W_t), \tag{8.23.1}$$

then the independent Gaussian increments $X_{t+\Delta} - X_t$ are Gaussian with mean $f(X_t; \theta)\Delta$ and variance $g(X_t; \theta)^2\Delta$. In this circumstance, the pseudo-transition density is

$$h(\Delta, X_{i-1}, X_i; \theta) = \frac{1}{\sqrt{2\pi\Delta g(X_{i-1}; \theta)}} e^{-\frac{\left(X_i - X_{i-1} - f(X_{i-1}; \theta)\Delta\right)^2}{2g(X_{i-1}; \theta)^2\Delta}}.$$

these three requirements, see Iacus (2008, pp. 122–125) and the references listed therein.)

### 8.3.1.2 The Ozaki Routine

Suppose we have a time-homogeneous SDE of the form

$$dX_t = f(X_t; \theta)dt + \sigma dW_t, t \geq 0, X(0) = X_0,$$

where $\sigma$ is constant. Given the Ozaki (1992, 1993) local linearization iteration scheme,

$$X_{t+\Delta} = A_t X_t + B_t W_{t+\Delta},$$

where

$$A_t = e^{K_t \Delta}, B_t = \sigma \left( \frac{e^{2K_t \Delta} - 1}{2K_t} \right)^{\frac{1}{2}},$$

and

$$K_t = \frac{1}{\Delta} \ln \left[ 1 + \frac{f(X_t)}{X_t f_X(X_t)} \left( e^{f_X(X_t)\Delta} - 1 \right) \right]$$

(see Equation (7.37)), the conditional distribution $X_{t+\Delta}|X_t$ is Gaussian with mean $E(X_{t+\Delta}) = e^{K_t \Delta} X_t$ and variance $V_t = \sigma^2((e^{2K_t \Delta} - 1)/2K_t)$. Then the pseudo-transition density function of the discretized process has the form

$$h(\Delta, X_{i-1}, X_i; \theta, \sigma) = \frac{1}{\sqrt{2\pi V_{i-1}}} e^{-\frac{1}{2} \frac{\left( X_i - e^{K_{i-1}\Delta} X_{i-1} \right)^2}{V_{i-1}}}, \tag{8.30}$$

and thus the logarithm of the pseudo-likelihood function is

$$\ln \mathcal{L}(\Delta, X_{i-1}, X_i; \theta, \sigma) = \ln h(X_0; \theta, \sigma) - \frac{1}{2} \sum_{i=1}^{n} \ln(2\pi V_{i-1})$$

$$- \frac{1}{2} \sum_{i=1}^{n} \frac{(X_i - e^{K_{i-1}\Delta} X_{i-1})^2}{V_{i-1}}$$

$$= \ln h(X_0; \theta, \sigma) - \frac{1}{2} \sum_{i=1}^{n} \ln(2\pi V_{i-1})$$

$$- \frac{1}{2} \sum_{i=1}^{n} \frac{(X_i - E_{i-1})^2}{V_{i-1}}, \tag{8.31}$$

where

$$E_i = X_i + \frac{f(X_i)}{f_X(X_t)}\left(e^{f_X(X_i)\Delta} - 1\right), V_i = \sigma^2\left(\frac{e^{2K_i\Delta} - 1}{2K_i}\right).$$

Then maximizing (8.31) with respect to $\theta$ and $\sigma$ yields the PML estimates $\hat{\theta}$ and $\hat{\sigma}$, respectively.

**Example 8.5** Given the SDE $dX_t = \theta X_t^3 dt + \sigma dW_t, t \geq 0$, determine the logarithmic expression for the pseudo-likelihood function via the Ozaki local linearization method. Proceeding as before, we need to first calculate

$$K_i = \frac{1}{\Delta}ln\left[1 + \frac{1}{3}\left(e^{3\theta X_i^2\Delta} - 1\right)\right].$$

Then

$$e^{K_i\Delta}X_i = X_t + \frac{X_i}{3}\left(e^{3\theta X_i^2\Delta} - 1\right) = E_i,$$

$$V_i = \sigma^2\left(\frac{\left[1 + (1/3)\left(e^{3\theta X_i^2\Delta} - 1\right)\right]^2 - 1}{(2/\Delta)ln\left[1 + (1/3)\left(e^{3\theta X_i^2\Delta} - 1\right)\right]}\right).$$

Next, a substitution of $E_{i-1}$ and $V_{i-1}$ into (8.31) gives the desired result. ∎

**Example 8.6** One of the key assumptions regarding the implementation of the Ozaki local linearization technique is that the diffusion term has a constant coefficient, that is, we have an SDE of the form (7.31.1). If this is not the case, then we can usually employ the Lamperti transformation

$$Y_t = F(X_t) = \int\frac{1}{g(u)}du\Big|_{u=X_t}$$

to achieve this form. This renders the transformed SDE

$$dY_t = f_F(Y_t, t)dt + dW_t, Y(0) = Y_0,$$

where the transformed drift coefficient is

$$f_F(Y_t, t) = \frac{f(F^{-1}(Y_t), t)}{g(F^{-1}(Y_t))} - \frac{1}{2}g_X\left(F^{-1}(Y_t)\right) = \frac{f(X_t, t)}{g(X_t)} - \frac{1}{2}g_X(X_t).$$

In this regard, suppose our objective is to apply the Ozaki local linearization method to the SDE

$$dX_t = (\mu_1 + \mu_2 X_t)dt + \sigma X_t dW_t, X(0) = X_0.$$

For our purposes, let $\sigma$ remain as the coefficient on $W_t$. Then set

$$Y_t = F(X_t) = \int \frac{1}{u} du \bigg|_{u = X_t} = \ln X_t$$

so that the transformed SDE appears as

$$dY_t = f_F(Y_t)dt + \sigma dW_t$$

$$= \left( \mu_1 X_t^{-1} + \mu_2 - \frac{1}{2} \right) dt + \sigma dW_t$$

$$= \left( \mu_1 e^{-Y_t} + \mu_2 - \frac{1}{2} \right) dt + \sigma dW_t.$$

To obtain the pseudo-likelihood function, we need to first find (in terms of the transformed variable $Y_t$)

$$K_i = \frac{1}{\Delta} \ln \left[ 1 + \frac{\mu_1 e^{-Y_i} + \mu_2 - (1/2)}{Y_i(-\mu_1 e^{-Y_1})} \left( e^{-\mu_1 e^{-Y_i \Delta}} - 1 \right) \right]$$

$$= \frac{1}{\Delta} \ln \left[ 1 - Y_i^{-1} \left\{ 1 + \frac{1}{\mu_1} \left( \mu_2 - \frac{1}{2} \right) e^{Y_i} \right\} \left( e^{-\mu_1 e^{-Y_i \Delta}} - 1 \right) \right]. \tag{8.32}$$

Then

$$e^{K_i \Delta} Y_i = Y_i - \left[ 1 + \frac{1}{\mu_1} \left( \mu_2 - \frac{1}{2} \right) e^{Y_i} \right] \left( e^{-\mu_1 e^{-Y_i \Delta}} - 1 \right) = E_i,$$

$$V_i = \sigma^2 \left( \frac{e^{2K_i \Delta} - 1}{2K_i} \right),$$

and thus the logarithm of the pseudo-likelihood function can be written as

$$\ln \mathcal{L}(\Delta, X_{i-1}, X_i; \mu_1, \mu_2, \sigma) = \ln h(Y_0; \mu_1, \mu_2, \sigma)$$

$$- \frac{1}{2} \sum_{i=1}^{n} \ln(2\pi V_{i-1}) - \frac{1}{2} \sum_{i=1}^{n} \frac{(Y_i - E_{i-1})^2}{V_{i-1}} \tag{8.33}$$

$$+ \sum_{i=0}^{n} \ln \left( \frac{1}{X_i} \right),$$

as determined earlier, where $Y_i = \ln X_i$. Why does the last term on the right-hand side of (8.33) need to be included in this expression? The answer to this question has to do with executing the Lamperti transformation and its effect on the (pseudo) log-likelihood function. Appendix 8.D covers the rationale for, and details of, this change-of-variable adjustment. ∎

### 8.3.1.3 The SO Routine

Given an SDE of the form

$$dX_t = f(X_t; \theta)dt + \sigma dW_t$$

(see the discussion underlying Equations (7.38) and (7.39)); the SO new or revised local linearization method that approximates this SDE at $X_t (\neq 0)$, with $L_t \neq 0$, is

$$X_{t+\Delta} = A(X_t)X_t + B(X_t)W_{t+\Delta},$$

where

$$A(X_t) = 1 + \frac{f(X_t)}{X_t L_t}\left(e^{L_t\Delta} - 1\right) + \frac{M_t}{X_t L_t^2}\left[\left(e^{L_t\Delta} - 1\right) - L_t\Delta\right],$$

$$B(X_t) = \sigma\left(\frac{e^{2L_t\Delta} - 1}{2L_t}\right)^{\frac{1}{2}}, L_t = f_X(X_t), \text{ and } M_t = \frac{\sigma^2}{2}f_{XX}(X_t)$$

(see Equation (7.46)). Here, the conditional distribution of $X_{t+\Delta}|X_t$ is Gaussian with mean $E(X_{t+\Delta}) = A(X_t)X_t$ and variance

$$V_t = \sigma^2\left(\frac{e^{2L_t\Delta} - 1}{2L_t}\right).$$

Let us set

$$E_i = A(X_i)X_i = X_i + \frac{f(X_i)}{L_i}\left(e^{L_i\Delta} - 1\right) + \frac{M_i}{L_i^2}\left[\left(e^{L_i\Delta} - 1\right) - L_i\Delta\right] \tag{8.34}$$

and

$$V_i = \sigma^2\left(\frac{e^{2L_i\Delta} - 1}{2L_i}\right). \tag{8.35}$$

Then the logarithm of the pseudo-likelihood function is

$$\ln\mathcal{L}(\Delta, X_{i-1}, X_i; \theta, \sigma) = \ln h(X_0; \theta, \sigma) - \frac{1}{2}\sum_{i=1}^{n}\ln(2\pi V_{i-1})$$
$$- \frac{1}{2}\sum_{i=1}^{n}\frac{(X_i - E_{i-1})^2}{V_{i-1}} \tag{8.36}$$

and may be maximized accordingly.

**Example 8.7** Suppose our objective is to specify the logarithmic expression for the pseudo-likelihood function using the SO local linearization routine given the SDE

$$dX_t = \theta X_t^3 dt + \sigma dW_t, t \geq 0.$$

Given $f(X_t; \theta) = \theta X_t^3$, it follows that

$$L_i = 3\theta X_i^2, M_i = \sigma^2 (3\theta X_i),$$

and thus

$$E_i = X_i + \frac{1}{3} X_i \left( e^{3\theta X_i^2 \Delta} - 1 \right) + \frac{\sigma^2}{3\theta X_i^3} \left( e^{3\theta X_i^2 \Delta} - 1 - 3\theta X_i^2 \Delta \right)$$

with

$$V_i = \sigma^2 \left( \frac{e^{6\theta X_i^2 \Delta} - 1}{6\theta X_i^2} \right).$$

Then substituting $E_{i-1}$ and $V_{i-1}$ into (8.36) yields the desired result. ∎

**Example 8.8**   We determined in Example 8.6 that, under the Lamperti transformation, the SDE

$$dX_t = (\mu_1 + \mu_2 X_t)dt + \sigma X_t dW_t, X(0) = X_0,$$

was converted to

$$dY_t = f_Y(Y_t)dt + \sigma dW_t$$

$$= \left( \mu_1 e^{-Y_t} + \mu_2 - \frac{1}{2} \right) dt + \sigma dW_t.$$

To obtain the logarithm of the pseudo-likelihood function for the SO routine, we need to calculate

$$L_i = \frac{\partial}{\partial Y} f_Y(Y_i), M_i = \frac{\sigma^2}{2} \frac{\partial^2}{\partial Y^2} f_Y(Y_i),$$

$$E_i = Y_i + \frac{f_Y(Y_i)}{L_i} \left( e^{L_i \Delta} - 1 \right) + \frac{M_i}{L_i^2} \left[ \left( e^{L_i \Delta} - 1 \right) - L_i \Delta \right], \tag{8.37}$$

and $V_i$ (determined from (8.35)). So for $f_Y(Y_i) = \mu_1 e^{-Y_i} + \mu_2 - (1/2)$, we have

$$L_i = -\mu_i e^{-Y_i}, M_i = \frac{\sigma^2}{2} \left( \mu_1 e^{-Y_i} \right),$$

and thus

$$E_i = Y_i - \left[ 1 + \frac{1}{\mu_1} \left( \mu_2 - \frac{1}{2} \right) e^{Y_i} \right] \left( e^{-\mu_1 e^{-Y_i} \Delta} - 1 \right)$$

$$+ \frac{\sigma^2 e^{Y_i}}{2\mu_1} \left( e^{-\mu_1 e^{-Y_i} \Delta} - 1 + \mu_1 e^{-Y_i} \Delta \right), \tag{8.38}$$

$$V_i = \sigma^2 \left( \frac{e^{-2\mu_1 e^{-Y_i \Delta}} - 1}{-2\mu_1 e^{-Y_i}} \right), \tag{8.39}$$

where $Y_i = \ln X_i$. Then, from (8.38) and (8.39), substituting $E_{i-1}$ and $V_{i-1}$ into (8.33) gives us the desired result. ∎

## Appendix 8.A   The ML Technique

Generally speaking, the **method of maximum likelihood** is a data reduction technique that yields statistics that are used to summarize sample information. As we shall soon see, this method requires knowledge of the form of the population probability density function.

To obtain an estimate of some unknown population parameter $\theta$, assume that the sample random variables $X_1, ..., X_n$ have been drawn from a population with probability density function $p(x; \theta)$. Given that the $X_i, i = 1, ..., n$ are independent and identically distributed, their joint probability density function has the form

$$L(x_1, ..., x_n; \theta, n) = \prod_{i=1}^{n} p(x_i; \theta). \tag{8.A.1}$$

Here, $\theta$ is fixed and the arguments are the variables $x_i, i = 1, ..., n$. But if the $x_i$'s are treated as realizations of the sample random variables and $\theta$ is taken to be a variable and no longer held constant, then (8.A.1) is termed the **likelihood function of the sample** and can be written as

$$\mathcal{L}(\theta; x_1, ..., x_n, n) = \prod_{i=1}^{n} p(x_i; \theta) \tag{8.A.2}$$

to highlight its dependence on $\theta$. For computational expedience, $\mathcal{L}$ will be transformed to the **logarithmic-likelihood function**

$$\ln \mathcal{L}(\theta; x_1, ..., x_n, n) = \sum_{i=1}^{n} \ln p(x_i; \theta). \tag{8.A.2.1}$$

Here, (A.8.2) depicts, in terms of $\theta$, the *a priori* probability of obtaining the observed random sample. In addition, as $\theta$ varies over some admissible range for fixed realization $x_i, i = 1, ..., n$, the said probability does likewise. In short, the logarithmic likelihood function expresses the probability of the observed random sample as a function of $\theta$. So with $\theta$ treated as a variable in $\ln \mathcal{L}$, the method of ML is based upon the **principle of maximum likelihood**: select as an estimate of $\theta$ that value of the parameter, $\hat{\theta}$, that maximizes the probability of observing the given random sample. So to find $\hat{\theta}$, we need only maximize $\ln \mathcal{L}$

with respect to $\theta$. In this regard, if $\mathcal{L}$ is a twice-differentiable function of $\theta$, then a necessary condition for $ln\,\mathcal{L}$ to attain a maximum at $\theta = \hat{\theta}$ is

$$\left.\frac{d\,ln\,\mathcal{L}}{d\theta}\right|_{\theta = \hat{\theta}} = 0.^5 \tag{8.A.3}$$

Hence, all we need to do is set $\partial ln\,\mathcal{L}/\partial\theta = 0$ and solve for the value of $\theta, \hat{\theta}$, which makes this derivative vanish. If $\hat{\theta} = g(x_1,...,x_n, n)$ is the value of $\theta$ that maximizes $ln\,\mathcal{L}$, then $\hat{\theta}$ is termed the **maximum likelihood estimate of** $\theta$; it is the realization of the **maximum likelihood estimator** $\hat{T} = g(x_1,...,x_n, n)$ and represents the parameter value *most likely* to have generated the sample realizations $x_1, i = 1,...,n$.

If $\theta$ is held fixed, then the population density $p(x; \theta)$ is completely specified. But if the $x_i$'s are held fixed and $\theta$ is variable, then we seek to determine from which density (indexed by $\theta$) the given set of sample values was most likely to have been drawn, that is, we want to determine from which density the likelihood is largest that the sample was obtained. This determination can be made by finding the value of $\theta, \hat{\theta}$, for which $\hat{\mathcal{L}} = \mathcal{L}\left(\hat{\theta}; x_1,...,x_n, n\right) = arg\,max_\theta\,\mathcal{L}(\theta; x_1,...,x_n, n) - -\hat{\theta}$ thus makes the probability of getting the observed sample greatest in the sense that it is the value of $\theta$ that would generate the observed sample *most often*.[6]

**Example 8.A.1**   Let $X_1, ..., X_n$ be a set of sample random variables drawn from a normal population with probability density function $p(x; \mu, \sigma) = \left(\sqrt{2\pi}\sigma\right)^{-1} exp\{-(x-\mu)^2/2\sigma^2\}$, $-\infty < x < \infty$. Let us find the ML estimates of $\mu$ and $\sigma^2$. To do so, it requires that we generalize (A.8.3). That is, if the likelihood function

---

5 In what follows, we shall usually deal with $ln\,\mathcal{L}$ rather than $\mathcal{L}$ itself since $ln\,\mathcal{L}$ is a strictly monotonic function of $\theta$, and thus the maximum of $\mathcal{L}$ occurs at the same $\theta$ as the maximum of $ln\,\mathcal{L}$. Thus, maximizing $ln\,\mathcal{L}$ is equivalent to maximizing $\mathcal{L}$ since $d\,ln\,\mathcal{L}/d\theta = (1/\mathcal{L})d\mathcal{L}/d\theta = 0$ implies $d\mathcal{L}/d\theta = 0$. In addition, it is assumed that $\hat{\theta}$ is an interior point of the set of admissible $\theta$s and that the sufficient condition for a maximum of $ln\,\mathcal{L}$ at $\hat{\theta}$ is satisfied, that is,

$$\left.\frac{d^2\,ln\,\mathcal{L}}{d\theta^2}\right|_{\theta = \hat{\theta}} < 0.$$

6 The notation **arg max** is an abbreviation of the phrase "the argument of the maximum." It is the set of points of a given argument for which the given function attains its greatest value. Hence, *arg max* $\mathcal{L}(\theta; x_1,...,x_n, n)$ is the set of points $\theta$ for which $\mathcal{L}$ attains its largest value. This set may be empty, a singleton, and have multiple elements. For instance, if $g(x) = 1 - |x|$, then

$$arg\,max_x\,(1 - |x|) = \{0\}.$$

The *arg max* operator is complementary to the **max operator** *max* $g(x)$ in that the latter returns the maximum value of $g(x)$ instead of the point(s) that produces that value, that is, $max_x g(x) = max_x (1 - |x|) = 1$.

depends on $h$ parameters $\theta_1, ..., \theta_h$, then the first-order conditions for $ln\mathcal{L}(\theta_1, ..., \theta_h; x_1, ..., x_n, n)$ to attain a maximum at $\theta_j = \hat{\theta}_j, j = 1, ..., h$, are

$$\frac{\partial ln\mathcal{L}}{\partial \theta_j}\bigg|_{\theta_j = \hat{\theta}_j} = 0, j = 1, ..., h. \tag{8.A.3.1}$$

Hence, we need only set $\partial ln\mathcal{L}/\partial \theta_j = 0, j = 1, ..., h$, and solve the resulting simultaneous equation system for the **maximum likelihood estimates** $\hat{\theta}_j = g^j(x_1, ..., x_n, n)$. Thus, the $\hat{\theta}_j$s are the realizations of the **maximum likelihood estimators** $\hat{T}_j = g^j(x_1, ..., x_n, n), j = 1, ..., h$.

The ML estimates of $\mu$ and $\sigma^2$ are, respectively, the values $\hat{\mu}$ and $\hat{\sigma}^2$ for which

$$\mathcal{L}(\mu, \sigma^2; x_1, ..., x_n, n) = \prod_{i=1}^{n} p(x_i; \mu, \sigma^2)$$

$$= (2\pi\sigma^2)^{-\frac{n}{2}} e^{-\sum_{i=1}^{n}(x_i - \mu)^2/2\sigma^2}$$

or

$$ln\mathcal{L} = -\frac{n}{2}ln(2\pi) - \frac{n}{2}ln\sigma^2 - \frac{1}{2\sigma^2}\sum_{i=1}^{n}(x_i - \mu)^2 \tag{8.A.4}$$

attains a maximum. Then from (A.8.3.1) and (8.A.4),

a. $\partial\dfrac{ln\mathcal{L}}{\partial\mu} = \dfrac{1}{\sigma^2}\sum_{i=1}^{n}(x_i - \mu) = 0,$

$$\tag{8.A.5}$$

b. $\partial\dfrac{ln\mathcal{L}}{\partial\sigma^2} = -\dfrac{n}{2\sigma^2} + \dfrac{1}{2\sigma^4}\sum_{i=1}^{n}(x_i - \mu)^2 = 0.$

From (8.A.5a), $\hat{\mu} = (1/n)\sum_{i=1}^{n} x_i = \bar{X}$ (the ML estimate of the mean $\mu$ of a normal population is the sample realization of the mean estimator $\bar{X} = (1/n)\sum_{i=1}^{n} x_i$). In addition, from (8.A.5b) and $\hat{\mu} = \bar{X}$, we obtain $\hat{\sigma}^2 = (1/n)\sum_{i=1}^{n}(x_i - \bar{X})^2$ (the ML estimate of the variance of a normal population is the realization of the variance estimator $s_0^2 = (1/n)\sum_{i=1}^{n}(x_i - \bar{X})^2$). (It is a well-known result that $\bar{X}$ is an unbiased estimator of $\mu$, and $s_0^2$ is a biased estimator of $\sigma^2$. An unbiased estimator of $\sigma^2$ is $s^2 = (1/n-1)\sum_{i=1}^{n}(x_i - \bar{X})^2$.) ∎

The ML technique will generally yield an efficient or minimum variance unbiased estimate of a parameter $\theta$, if one exists. That is, if a ML estimator $\hat{\theta}$ can be found and $\hat{\theta}$ is unbiased for $\theta$, then $\hat{\theta}$ will typically be an efficient or best unbiased estimator of $\theta$. In addition, ML estimators are consistent estimators of population parameters.

## Appendix 8.B    The Log-Normal Probability Distribution

Suppose a continuous random variable $X$ is **normally distributed** with mean $\mu$ and standard deviation $\sigma$ (i.e., $X \sim N(\mu, \sigma)$). Then $X$'s probability density function has the form

$$p(x; \mu, \sigma) = \frac{1}{\sqrt{2\pi}\sigma} e^{-\frac{1}{2}\left(\frac{x-\mu}{\sigma}\right)^2}, -\infty < x < +\infty. \tag{8.B.1}$$

Let us now consider the notion of "functions of random variables." That is, given a continuous random variable $X$ with probability density function $f(x)$, suppose another random variable $Y$ can be written as a function of $X$ or $Y = u(X)$. How can we determine the probability density function of $Y$? If we can find $Y$'s cumulative distribution function

$$G(y) = P(Y \le y) = P(u(x) \le y)$$

$$= \int_{-\infty}^{y} f(x) \, dx = \int_{-\infty}^{u(x)} f(x) \, dx, \tag{8.B.2}$$

then its probability density is given by

$$g(y) = G'(y) = f(u(x))u'(x) \tag{8.B.3}$$

(provided $u'(x)$ exists).

In this regard, let $X \sim N(\mu, \sigma)$ and let $y = e^X$. Then the cumulative distribution of $Y$ is

$$G(y) = P(Y \le y) = P(e^X \le y) = P(X \le \ln y), y > 0.$$

Hence, from (8.B.2),

$$G(y) = \int_{-\infty}^{\ln x} \frac{1}{\sqrt{2\pi}\sigma} e^{-\frac{1}{2}\left(\frac{x-\mu}{\sigma}\right)^2} dx, y > 0,$$

and thus the probability density function of $Y$ is

$$g(y) = G'(y) = f(u(x))u'(x) = f(\ln y)\frac{1}{y}$$

$$= \frac{1}{\sqrt{2\pi}\sigma y} e^{-\frac{1}{2}\left(\frac{\ln y - \mu}{\sigma}\right)^2}, y > 0, \tag{8.B.4}$$

the **log-normal distribution**. Thus, the random variable $Y$ has a log-normal distribution if $X = \ln Y$ has a normal distribution. Stated alternatively, if $X$ is a random variable with a normal distribution, then $Y = e^X$ has a log-normal distribution. The parameters of the log-normal distribution are $\mu$ and $\sigma$—the mean of $ln\,Y$ is $\mu$, and the standard deviation of $ln\,Y$ is $\sigma$. In addition, for $X = ln\,Y \sim N(\mu, \sigma)$, the mean and variance of the log-normal variable $Y = e^X$ are, respectively,

$$E(Y) = E\left(e^X\right) = e^{\mu + \frac{1}{2}\sigma^2} \text{ and}$$

$$V(Y) = V\left(e^X\right) = e^{2\mu + \sigma^2}\left(e^{\sigma^2} - 1\right).$$

## Appendix 8.C  The Markov Property, Transitional Densities, and the Likelihood Function of the Sample

Let $X = \{X_t\}_{t \in [t_0, T]}$ be the path of a Markov process defined on the probability space $(\Omega,\ \mathcal{A},\ P)$ and which assumes continuous values in $R$. (Here, $\mathcal{A}$ is the $\sigma$-algebra of Borel sets $B$ in $R$.) We know that a process such as this is called a **diffusion process**. Also, for $X(t_i) \equiv X_i, i = 1, \ldots, n$, let $P(X_1, \ldots, X_n; \theta)$ be the joint probability of observing the path when $\theta$ is the true parameter, where it is assumed that $X_i \in B$ for all $i = 1, \ldots, n$ and all Borel sets $B \in R$. From the definition of conditional probability,

$$P(X_1, \ldots, X_n; \theta) = P(X_n | X_{n-1}, \ldots, X_n; \theta) \cdot P(X_1, \ldots, X_{n-1}; \theta). \tag{8.C.1}$$

With $X$ as Markov process, let us write the **Markov property** in terms of transition probabilities as $P(X_n | X_{n-1}, \ldots, X_1; \theta) = P(X_n | X_{n-1}; \theta)$ (once the present is known, the past and future are independent). Hence, (8.C.1) simplifies to

$$P(X_1, \ldots, X_n; \theta) = P(X_n | X_{n-1}; \theta) \cdot P(X_1, \ldots, X_{n-1}; \theta). \tag{8.C.2}$$

Applying the Markov property to the term $P(X_1, \ldots, X_{n-1}; \theta)$ in (8.C.2) yields

$$P(X_1, \ldots, X_{n-1}; \theta) = P(X_{n-1} | X_{n-2}; \theta) \cdot P(X_1, \ldots, X_{n-2}; \theta). \tag{8.C.3}$$

Then inserting (8.C.3) into (8.C.2) gives

$$P(X_1, \ldots, X_n; \theta) = P(X_n | X_{n-1}; \theta) \cdot P(X_{n-1} | X_{n-2}; \theta) \cdot P(X_1, \ldots, X_{n-3}; \theta). \tag{8.C.4}$$

If we repeat this process of applying the Markov property $n$ times in succession, we ultimately obtain

$$P(X_1, \ldots, X_n; \theta) = P(X_n | X_{n-1}; \theta) \cdot P(X_{n-1} | X_{n-2}; \theta) \cdots P(X_2 | X_1; \theta) \cdot P(X_0; \theta). \tag{8.C.5}$$

We know that for $i = 1, \ldots, n$, every joint probability $P(X_1, \ldots, X_n; \theta)$ of the stochastic process $X = \{X_t\}_{t \in [t_0, T]}$ has a probability density $p(X_1, \ldots, X_n; \theta)$. Hence, we can express the Markov property in terms of the transition density function as $p(X_n|X_1,...,X_{n-1};\theta) = p(X_n|X_{n-1};\theta)$. From the definition of the conditional density function, we can write the joint density as

$$p(X_1,...,X_n;\theta) = p(X_n|X_{n-1},...,X_1;\theta) \cdot p(X_1,...,X_{n-1};\theta)$$
$$= p(X_n|X_{n-1};\theta) \cdot p(X_1,...,X_n;\theta)$$

via the Markov property applied to transition densities. (The reader can now see where all this is going.) By continuing this process we ultimately obtain

$$p(X_1,...,X_n;\theta) = p(X_n|X_{n-1};\theta) \cdot p(X_{n-1}|X_{n-2};\theta) \cdots p(X_2|X_1;\theta) \cdot p(X_0;\theta),$$
(8.C.6)

where $p(X_0;\theta)$ is the density of the initial state. Thus, the conditional densities on the right-hand side of (8.C.6) are the transition probability densities of the Markov process.

With the $X_i, i = 1,...,n$, independent and identically distributed, their joint probability density function has the form

$$L(X_1,...,X_n;\theta,n) = p(X_0;\theta)\prod_{i=1}^{n}p(X_i|X_{i-1};\theta).$$

If we now treat the $X_i$s as realizations of the sample random variables and $\theta$ is deemed variable, then the likelihood function of the sample is

$$\mathcal{L}(\theta;x_1,...,x_n,n) = p(x_0;\theta)\prod_{i=1}^{n}p(x_i|x_{i-1};\theta).$$
(8.C.7)

We noted in Sections 3.5.2 and 5.3.2 that a continuous-time Markov process is homogeneous (with respect to time) if all of its transition densities depend only on the time difference $t-s$ rather than on specific values of $s$ and $t$, that is, $p(x,s;t,y) = p(t-s,x,y), 0 \le s \le t$. For instance, a standard Brownian motion (SBM) or Wiener process $\{W_t\}_{t\ge 0}$ is a homogeneous Markov process with transition density

$$p(x,s;t,y) = \frac{1}{\sqrt{2\pi(t-s)}}e^{-\frac{1}{2}(y-x)^2/(t-s)}, 0 \le s \le t,$$

given that $\{W_t\}_{t\ge 0}$ is Gaussian with independent increments for which $W_0 = 0$ a.s., $E(W_t) = 0$, and $V(W_t) = t-s, 0 \le s \le t$.

In terms of our immediately preceding notation, we can write the transition density as $p(x,t_{i-1},t_i,y) = p(\Delta_i,x,y)$. If $\Delta_i = t_i - t_{i-1} = \Delta =$ constant, then this density is written as $p(\Delta,x,y)$ or, for our purposes, as $p(\Delta,X_{i-1},X_i)$. Under this convention, (8.C.7) becomes

$$\mathcal{L}(\theta, x_1, \ldots, x_n, n) = p(x_0; \theta) \prod_{i=1}^{n} p(\Delta, x_{i-1}, x_i; \theta) \qquad (8.C.8)$$

or, more generally, Equation (8.16).

## Appendix 8.D   Change of Variable

Suppose we are interested in obtaining the probability density function of a continuous random variable $Y$ given that we have the probability density function of a random variable $X$ and a function $y = h(x)$ connecting the variables $X$ and $Y$. Suppose further that $h(x)$ is differentiable and either increasing or decreasing for all values within the range of $X$ so that the inverse function $x = h^{-1}(y)$ exists for all of the corresponding values of $Y$. Furthermore, $h^{-1}$ is taken to be differentiable except when $h'(x) = 0$. Under these conditions, we state

**Theorem 8.D.1**   If the probability density function of the random variable $X$ is given by $p(x)$ and $y = h(x)$ is differentiable and either increasing or decreasing for all values within the range of $X$ for which $p(x) \neq 0$, then the probability density function of $Y$ is given by

$$p(y) = p(x) \cdot \left| \frac{dx}{dy} \right|$$

$$= p(h^{-1}(y)) \cdot \left| \frac{dx}{dy} \right|, \frac{dx}{dy} \neq 0, \qquad (8.D.1)$$

where $dx/dy$ is the derivative of the inverse function $x = h^{-1}(y)$.

**Example 8.D.1**   Let the probability density function of the random variable $X$ be specified as

$$p(x) = \begin{cases} e^{-x} & , x > 0; \\ 0 & , x \leq 0 \end{cases}$$

and let the random variable $Y$ assume values that are given by $y = h(x) = +x^{1/2}$. Clearly, $x = h^{-1}(y) = y^2$. Since $dx/dy = 2y$, we obtain, from (8.D.1),

$$p(y) = \begin{cases} e^{-y^2} \cdot |2y| = 2ye^{-y^2} & , y > 0; \\ 0 & , y \leq 0. \end{cases} \blacksquare$$

Next, let us consider the transformation from two continuous random variables to two new random variables. More specifically, suppose we are given two continuous random variables $Y_1$ and $Y_2$ with joint probability density function $p(y_1, y_2)$. Moreover, these random variables are taken to be related (on a one-to-one basis) to the two random variables $X_1$ and $X_2$ by the known functions $x_1 = u_1(y_1, y_2)$ and $x_2 = u_2(y_1, y_2)$ that have unique (single-valued) inverses $y_1 = v_1(x_1, x_2)$ and $y_2 = v_2(x_1, x_2)$. If we denote the joint probability density function of the random variables $X_1$ and $X_2$ by $p(x_1, x_2)$, then

$$p(x_1, x_2) = p(y_1, y_2)|J|$$

$$= p(v_1(x_1, x_2), v_2(x_1, x_2))|J|,$$

(8.D.2)

where $|J|$ is the absolute value of the **Jacobian determinant**

$$J = \left| \frac{\partial(y_1, y_2)}{\partial(x_1, x_2)} \right| = \begin{vmatrix} \dfrac{\partial y_1}{\partial x_1} & \dfrac{\partial y_1}{\partial x_2} \\ \dfrac{\partial y_2}{\partial x_1} & \dfrac{\partial y_2}{\partial x_2} \end{vmatrix}.$$

(8.D.3)

**Example 8.D.2**  Let

$$p(y_1, y_2) = e^{-(y_1 + y_2)}, 0 \le y_1, y_2 < +\infty,$$

with

$$x_1 = u_1(y_1, y_2) = y_1 + y_2 \quad \text{and} \quad x_2 = u_2(y_1, y_2) = \frac{y_1}{y_2}$$

or, after a bit of algebra,

$$y_1 = v_1(x_1, x_2) = x_1 x_2 (1 + x_2)^{-1} \quad \text{and} \quad y_2 = v_2(x_1, x_2) = x_1 (1 + x_2)^{-1}.$$

Then, from (8.D.3),

$$J = \begin{vmatrix} \dfrac{x_2}{1 + x_2} & x_1(1 + x_2)^{-1}\left[1 - x_2(1 + x_2)^{-1}\right] \\ \dfrac{1}{1 + x_2} & -x(1 + x_2)^{-2} \end{vmatrix} = -\frac{x_1}{(1 + x_2)^2},$$

and thus, from (8.D.2),

$$p(x_1, x_2) = e^{-x_1}\left[ \frac{x_1}{(1 + x_2)^2} \right], 0 \le x_1, x_2 < +\infty. \blacksquare$$

In general, we can now turn to the transformation from $n$ continuous random variables to $n$ new random variables. That is, suppose we have $n$ continuous random variables $y_1, y_2, \ldots, y_n$ with joint probability density function $p(y_1, y_2, \ldots, y_n)$.

As required, these random variables are related (in a one-to-one fashion) to the $n$ random variables $x_1, x_2, ..., x_n$ by the known functions

$$x_i = u_i(y_1, y_2, ..., y_n), i = 1, ..., n,$$

which have unique (single-valued) inverses

$$y_i = v_i(x_1, x_2, ..., x_n), i = 1, ..., n.$$

Let $p(x_1, x_2, ..., x_n)$ represent the joint probability density function of the random variables $X_i, i = 1, ..., n$. Then

$$p(x_1, x_2, ..., x_n) = p(y_1, y_2, ..., y_n)|J|$$

$$= p(v_1(x_1, x_2, ..., x_n), ..., v_n(x_1, x_2, ..., x_n))|J|, \qquad (8.D.4)$$

where $|J|$ is the absolute value of the **$n$th-order Jacobian determinant**

$$J = \left| \frac{\partial(y_1, y_2, ..., y_n)}{\partial(x_1, x_2, ..., x_n)} \right| = \begin{vmatrix} \dfrac{\partial y_1}{\partial x_1} \dfrac{\partial y_1}{\partial x_2} \cdots \dfrac{\partial y_1}{\partial x_n} \\ \dfrac{\partial y_2}{\partial x_1} \dfrac{\partial y_2}{\partial x_2} \cdots \dfrac{\partial y_2}{\partial x_n} \\ \vdots \quad \vdots \qquad \vdots \\ \dfrac{\partial y_n}{\partial x_1} \dfrac{\partial y_n}{\partial x_2} \cdots \dfrac{\partial y_n}{\partial x_n} \end{vmatrix}. \qquad (8.D.5)$$

Now, how does all this relate to the structure of Equation (8.30)? Going back to Equation (8.D.4), we can write

$$ln\, p(x_1, x_2, ..., x_n) = ln\, p(y_1, y_2, ..., y_n) + ln|J|.$$

Since $y_i = ln\, x_i, i = 1, ..., n$, in Example 8.5

$$J = \begin{vmatrix} \dfrac{1}{x_1} & 0 & \cdots & 0 \\ 0 & \dfrac{1}{x_2} & \cdots & 0 \\ \vdots & \vdots & & \vdots \\ 0 & 0 & \cdots & \dfrac{1}{x_n} \end{vmatrix} = \prod_{i=1}^{n} \left( \frac{1}{x_i} \right)$$

since the determinant of a diagonal matrix equals the product of the elements on the main diagonal. Then,

$$ln|J| = ln \prod_{i-1}^{n} \left( \frac{1}{x_i} \right) = \sum_{i=1}^{n} ln \left( \frac{1}{x_i} \right)$$

as required.

# Appendix A

# A Review of Some Fundamental Calculus Concepts

## A.1 Limit of a Function

Let $y$ be a real-valued function of $x$ and written $y = f(x)$, with $f$ defined at all points $x$ within some interval about point $x_0$. $f$ is said to approach a **limit** $L$ as $x \to x_0$ if

$$\lim_{x \to x_0} f(x) = L. \tag{A.1}$$

Under what conditions does $L$ exist? Let us denote the **left-hand limit** of $x$ at $x_0$ (the limit of $f$ as $x \to x_0$ from below) as

$$\lim_{x \to x_0^-} f(x);$$

and the **right-hand limit** of $f$ at $x_0$ (the limit of $f$ as $x \to x_0$ from above) as

$$\lim_{x \to x_0^+} f(x).$$

Now, if $\lim_{x \to x_0^-} f(x) = \lim_{x \to x_0^+} f(x) = L$, then the limit $L$ exists.[1]

## A.2 Continuity

If $L = f(x_0)$ in (A.2), then $f$ is said to be **continuous at a point** $x = x_0$, that is,

$$\lim_{x \to x_0} f(x) = f(x_0).$$

---

1 A function $y = f(x)$ is said to be **bounded** on a set $A$ if there is a number $M$ such that $|f(x)| \le M$ for all $x \in A$. $f$ is **monotone increasing** (resp. **monotone decreasing**) on $A$ if $f(x_1) \le f(x_2)$ (resp. $f(x_1) \ge f(x_2)$) for $x_1 < x_2$ and $x_1, x_2 \in A$. If $f$ is bounded and monotone increasing on $(a, b) = \{x | a < x < b\}$, then the limit $\lim_{x \to x_0^-} f(x)$ exists. If $f$ is bounded and monotone decreasing on $(a, b)$, then $\lim_{x \to x_0^+} f(x)$ exists.

*Stochastic Differential Equations: An Introduction with Applications in Population Dynamics Modeling*, First Edition. Michael J. Panik.
© 2017 John Wiley & Sons, Inc. Published 2017 by John Wiley & Sons, Inc.

Furthermore, a function $f$ is termed **left-continuous** at $x_0$ if

$$\lim_{x \to x_0-} f(x) = f(x_0)$$

($f(x) \to f(x_0)$ as $x \to x_0$ from below); and **right-continuous** at $x_0$ if

$$\lim_{x \to x_0+} f(x) = f(x_0)$$

($f(x) \to f(x_0)$ as $x \to x_0$ from above). So if $f$ is continuous at $x_0$, then it is both left-continuous and right-continuous at that point.

Looked at in another fashion, a real-valued function $y = f(x)$ is continuous at a point $x = x_0$ if the increment in $f$ over a "small" interval is "small," that is,

$$\Delta f(x) = f(x) - f(x_0) \to 0 \text{ as } \Delta x = x - x_0 \to 0$$

or, for $h = x - x_0$,

$$\Delta f(x) = f(x_0 + h) - f(x_0) \to 0 \text{ as } h \to 0.$$

$f$ is said to be **continuous** if it is continuous at each point of its domain.

A function is termed **regular** if it has both left- and right-hand limits at any point of its domain, and has one-sided limits at its boundary. **A regular right-continuous function** (RRC) is one that is right-continuous with left-handed limits.

A function may fail to be continuous at $x = x_0$ if either $f(x)$ does not approach any limit as $x \to x_0$, or because it approaches a limit that is different from $f(x_0)$. In this regard, a **discontinuity** refers to a point at which $f$ is defined but not continuous, or to a point at which $f$ is not defined (e.g., $f(x) = x^{-1}$ is discontinuous at $x = 0$). So if $f(x_0)$ is defined and $L = \lim_{x \to x_0} f(x)$ exists but $L \neq f(x_0)$, then $f$ has a discontinuity at $x_0$. When $\lim_{x \to x_0} f(x)$ does not exist, $x_0$ is termed an **essential discontinuity** for $f$.

## A.3  Differentiable Functions

A function $y = f(x), x \in (a, b)$, is said to be **differentiable at a point** $x = x_0 \in (a, b)$ if $f$ is defined on a neighborhood of $x_0$ and has a derivative at $x_0$ defined by

$$\lim_{\Delta x \to 0} \frac{\Delta y}{\Delta x} = \lim_{x \to x_0} \frac{f(x) - f(x_0)}{x - x_0} = \lim_{h \to 0} \frac{f(x_0 + h) - f(x_0)}{h}$$

$$= f'(x_0) \left( \text{or } = \frac{dy}{dx} \bigg|_{x = x_0} \right). \tag{A.2}$$

Under what conditions does $f'(x_0)$ exist? Let us denote the **left-hand derivative** of $f$ at $x_0$ as

$$\lim_{x \to x_0-} \frac{f(x)-f(x_0)}{x-x_0} = f'_-(x_0);$$

the **right-hand derivative** of $f$ at $x_0$ is represented as

$$\lim_{x \to x_0+} \frac{f(x_0)-f(x_0)}{x-x_0} = f'_+(x_0).$$

Then if $f$ is differentiable at $x_0$, we must have $f'_-(x_0) = f'_+(x_0)$. Note: if $f$ is differentiable at $x_0$, then it is also continuous there; but if $f$ is continuous at $x_0$, this does not necessarily imply that $f$ is differentiable at that point. $f$ is said to be **differentiable** if it is differentiable at each point of its domain.

## A.4  Differentials

Let the real-valued function $y = f(x)$ be differentiable at $x = x_0$. If $dx$ is taken to be an independent variable whose value is arbitrary, then the expression

$$dy = f'(x_0)dx \tag{A.3}$$

is called the **differential of $f$ at $x_0$**, with $dy$ serving as the dependent variable. Thus, the differential of $f$ is a homogeneous linear function of the independent variable $dx$. So for $x$ held fixed, $dy$ is the value of the differential for any chosen $dx$. (We can also regard the differential as a function of $x$ in that it is defined at each point where the derivative of $f$ exists.)

Next, suppose that the real-valued function of $x = g(t)$ is defined for $\alpha < t < \beta$, and that $a < g(t) < b$. Also, let $y = f(x)$ be defined for $a < x < b$. Then if we replace $x$ by $g(t)$ in $f$, we obtain the **composite function** $y = F(t) = f(g(t))$. How do we differentiate a composite function? The answer is provided by Theorem A.1.

**Theorem A.1**  Composite Function (Chain) Rule
Suppose $g$ is differentiable at the point $t_0 \in \{t | \alpha < t < \beta\}$, with $x_0 = g(t_0)$. Suppose also that $f$ is differentiable at $x_0$. Then the composite function $F(t) = f(g(t))$ is differentiable at $t_0$, and

$$F'(t_0) = f'(x_0)g'(t_0).$$

Given this result, we can readily specify the differential of a composite function as follows. If $y = f(x)$ and $x = g(t)$ so that $y = f(g(t)) = F(t)$, then

$$dy = F'(t)dt = f'(x)g'(t)dt \,(= f'(x)dx). \tag{A.4}$$

(Note: in (A.4), $t$ is the independent variable; $x$ and $dx$ are not independent—as they were in (A.3).)

Looking to the differential of a multivariate function, suppose $v = f(x, y)$, where $x, y \in R^2$. Think of the **total differential of** $f$, $df$, as a linear combination of $dx$ and $dy$, where $dx$ and $dy$ are independent variables. In practice, we write

$$df = f_x\, dx + f_y\, dy, \tag{A.5}$$

where $f_x$ and $f_y$ are assumed to exist at the point $(x, y)$. (We can also regard $df$ as a function of $(x, y)$ if $(x, y)$ is a point at which $f$ is differentiable.) However, the existence of the partial derivatives at $(x, y)$ is not sufficient to guarantee the existence of the total differential $df$. Indeed, there are functions for which $f_x$ and $f_y$ both exist at a point $(x, y)$, and yet $df$ does not exist. However, if $f_x$ and $f_y$ are defined throughout a neighborhood of the point $(x_0, y_0)$, and are continuous at that point, then $df$ exists (or $f$ is differentiable) at $(x_0, y_0)$. In addition, at each point where $df$ does exist, it is defined for all values of $dx$ and $dy$. In general, if $v = f(x, y)$ has continuous first partial derivatives in a set $A$ (the domain of $f$), then $f$ has a total differential $df = f_x\, dx + f_y\, dy$ at every point $(x, y)$ of $A$.

An important application of the derivative concept is Theorem A.2, the *law of the mean*.

**Theorem A.2**   Mean Value Theorem for Derivatives
Suppose a real-valued function $y = f(x)$ is continuous over $[a, b] = \{x \mid a \le x \le b\}$ and differentiable over $(a, b)$. Then there is a point $x = x_0 \in (a, b)$ such that

$$f(b) - f(a) = f'(x_0)(b - a). \tag{A.6}$$

## A.5   Integration

### A.5.1   Definite Integral

Suppose a function $y = f(x)$ is defined and continuous over $[a, b]$. The **definite integral** of $f$ over $[a, b]$ is the number

$$\int_a^b f(x)\, dx,$$

with $f(x)$ termed the **integrand**. How is the value of this number determined? To answer this question, we shall perform the following stepwise construction:

1. Select an integer $n \ge 1$ and subdivide $[a, b]$ into $n$ subintervals by picking the points $x_0, x_1, ..., x_n$, with $a = x_0 < x_1 < \cdots < x_n = b$, and setting $\Delta x_i = x_i - x_{i-1}, i = 1, ..., n$. (Here, we have formed what is commonly called a **partition** $P = \{a = x_0, x_1, ..., x_n = b\}$ of $[a, b]$.)
2. Within the $i$th subinterval $\Delta x_i$, select an arbitrary point $x_i'$ such that $x_{i-1} \le x_i' \le x_i, i = 1, ..., n$ (Figure A.1a).

(a)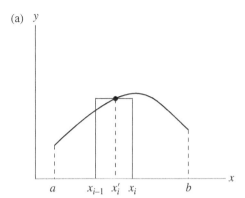

(b)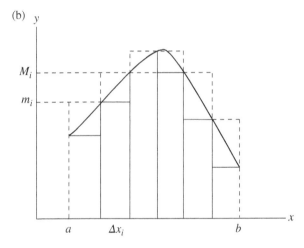

**Figure A.1** (a) An approximating rectangle. (b) Largest and smallest values of $f$ in $\Delta x_i$.

3. Determine the value of $f(x)$ at each of the $x_i'$s and form the **approximating sum**

$$f(x_1')\Delta x_1 + f(x_2')\Delta x_2 + \cdots + f(x_n')\Delta x_n.$$

4. Take the limit of the approximating sum as $n$ increases without bound and the largest of the $\Delta x_i$s approaches zero, that is, find

$$L = \int_a^b f(x)\, dx = \lim_{|P| \to 0} \sum_{i=1}^n f(x_i')\, \Delta x_i, \tag{A.7}$$

where the fineness or **mesh** of $P$ is $|P| = max\{\Delta x_i, i = 1, ..., n\}$. When the indicated limit exists, the area under $f(x)$ from $a$ to $b$ is $L$. The process of determining this limit is termed **Riemann integration,** and the actual limit is called the **definite integral of $f$ from $a$ to $b$.**

Under what conditions is a function integrable over $(a, b)$? The answer depends upon the behavior of the limit in (A.7). Let $m_i$ be the smallest value (a greatest lower bound) of $f(x)$ in the subinterval $\Delta x_i$. If $x_i'$ is chosen as the point at which $f(x)$ equals $m_i$ for each $\Delta x_i$, then we can define the **lower approximating sum** or **lower Darboux sum** as

$$s = \sum_{i=1}^{n} m_i \Delta x_i.$$

In similar fashion, for $M_i$ the largest value (or least upper bound) of $f(x)$ in $\Delta x_i$, and $x_i'$ is selected as the point at which $f(x)$ equals $M_i$ for each $\Delta x_i$, then the **upper approximating sum** or **upper Darboux sum** is

$$S = \sum_{i=1}^{n} M_i \Delta x_i.$$

Clearly, $m_i \leq f\left(x_i'\right) \leq M_i, i = 1,\dots,n$, and thus $s \leq \sum_{i=1}^{n} f\left(x_i'\right) \Delta x_i \leq S$ (Figure A.1b). Now, if we can demonstrate that both $s$ and $S$ approach the same limit $L$ as $|P| \to 0$, then $\sum_{i=1}^{n} f\left(x_i'\right) \Delta x_i \to L$.

To this end, let $m$ and $M$ denote, respectively, the greatest lower bound and least upper bound of $f$ on $(a, b)$. Then $m \leq m_i$ and $M_i \leq M$. With $s$ and $S$ contained between fixed bounds, we can conclude that $s$ has a least upper bound $I$ (called the **lower integral of $f$ over $(a, b)$**), and $S$ has a greatest lower bound $J$ (termed **upper integral of $f$ over $(a, b)$**), or $s \leq I \leq J \leq S$. In fact, $I$ and $J$ always exist if $f$ is bounded on $(a, b)$. Moreover, if we consider $s$ and $S$ for all possible choices of subintervals $\Delta x_i$ over $(a, b)$, then

$$\lim_{|P| \to 0} s = I \text{ and } \lim_{|P| \to 0} S = J.$$

Hence, our answer to the preceding question pertaining to the integrability of $f$ over $(a, b)$ is provided by Theorem A.3.

**Theorem A.3** A bounded function $f(x)$ is integrable over $(a, b)$ if and only if its lower and upper integrals are equal or

$$\int_{a}^{b} f(x)\, dx = I = J.$$

We also note briefly the following

i. A function $f(x)$ is integrable over $[a, b]$ if it is continuous over this interval.
ii. A bounded function is integrable in any interval in which it is monotonic (e.g., $f(x_1) \leq f(x_2)$ if $x_1 < x_2$).

iii. A bounded function is integrable in any interval in which it has only a finite number of discontinuities.

iv. A bounded function with an infinite number of discontinuities is integrable if and only if its points of discontinuity form a **set of measure zero**. (A set of points has a zero measure if it is possible to enclose them in a countable set of intervals whose total length is arbitrarily small.)

v. If $f(x)$ and $g(x)$ are integrable on $[a, b]$, $a < b$, and $f(x) \leq g(x)$, then

$$\int_a^b f(x)\, dx \leq \int_a^b g(x)\, dx.$$

vi. If $f(x)$ is integrable in $(a, b)$, $a < b$, then so is $|f(x)|$ and

$$\left| \int_a^b f(x)\, dx \right| \leq \int_a^b |f(x)|\, dx.$$

### A.5.2 Variable Limits of Integration

Let the real-valued function $f(t)$ be integrable over $[a, b]$ and let $x \in [a, b]$. Then an integral can be expressed as a function of its upper limit as

$$F(x) = \int_a^x f(t)\, dt. \tag{A.8}$$

Here, $F(x)$ is differentiable and continuous over any interval $(a, b)$ in which $f(x)$ is integrable. Looking to the derivative of $F(x)$ we have Theorem A.4, an existence theorem.

**Theorem A.4**   If $f(x)$ is integrable in $(a, b)$, then

$$\frac{dF}{dx} = \frac{d}{dx} \int_a^b f(t)\, dt = f(x), \quad x \in [a, b], \tag{A.9}$$

for all $x$ where $f(x)$ is continuous. (Similarly,

$$\frac{d}{dx} \int_x^b f(t)\, dt = -f(x).)$$

Note: this theorem legitimizes the existence of a function $F(x)$ whose derivative is a given function $f(x)$ at each of its points of continuity.

The importance of Theorem A.4 is the following. We found that if $f(x)$ is continuous in $(a, b)$, then $dF/dx = f(x)$ for all $x \in (a, b)$. Hence, the differential equation $dF = f(x)\, dx$ has the general solution $y = F(x) + C$, where $C$ is an *arbitrary* constant and $F(x) = \int_a^x f(t)\, dt$ for fixed $a$ and $x$ a member of $[a, b]$, that is, (A.9) involves all functions for which $F'(x) = f(x)$. Here, $F(x)$ is termed the **indefinite integral** of $f(x)$.

We next have Theorem A.5, which provides us with a systematic process or standard for evaluating definite integrals.

**Theorem A.5**  Fundamental Theorem of Integral Calculus
If $f(x)$ is integrable on $(a, b)$ and $F(x)$ is any function having $f(x)$ as a derivative, then

$$\int_a^b f(x)\,dx = F(x)\big|_a^b = F(b) - F(a). \tag{A.10}$$

Finally, Theorem A.6 offers a procedure for transforming definite integrals.

**Theorem A.6**  Given that

i. $f(x)$ is continuous on $[a, b]$;
ii. $\varnothing(t)$ and $\varnothing'(t)$ are continuous and $\varnothing'(t) \in [a, b]$ whenever $t \in [\alpha, \beta]$; and
iii. $\varnothing(\alpha) = a$ and $\varnothing(\beta) = b$,

it follows that, for $g(t) = f(\varnothing(t))\varnothing'(t)$,

$$\int_a^b f(x)\,dx = \int_\alpha^\beta g(t) = \int_\alpha^\beta f(\varnothing(t)\,\varnothing'(t))\,dt. \tag{A.11}$$

### A.5.3  Parametric Integrals

Suppose we have a situation in which a parameter $\theta$ is present: (1) under the integral sign ($\theta$ is an argument of the integrand $f$); or (2) as a limit of integration; or (3) $\theta$ possibly appears in both places.

#### A.5.3.1  Parameter in the Integrand

Suppose the real-valued function $f$ is defined at a point $(x, \theta)$ in the rectangular region $\mathcal{K} = \{(x, \theta)|a \le x \le b, c \le \theta \le d\}$, and let the integral appear as

$$F(\theta) = \int_a^b f(x, \theta)\,dx.$$

What properties of $F$ can be gleaned from those of $f$? Specifically, if $f$ is continuous at each point of $\mathcal{K}$, then $F$ is continuous for each $\theta \in [c, d]$. In addition, if $f(x, \theta)$ is integrable with respect to $x$ for each value of $\theta$, and $\partial f/\partial \theta$ exists and is continuous in $x$ and $\theta$ throughout $\mathcal{K}$, then $F$ is differentiable with respect to $\theta$ and

$$F'(\theta) = \int_a^b \left(\frac{\partial f}{\partial \theta}\right)\,dx. \tag{A.12}$$

### A.5.3.2 Parameter in the Limits of Integration

Suppose

$$G(u, v) = \int_u^v f(x)\, dx.$$

If $u = u(\theta)$ and $v = v(\theta)$, then

$$G(u(\theta), v(\theta)) = \int_{u(\theta)}^{v(\theta)} f(x)\, dx = F(\theta).$$

Then from the chain rule and Theorem (A.4),

$$
\begin{aligned}
F'(\theta) &= \frac{\partial G}{\partial u}\frac{\partial u}{\partial \theta} + \frac{\partial G}{\partial v}\frac{\partial v}{\partial \theta} \\
&= -f(u)\frac{\partial u}{\partial \theta} + f(v)\frac{\partial v}{\partial \theta}.
\end{aligned}
\tag{A.13}
$$

### A.5.3.3 Parameter in the Integrand and in the Limits of Integration

Let

$$F(\theta) = \int_a^u f(x, \theta)\, dx.$$

For $u = u(\theta)$, let

$$G(u, v) = \int_a^{u(\theta)} f(x, v)\, dx,$$

where $u = u(\theta)$ and $v = \theta$. Then, via the chain rule and Theorem (A.4),

$$\frac{\partial G}{\partial u} = f(u, v), \quad \frac{\partial G}{\partial v} = \int_a^u \left( \frac{\partial f(x, v)}{\partial v} \right) dx,$$

and

$$
\begin{aligned}
F'(\theta) &= \frac{\partial G}{\partial u}\frac{\partial u}{\partial \theta} + \frac{\partial G}{\partial v}\frac{\partial v}{\partial \theta} \\
&= f(u, v)\frac{\partial u}{\partial \theta} + \int_a^u \left( \frac{\partial f(x, v)}{\partial v} \right) dx,
\end{aligned}
\tag{A.14}
$$

where $\partial v / \partial \theta = 1$.

### A.5.4 Riemann–Stieltjes Integral

As we shall now see, the Riemann–Stieltjes (RS) integral is a generalization of the ordinary Riemann integral. Suppose we have two real-valued functions $f(x)$

and $g(x)$, both defined on $[a, b]$. Then the **RS integral of $f$ with respect to $g$** can be expressed as

$$\int_a^b f(x)\, dg(x).\qquad\qquad\text{(A.15)}$$

Here, $f$ is the **integrand** and $g$ is termed the **integrator**. In fact, if $g(x) = x$, then (A.15) becomes the Riemann integral (A.7). How is the RS integral defined? To answer this question, let us form a partition $P$ of $[a, b]$ as $\{a = x_0, x_1, ..., x_n = b\}$. This partition of $[a, b]$ renders $n$ subintervals $\Delta x_i = x_i - x_{i-1}, i = 1, ..., n$, where the fineness or **mesh** of $P$ is $|P| = max\{\Delta x_i, i = 1, ..., n\}$. In addition, select a set of $n$ points $\{x_1', x_2', ..., x_n'\}$ and place one of them in each of the aforementioned subintervals. Next, define

$$\int_a^b f(x)\, dx = \lim_{|P|\to 0} \sum_{i=1}^n f(x_i')[g(x_i) - g(x_{i-1})]\qquad\qquad\text{(A.16)}$$

provided that a unique limit exists. (Note: given $f$, if $g$ is constant on $[a, b]$, then $g(x_i) - g(x_{i-1}) = 0$ for all $i$ and thus (A.16) is zero.)

To address the issue of the existence of the limit in (A.16), let us employ the following generalizations of the lower and upper approximating sums (Darboux sums). Given a partition $P$ and, say, a nondecreasing function $g$ on $[a, b]$, define the **upper Darboux sum of $f$ with respect to $g$** as

$$U = \sum_{i=1}^n \sup_{x_i' \in \Delta x_i} f(x_i')[g(x_i) - g(x_{i-1})].$$

In addition, define the **lower Darboux sum of $f$ with respect to $g$** as

$$L = \sum_{i=1}^n \inf_{x_i' \in \Delta x_i} f(x_i')[g(x_i) - g(x_{i-1})].$$

Then $f$ is said to be **RS integrable** with respect to $g$ if

$$\lim_{|P|\to 0} (U - L) = 0.$$

Under what conditions on $f$ and $g$ will the immediately preceding limit (or the limit in (A.16)) exist? Before answering this question, we need to consider the following definition pertaining to the function $g$. For a given partition $P$ of $[a, b]$, the **variation** of $g$ on $[a, b]$ is specified as

$$V_g(P) = \sum_{i=1}^n |g(x_i) - g(x_{i-1})|.$$

Then the **total variation** of $g$ on $[a, b]$ is defined as

$$V_g([a, b]) = \sup_P \{V_g(P)\}$$

(i.e., we are taking the least upper bound of the set of all sums $V_g(P)$) or

$$V_g([a,b]) = \lim_{|P| \to 0} V_g(P).$$

If $V_g([a,b]) < +\infty$, then $g$ is said to be of **finite (bounded) variation** on $[a, b]$. Given this concept of bounded variation, we now look to Theorem A.7.

**Theorem A.7**   Existence Theorem for RS Integrals
For real-valued functions $f(x)$ and $g(x)$ defined on $[a, b]$, if $f$ is continuous and $g$ is of bounded variation on $[a, b]$, then the RS integral exists.

Note: a real-valued function that is monotone on a given interval satisfies the condition of being of bounded variation on that interval, for example, for $g$ monotone on $[a, b]$, $V_g([a, b]) = |g(a) - g(b)|$. So for the existence of the RS integral, it is sufficient to define the integral with respect to monotone functions. To summarize, given that $f$ is continuous on $[a, b]$, conditions on $g$ that are sufficient to guarantee the existence of the RS integral are the following:

i. $g$ is nondecreasing on $[a, b]$.
ii. $g$ is nonincreasing on $[a, b]$.
iii. $g$ is the sum of a nondecreasing function and a nonincreasing function on $[a, b]$.

A few of the important features of the RS integral are the following:

1. If $f$ is Riemann integrable and $g$ has a continuous derivative, then

$$\underbrace{\int_s^b f(x)\, dg(x)}_{\text{(RS integral)}} = \underbrace{\int_a^b f(x)\, g'(x)\, dx}_{\text{(Riemann integral)}}.$$

2. The RS integral allows integration by parts, or

$$\int_a^b f(x)\, dg(x) = f(b)g(b) - f(a)g(a) - \int_a^b g(x)\, df(x).$$

3. Suppose $f$ is continuous and $g$ is discontinuous with a finite number of discontinuities at which $g$'s function value "jumps" and then remains constant in the open intervals between the points of discontinuity. To see this, suppose $[a, b]$ is divided into $n$ parts by the partition $\{a = x_0, x_1, ..., x_n = b\}$ and let $g(x)$ assume the constant value $c_i$ in the *interior* of the $i$th subinterval (Figure A.2). For $f$ continuous on $[a, b]$, it can be demonstrated that

$$\int_a^b f(x)\, dg(x) = f(x_0)[c_1 - g(x_0)] + f(x_1)(c_2 - c_1)$$

$$+ \cdots + f(x_{n-1})(c_n - c_{n-1}) + f(x_n)[g(x_n) - c_n].$$

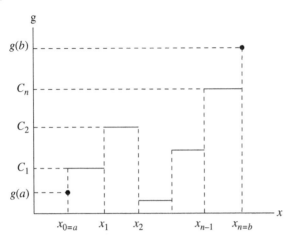

**Figure A.2** The function $g$ exhibits a finite number of discontinuities.

## A.6 Taylor's Formula

### A.6.1 A Single Independent Variable

**Theorem A.8**  Taylor's Theorem
Let the real-valued function $y = f(x)$ and its first $n + 1$ $(n \geq 0)$ derivatives be continuous throughout a closed interval $[a, b]$ containing the point $x_0$. Then the value of $f$ at any point $x$ near $x_0$ is

$$f(x) = f(x_0) + f'(x_0)(x - x_0) + \frac{1}{2!}f''(x_0)(x - x_0)^2$$

$$+ \cdots + \frac{1}{n!}f^{(n)}(x_0)(x - x_0)^n + R_{n+1}, \tag{A.17}$$

where the remainder term $R_{n+1}$ has the form

$$R_{n+1} = \frac{1}{n!}\int_{x_0}^{x} (x - t)^n f^{(n+1)}(t)\, dt. \tag{A.18}$$

Equation (A.17) is known as **Taylor's formula with integral remainder**. While the remainder term can assume various forms, we shall employ **Lagrange's form of the remainder,** or

$$R_{n+1} = \frac{1}{(n+1)!}f^{(n+1)}(\xi)(x - x_0)^{n+1}, x_0 < \xi < x.^2 \tag{A.19}$$

---

2 This result is a consequence of the mean value theorem for integrals: let $g(t)$, $h(t)$ be continuous on $[\alpha, \beta]$ with $h(t) \geq 0$ for all $t \in [\alpha, \beta]$. Then for some $t = \xi$ such that $\alpha < \xi < \beta$,

$$\int_{\alpha}^{\beta} g(t)h(t)\, dt = g(\xi)\int_{\alpha}^{\beta} h(t)\, dt.$$

Then combining (A.17) and (A.19) renders **Taylor's formula with Lagrange's form of the remainder,** or

$$f(x) = f(x_0) + f'(x_0)(x - x_0) + \frac{1}{2!}f''(x_0)(x - x_0)^2$$

$$+ \cdots + \frac{1}{n!}f^{(n)}(x_0)(x - x_0)^n + \frac{1}{(n+1)!}f^{(n+1)}(\xi)(x - x_0)^{n+1}, \; x_0 < \xi < x.$$

$$(A.20)$$

(Note: Taylor's formula with Lagrange's form of the remainder is an extension of the mean value theorem for derivatives, that is, (A.20) coincides with (A.6) if $n = 0$.) Suppose we set $x = x_0 + h$. Then $x_0 < \xi < x_0 + h$ can be expressed as $\xi = x_0 + \theta h, 0 < \theta < 1$, so that (A.20) becomes

$$f(x) = f(x_0) + f'(x_0)(x - x_0) + \frac{1}{2!}f''(x_0)(x - x_0)^2$$

$$+ \cdots + \frac{1}{n!}f^{(n)}(x_0)(x - x_0)^n + \frac{1}{(n+1)!}f^{(n+1)}(x_0 + \theta h)^{n+1}, 0 < \theta < 1.$$

$$(A.21)$$

### A.6.2 Generalized Taylor's Formula with Remainder

Suppose $h = x - x_0$ is an $(n \times 1)$ vector with components $h_i = x_i - x_i^0, i = 1, \ldots, n$. Then the **jth order differential** of the real-valued function $y = f(x)$ at point $x_0$ interior to a closed region $K \subset R^n$ may be written as

$$d^j f(x_0, h) = \left( h_1 \frac{\partial}{\partial x_1} + \cdots + h_n \frac{\partial}{\partial x_n} \right)^j_{x_0} f, j = 0, 1, \ldots, n.$$

For instance,

$$d^0 f(x_0, h) = f(X_0),$$

$$d^1 f(x_0, h) = \sum_{r=1}^{n} h_r \frac{\partial f(x_0)}{\partial x_r},$$

$$d^2 f(x_0, h) = \sum_{r=1}^{n} \sum_{s=1}^{n} h_r h_s \frac{\partial^2 f(x_0)}{\partial x_r \partial x_s}, \ldots.$$

In this regard, we have

**Theorem A.9** Generalized Taylor's Formula with Remainder
Let the real-valued function $y = f(x), x \in R^n$, and its first $n + 1$ $(n \geq 0)$ derivatives be continuous throughout a closed region $K$ of $R^n$ containing the point $x_0$ Then the value of $f$ at any point $x$ near $x_0$ is

$$f(x) = f(x_0) + df(x_0, h) + \frac{1}{2!}d^2f(x_0, h)$$

$$+ \cdots + \frac{1}{n!}d^nf(x_0, h) + R_{n+1},$$

(A.22)

where $R_{n+1}$ is a remainder term of the form

$$R_{n+1} = \frac{1}{(n+1)!}d^{n+1}f(x_0 + \theta h, h), 0 < \theta < 1.$$

(A.23)

An important special case of (A.22) occurs when $x \in R^2$, that is,

$$f(x_1, x_2) = \sum_{j=0}^{n} \frac{1}{j!}\left(h_1\frac{\partial}{\partial x_1} + h_2\frac{\partial}{\partial x_2}\right)^j f \Big|_{(x_1^0, x_2^0)}$$

$$+ \frac{1}{(n+1)!}\left(h_1\frac{\partial}{\partial x_1} + h_2\frac{\partial}{\partial x_2}\right)^{n+1} f\Big|_{(x_1^0 + \theta h_1, x_2^0 + \theta h_2)}, \quad 0 < \theta < 1.$$

For $n = 2$,

$$f(x_1, x_2) = \sum_{j=0}^{2} \frac{1}{j!}\left(h_1\frac{\partial}{\partial x_1} + h_2\frac{\partial}{\partial x_2}\right)^j f\Big|_{(x_1^0, x_2^0)}$$

$$+ \frac{1}{3!}\left(h_1\frac{\partial}{\partial x_1} + h_2\frac{\partial}{\partial x_2}\right)^3 f\Big|_{(x_1^0 + \theta h_1, x_2^0 + \theta h_2)}$$

$$= f(x_1^0, x_2^0) + f_1(x_1^0, x_2^0)h_1 + f_2(x_1^0, x_2^0)h_2$$

$$+ \frac{1}{2!}\left[f_{11}(x_1^0, x_2^0)h_1^2 + 2f_{12}(x_1^0, x_2^0)h_1, h_2\right.$$

$$\left. + f_{22}(x_1^0, x_2^0)h_2^2\right]$$

$$+ \frac{1}{3!}\left(h_1\frac{\partial}{\partial x_1} + h_2\frac{\partial}{\partial x_2}\right)^3 f\Big|_{(x_1^0 + \theta h_1, x_2^0 + \theta h_2)}, 0 < \theta < 1.$$

(Note: the differential operator

$$\left(h_1\frac{\partial}{\partial x_1} + h_2\frac{\partial}{\partial x_2}\right)^j$$

is applied to $f$ via the binomial expansion

$$\left(h_1\frac{\partial}{\partial x_1} + h_2\frac{\partial}{\partial x_2}\right)^j f = \sum_{i=0}^{j}\binom{j}{i}h_1^i h_2^{j-i}\frac{\partial^j f}{\partial x_1^i \partial x_2^{j-i}}.$$

# Appendix B

## The Lebesgue Integral

Suppose $f : [a,b] \to R$ is a bounded Lebesgue measurable function on the bounded (Lebesgue) measurable set $[a,b] \subset R$. Consider the partition of the vertical-axis $P = \{l = y_0 < y_1 < y_2 < \cdots < y_n = u\}$, where

$$l = inf\{f(x)|x \in [a,b]\},$$
$$u = sup\{f(x)|x \in [a,b]\},$$

and the mesh of $P$ is

$$|P| = max\{\Delta y_i = y_i - y_{i-1}, i = 1,\ldots,n\}.$$

Next, for $y_i' \in [y_{i-1}, y_i], i = 1,\ldots,n$, let us define the **Lebesgue sum** of $f$ with respect to the partition $P$ on the interval $[l,u], l < u$, as

$$S(f,P) = \sum_{i=1}^{n} y_i' \mu(E_i), \tag{B.1}$$

where

$$E_i = \{x \in [a,b] | y_{i-1} < f(x) < y_i\}, i = 1,\ldots,n,$$

that is, set $E_i$ is composed of all $x$ values for which the corresponding $f(x)$ values lie between $y_{i-1}$ and $y_i$.

To now pass to the Lebesgue integral, we need only apply the limit concept to (B.1). Specifically, a bounded Lebesgue measurable function $f : [a,b] \to R$ is **Lebesgue integrable** on $[a,b]$ if there is a real number $L$ such that, given $\varepsilon > 0$, there exists a $\delta > 0$ such that

$$|S(f,P) - L| < \varepsilon \text{ whenever } |P| < \delta.$$

Here, $L$ is termed the **Lebesgue integral of $f$** on $[a,b]$ and denoted $\int_{[a,b]} f d\mu$.

*Stochastic Differential Equations: An Introduction with Applications in Population Dynamics Modeling*, First Edition. Michael J. Panik.
© 2017 John Wiley & Sons, Inc. Published 2017 by John Wiley & Sons, Inc.

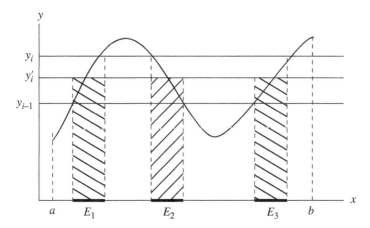

**Figure B.1** Lebesgue integration.

Looking to Figure B.1, $y_i'$ is the height of the $i$th rectangle and $\mu(E_i)$ is the length of its base. For the particular subinterval $\Delta y_i$ considered, $B = E_1 \cup E_2 \cup E_3 = \{x \in [a, b] | y_{i-1} < f(x) < y_i\}$. Then $\mu(B) = \mu(E_1) + \mu(E_2) + \mu(E_3)$ and the area associated with $y_i', y_{i-1} < y_i' < y_i$, is $y_i'\mu(B)$—the amount contributed to the total integral.

An alternative procedure for defining the Lebesgue integral (one that mirrors somewhat the development of the Riemann integral) is the following. Given the preceding partition $P$ along with the definitions of $l$, $u$, and $E_i, i = 1,...,n$, set $E_0 = \{x \in [a, b] | f(x) = l\}$ and $y_{-1} = l$. Then we can define the **upper and lower Lebesgue sums** as

$$S_p = \sum_{i=0}^{n} y_i \mu(E_i) \text{ and } s_p = \sum_{i=0}^{n} y_{i-1}\mu(E_i),$$

respectively. Then the **upper Lebesgue integral** is defined as $U = \inf_P S_p$ and the **lower Lebesgue integral** is specified as $L = \sup_P s_p$. Hence, a bounded Lebesgue measurable function $f$ on $[a, b]$ is **Lebesgue integrable** if

$$\lim_{|P| \to 0} |L - U| = 0,$$

In which case $L = U$ and the common value is termed the **Lebesgue integral** of $f$ on $[a, b]$ and written $\int_{[a, b]} f d\mu$.

It is important to note (and the reader might already have guessed) that if $f : [a, b] \to R$ is continuous on a finite closed interval, then the Lebesgue integral $\int_{[a, b]} f d\mu$ exists and has the same value as the Riemann integral $\int_a^b f(x) dx$.

# Appendix C

# Lebesgue–Stieltjes Integral

As we shall now see, the Lebesgue–Stieltjes (LS) integral is based on the general definition of the Lebesgue integral; it is the ordinary Lebesgue integral with respect to the LS measure.

Let $F(x):R \to R$ be a monotone, nondecreasing real-valued function (or, equivalently, $F$ is of bounded variation on $R$) which is everywhere right-continuous. In addition, let $\mu_F((a, b]) = F(b) - F(a)$ for each semi-open interval $(a, b]$, that is, $\mu_F$ is a non-negative and finitely additive measure function for the class of intervals of the form $(a, b]$. Since $\mu_F$ can be extended to the $\sigma$-ring $\mathcal{B}$ of Borel sets in $R$, this extension defines $\mu_F$ on the completion of $\mathcal{B}$ with respect to $\mu_F$. Hence, the finite Borel measure corresponding to $F$, $\mu_F$, is called the **LS measure corresponding to $F$**, and the resulting class of sets is termed **LS measurable for $F$**.

We know that a general function $F(x) : R \to R$ is a **distribution function** (see Equation (2.2)) if

i. $F$ is monotone increasing and continuous from the right; and
ii. $F(x) \to 0$ as $x \to -\infty$ and $F(x) \to 1$ as $x \to +\infty$.

Clearly, a distribution function $F$ can be used to define an LS measure $\mu_F$ on a $\sigma$-algebra $\mathcal{I}_F$ within the measure space $(R, \mathcal{I}_F, \mu_F)$, where $\mathcal{I}_F$ is the class of sets that are LS measurable for $F$.

Given the preceding discussion, suppose $F$ is a distribution function with corresponding LS measure $\mu_F$ and let $f$ be a bounded Borel-measurable function. (Remember that $f$ is Borel measurable if $\{x|f(x) \in B\}$ is a Borel set for each Borel set $B \in R$.) Then the **LS integral of $f$ with respect to $F$** is

$$\int f(x)\, dF(x) = \int f(x)\, d\mu_F(x). \tag{C.1}$$

(Note: we can also view the LS integral as an ordinary Lebesgue integral with respect to the LS measure that is associated with any function $F$ of bounded

*Stochastic Differential Equations: An Introduction with Applications in Population Dynamics Modeling*, First Edition. Michael J. Panik.
© 2017 John Wiley & Sons, Inc. Published 2017 by John Wiley & Sons, Inc.

variation on $R$. Stated alternatively, in defining the LS integral we have shifted our focus from a nondecreasing, right-continuous function to a measure function on $R$, that is, we have adopted the view that if $F$ is nondecreasing and right-continuous, then $\int f\,dF(x)$ represents the Lebesgue integral of $f$ using the measure generated by $F$.)

When $f(x)$ is a continuous, bounded real-valued function and $F(x)$ is a nondecreasing real-valued function, the LS integral is equivalent to the Riemann–Stieltjes (RS) integral.

The reason why the LS integral is important for our purposes is that it can be used to evaluate expectations via the distribution function. Suppose we have a probability space of the form $(R, \mathcal{B}, P)$, where $\mathcal{B}$ is the Borel $\sigma$-algebra on $R$ and $P$ is a probability measure on $\mathcal{B}$. As a practical matter, the characterization of $P$ on $\mathcal{B}$ can be made by a distribution function $F(x) = P((-\infty, x])$, with $dF(x)$ representing the LS probability measure associated with $F$. This is because there exists a one-to-one correspondence between the distribution function $F$ on $R$ and the probability measure $P$ on $\mathcal{B}$. In fact, for $(a, b] \in \mathcal{B}, P((a, b]) = F(b) - F(a)$.

Since a random variable on $(R, \mathcal{B}, P)$ can be viewed as a measurable function $f(x)$, its expectation can be written as $E(f(x)) = \int_R f(x)\,dF(x)$, which is the **LS integral of $f$ with respect to $F$**. More formally, if $X$ is a random variable with distribution function $F(x)$ and $f(x)$ is a measurable function on $R$, with $f(x)$ integrable, then

$$E(f(x)) = \int_R f(x)\,dP(x) = \int_R f(x)\,dF(x). \tag{C.2}$$

Note that for $f(X) = X, E(X) = \int_R x\,dF(x)$, provided this integral is finite or exists.

# Appendix D

# A Brief Review of Ordinary Differential Equations

## D.1 Introduction

A basic problem encountered in the integral calculus is the determination of a function $y = f(x)$ when its derivative is a given function of $x$, $g(x)$. In this regard, given that

$$\frac{dy}{dx} = g(x) \text{ or } dy = g(x)dx, \tag{D.1}$$

the function $y$ can be expressed as

$$y = \int g(x)\, dx = c, \quad c \text{ an arbitrary constant.}$$

Equation (D.1) is an elementary example of an **ordinary differential equation** (ODE).

## D.2 The First-Order Differential Equation

Consider the ODE of the form

$$F\left(x, y, \frac{dy}{dx}\right) = 0, \tag{D.2}$$

where $F$ is a given function of the arguments $x$, $y$, and $dy/dx$. A solution to (D.2) will typically be a relation of the form

$$G(x, y, c) = 0, \quad c \text{ an arbitrary constant.} \tag{D.3}$$

Given (D.3), we may obtain an ODE that is solved by (D.3) by differentiating this equation with respect to $x$,

$$\frac{\partial G}{\partial x} + \frac{\partial G}{\partial y}\frac{dy}{dx} = 0, \tag{D.4}$$

*Stochastic Differential Equations: An Introduction with Applications in Population Dynamics Modeling*, First Edition. Michael J. Panik.
© 2017 John Wiley & Sons, Inc. Published 2017 by John Wiley & Sons, Inc.

and eliminating $c$ from (D.3) and (D.4).

For instance, suppose

$$G(x, y, c) = y - cx = 0 \tag{D.5}$$

so that

$$\frac{dy}{dx} - c = 0. \tag{D.6}$$

If we eliminate $c$ from the preceding two equations, we obtain the ODE

$$y - x\frac{dy}{dx} = 0. \tag{D.7}$$

Does (D.5) represent *all* solutions to (D.7)? A moment's reflection on this issue will elicit a *no* answer.

A general theorem pertaining to the question of all solutions of the ODE $dy/dx = f(x, y)$ is provided by Theorem D.1.

**Theorem D.1**   Existence and Uniqueness
Let the functions $f$ and $\partial f/\partial y$ be continuous in some rectangular region $R = \{(x, y) | a < x < b, c < y < d\}$ of the $xy$-plane containing the point $(x_0, y_0)$. Then within some subinterval $x_0 - h < x < x_0 + h$ of $a < x < b$, there is a unique solution $y = y(x)$ of the ODE $dy/dx = f(x, y)$ with the initial condition $y_0 = y(x_0)$.

The conditions in this theorem are *sufficient* to guarantee the existence of a unique solution of $dy/dx = f(x, y), y_0 = y(x_0)$. (Note: if $f$ does not satisfy the hypotheses of the theorem, a unique solution may still exist.) The expression $G(x, y, c) = 0$ that contains all solutions of $F(x, y, (dy/dx)) = 0$ in $R$ will be termed the **general solution** of this ODE in $R$. A solution obtained from the general solution of an ODE via the assignment of a particular value of $c$ will be called a **particular solution**.

## D.3   Separation of Variables

The simplest class of ODEs $dy/dx = f(x, y)$ that can be solved in a straightforward fashion are the equations in which the variables are separable in a way such that

$$N(y)\frac{dy}{dx} + M(x) = 0$$

or

$$M(x)dx + N(y)dy = 0. \tag{D.8}$$

Then the general solution of (D.8) is

$$\int M(x)\, dx + \int N(y)\, dy = c, \quad c \text{ an arbitrary constant.}$$

For instance, suppose

$$\frac{dy}{dx} + xe^{-y} = 0.$$

Then $e^{y} dy + x dx = 0$ (see (D.8)),

$$\int e^{y} dy + \int x dx = c$$

or

$$e^{y} + \frac{1}{2}x^{2} = c.$$

Then solving for $y$ yields

$$y = \ln\left(c - \frac{1}{2}x^{2}\right).$$

(Any restriction on $c$?)

A class of first-order ODEs that can be rendered separable by a change of variable is

$$\frac{dy}{dx} = g\left(\frac{y}{x}\right). \tag{D.9}$$

Here, $g$ is said to be homogeneous in that it depends only on the ratio $y/x$ (or $x/y$) and not on $x$ and $y$ separately. If we set $v = y/x$ so that $dy/dx = g(v)$, where $x$ is the independent variable and $v$ is taken to be the new dependent variable, then, if we can find $v$ as a function of $x$, $y = vx$ will ultimately be a function of $x$ and a solution to (D.9) follows. To find $v$, use $y = v(x)$ to obtain

$$\frac{dy}{dx} = v + x\frac{dv}{dx}.$$

Then (D.9) becomes

$$v + x\frac{dv}{dx} = g(v),$$

and thus the variables $v$ and $x$ are separable as

$$\frac{dv}{v - g(v)} + \frac{dx}{x} = 0. \quad \text{(See (D.8))}$$

For example, let us solve

$$x\frac{dy}{dx} - xe^{-\frac{y}{x}} - y = 0.$$

(Clearly this expression is homogeneous once both sides are divided by $x$.) Setting $y = vx$, we obtain

$$x\left(v + x\frac{dv}{dx}\right) - xe^{-v} - vx = 0$$

or

$$x\frac{dv}{dx} - e^{-v} = 0,$$

and thus

$$e^v dv - dx/x = 0$$

(separability holds). Then integrating yields

$$e^v - lnx = ln\,c,$$
$$e^v = ln\,cx,$$
$$v = ln(ln\,cx)$$

or

$$y = vx = x\,ln(ln\,cx), cx > 1.$$

## D.4  Exact ODEs

Given the function

$$g(x, y) = c, \tag{D.10}$$

the associated ODE appears as

$$dg = \frac{\partial g}{\partial x}dx + \frac{\partial g}{\partial y}dy = 0.$$

This said, suppose we have an ODE of the form

$$M(x, y)dx + N(x, y)dy = 0. \tag{D.11}$$

If $M$ and $N$ are the partial derivatives of some function $g(x, y)$, with

$$M(x, y) = \frac{\partial g}{\partial x}, N(x, y) = \frac{\partial g}{\partial y}, \tag{D.12}$$

then we can argue that if (D.10) and (D.11) hold, it must be the case that (D.12) also holds. That is, if there exists a function $g(x, y)$ for which (D.12) is true, then the solution of the ODE (D.11) is provided by (D.10). In this instance, the expression $M(x,y)dx + N(x,y)dy$ is an exact differential, $dg$, and the ODE (D.11) is called an **exact ODE**.

A necessary and sufficient condition on the functions $M$ and $N$ which ensures that (D.11) be an exact ODE is that

$$\frac{\partial M(x,y)}{\partial y} = \frac{\partial N(x,y)}{\partial x}. \tag{D.13}$$

An obvious question that presents itself at this point in our discussion is "When (D.13) holds, can we always find a function $g$ for which (D.12) also holds?" Let's start with determining a $g$ such that

$$\frac{\partial g(x,y)}{\partial x} = M. \tag{D.14}$$

Given $g$ such that (D.14) holds, can we also find $N = \partial g / \partial y$? From (D.14),

$$g(x,y) = \int^x M(t,y)\,dt + h(y) \tag{D.15}$$

(we integrate partially with respect to $x$), where $h(y)$ is an *arbitrary function of integration*. Then from (D.15),

$$\begin{aligned}\frac{\partial g}{\partial y} &= \frac{\partial}{\partial y}\int^x M(t,y)\,dt + h'(y) \\ &= \int^x \frac{\partial M(t,y)}{\partial y}\,dt + h'(y).\end{aligned} \tag{D.16}$$

Substituting $\partial g / \partial y = N(x,y)$ into (D.16) yields

$$h'(y) = N(x,y) - \int^x \frac{\partial M(t,y)}{\partial y}\,dt. \tag{D.16.1}$$

Our next step is to determine $h(y)$. While not obvious, the right-hand side of (D.16.1) is a function of $y$ only. To verify this, let us differentiate the right-hand side of this expression with respect to $x$ so as to obtain

$$N_x(x,y) - M_y(x,y) = 0$$

via (D.13). Since the right-hand side of (D.16.1) is independent of $x$, a single integration with respect to $y$ gives $h(y)$ or $h(y)$ can be obtained from

$$h'(y) = N - \frac{\partial}{\partial y}\int M\partial x.$$

Finally, substituting form $h(y)$ in (D.15) gives

$$g(x, y) = \int^x M(t, y)\, dt + \int^y \left[ N(x, s) - \int^x M_s(t, s)\, dt \right] ds. \tag{D.17}$$

For example, we can readily determine that the ODE

$$\underbrace{\left( 2xy^2 + 2y \right)}_{M(x,y)} dx + \underbrace{\left( 2x^2 y + 2x \right)}_{N(x,y)} dy = 0$$

is exact since

$$\frac{\partial M}{\partial y} = 4xy + 2 = \frac{\partial N}{\partial x}.$$

To obtain a solution to this exact ODE, let us start with

$$\frac{\partial g}{\partial x} = M = 2xy^2 + 2y,$$

$$g = \int \left( 2xy^2 + 2y \right) \partial x + h(y)$$

$$= x^2 y^2 + 2xy + h(y),$$

and thus

$$\frac{\partial g}{\partial y} = 2x^2 y + 2x + h'(y).$$

Since also

$$\frac{\partial g}{\partial y} = N = 2x^2 y + 2x,$$

it follows that

$$h'(y) = 2x^2 y + 2x - \frac{\partial}{\partial y} \int M \partial x$$

$$= 2x^2 y + 2x - \frac{\partial}{\partial y} \left( x^2 y^2 + 2xy \right)$$

$$= 2x^2 y + 2x - \left( 2x^2 y + 2x \right) = 0$$

or $h(y) = c_0 =$ constant. The solution to the original ODE is thus $g = c$ or

$$g = x^2 y^2 + 2xy = c,$$

where $c_0$ has been absorbed into the constant $c$.

How about the ODE

$$\underbrace{(2x+4y)\,dx}_{M(x,y)} + \underbrace{(2x-2y)\,dy}_{N(x,y)} = 0?$$

Is it exact? Since

$$\frac{\partial M}{\partial y} = 4 \neq \frac{\partial N}{\partial x} = 2,$$

the answer is no.

## D.5 Integrating Factor

Suppose an ODE is not exact. In this circumstance, our objective is to transform a non-exact ODE to an exact one by multiplying the non-exact one by a suitable function of $x$ and $y$. That is, given

$$M(x,y)\,dx + N(x,y)\,dy = 0, \tag{D.18}$$

with $\partial M/\partial y \neq \partial N/\partial x$, can (D.18) be made exact by forming

$$\mu M dx + \mu N dy = 0, \tag{D.19}$$

where $\mu = \mu(x,y)$. If any such $\mu$ exists, then it is called an **integrating factor** of (D.18). The **requirement for exactness** of (D.19) is

$$\frac{\partial(\mu M)}{\partial y} = \frac{\partial(\mu N)}{\partial x},$$

$$\left(\frac{\partial \mu}{\partial y}\right)M + \left(\frac{\partial M}{\partial y}\right)\mu = \left(\frac{\partial \mu}{\partial x}\right)N + \left(\frac{\partial N}{\partial y}\right)\mu,$$

or

$$\left(\frac{\partial \mu}{\partial y}\right)M - \left(\frac{\partial \mu}{\partial x}\right)N = \left(\frac{\partial N}{\partial x} - \frac{\partial M}{\partial y}\right)\mu. \tag{D.20}$$

So if a function $\mu$ satisfying (D.20) can be found, then (D.19) will be exact. The solution of (D.19) can then be obtained by the method of Section D.4 and will appear implicitly as $g(x,y) = c$. Note that this expression also defines the solution to (D.18) since $\mu$ can be eliminated from all terms of (D.19).

Rather than trying to solve the partial differential equation (D.20) for $\mu$, let us, as a compromise, take an alternative tack. Specifically, we shall consider the conditions on $M$ and $N$ that will enable us to obtain an integrating factor $\mu$ of a particular form. That is, we shall restrict the equation for $\mu$ to one that depends on $x$ alone. In this instance, we set $\partial \mu/\partial y = 0$ and rewrite (D.20) as

$$\frac{1}{\mu}\frac{d\mu}{dx} = \frac{(\partial M/\partial y) - (\partial N/\partial x)}{N}. \tag{D.20.1}$$

Since the left-hand side of this equation depends only on $x$, we should be able to find an integrating factor $\mu = \mu(x)$ such that the right-hand side of (D.20.1) also depends on $x$ alone. If we set the right-hand side of (D.20.1) equal to $r(x)$, then our integrating factor has the form

$$\frac{1}{\mu}\frac{d\mu}{dx} = r(x),$$

and thus

$$\mu = e^{\int r(x)\,dx}. \tag{D.21}$$

For instance, let us solve the ODE

$$\underbrace{\left(3xy + y^2\right)}_{M(x,y)} dx + \underbrace{\left(x^2 + xy\right)}_{N(x,y)} dy = 0.$$

Since

$$\frac{\partial M}{\partial y} = 3x + 2y \neq \frac{\partial N}{\partial x} = 2x + y,$$

this ODE is not exact. From (D.20.1),

$$\frac{1}{\mu}\frac{d\mu}{dx} = \frac{3x + 2y - (2x + y)}{x^2 + xy} = \frac{1}{x}$$

or $d\mu/\mu = dx/x$, and thus $\ln\mu = \ln x$ or $\mu(x) = x$. Then from (D.19),

$$\underbrace{\left(3x^2 y + xy^2\right)}_{M(x,y)} dx + \underbrace{\left(x^3 + x^2 y\right)}_{N(x,y)} dy = 0.$$

Clearly, this ODE is exact since

$$\frac{\partial M}{\partial y} = 3x^2 + 2xy = \frac{\partial N}{\partial x} = 3x^2 + 2xy.$$

Then it can be shown that the solution to this exact ODE is

$$g(x,y) = x^3 y + \frac{1}{2}x^2 y^2 = c.$$

An important example of an ODE in which an integrating factor $\mu$ can be found which depends only on $x$ is that of a linear ODE of the form

$$\frac{dy}{dx} + R(x)y = Q(x) \tag{D.22}$$

or

$$[R(x)y - Q(x)]\underbrace{dx}_{M(x,y)} + \underbrace{dy}_{(N(x,y)=1)} = 0. \tag{D.22.1}$$

(Note: this ODE is linear since $y$ and $dy/dx$ appear linearly in (D.22).) Then we can determine

$$\frac{(\partial M/\partial y) - (\partial N/\partial x)}{N} = R(x)$$

so that the integrating factor has the form

$$\mu = e^{\int R(x)\,dx}. \tag{D.23}$$

Now, multiplying both sides of (D.22) by this integrating factor gives

$$e^{\int R(x)\,dx}\left[\frac{dy}{dx} + R(x)y\right] = e^{\int R(x)\,dx}Q(x)$$

or

$$\frac{d}{dx}\left(e^{\int R(x)\,dx}y\right) = e^{\int R(x)\,dx}Q(x).$$

Then from this latter expression, we get the solution to (D.22) or

$$e^{\int R(x)\,dx}y = \int e^{\int R(x)\,dx}Q(x)\,dx + c. \tag{D.24}$$

For example, consider the linear ODE

$$\frac{dy}{dx} + y = x.$$

(Here, $R = 1$ and $Q = x$.) From (D.23), an integrating factor is

$$\mu = e^{\int dx} = e^x.$$

Thus, via (D.24), the solution is

$$e^x y = \int xe^x\,dx + c.$$

Employing integration by parts $\left( \int u\,dv = uv - \int v\,du \right)$, we ultimately obtain

$$y = x - 1 + ce^{-x}.$$

## D.6   Variation of Parameters

Let us return to the linear ODE (D.22) or

$$\frac{dy}{dx} + R(x)y = Q(x) \tag{D.22}$$

and express its homogeneous version as

$$\frac{dy_h}{dx} + R(x)y_h = 0 \tag{D.25}$$

or

$$\frac{dy_h}{y_h} + R(x)dx = 0.$$

The solution to this homogeneous ODE is

$$y_h = c_0 e^{-\int R(x)dx} \tag{D.26}$$

$$c_0 u(x), c_0 \text{ a parameter,}$$

where

$$u(x) = e^{-\int R(x)dx}. \tag{D.27}$$

Suppose we now *guess* that the *full solution* to (D.22) has the form

$$y = v(x)u(x). \tag{D.28}$$

(Clearly, we have replaced the arbitrary parameter $c_0$ in (D.26) by a term dependent on $x$, that is, we are *varying the parameter* $c_0$.) If we substitute (D.28) into (D.22), we obtain an ODE for $v(x)$ as follows:

$$\frac{dv(x)}{dx}u(x) + \underbrace{v(x)\frac{du(x)}{dx} + R(x)v(x)u(x)}_{\frac{dy_h}{dx} + R(x)y_h = 0} = Q(x),$$

$$\frac{dv(x)}{dx}u(x) = Q(x), \tag{D.29}$$

$$dv(x) = \frac{Q(x)}{u(x)}dx,$$

$$v(x) = \int \frac{Q(x)}{u(x)}dx + c.$$

Then inserting (D.27) and (D.29) into (D.28) ultimately yields

$$y = u(x)v(x)$$

$$e^{-\int R(x)dx}\left(\int Q(x)e^{\int R(x)dx}dx + c\right). \tag{D.30}$$

For instance, suppose we have the ODE

$$\frac{dy}{dx} - y = e^x.$$

Here, $R(x) = -1, Q(x) = e^x$, and thus, from (D.27),

$$u(x) = e^{-\int R(x)dx} = e^{\int dx} = e^x.$$

Then (D.30) becomes

$$y = e^x\left(\int e^x e^{-\int dx}dx + c\right)$$

$$= e^x\left(\int e^x e^{-x}dx + c\right)$$

$$= e^x(x + c) = xe^x + ce^x.$$

# References

Allee, W. C. *Animal Aggregations: A Study in General Sociology*. Chicago, IL: University of Chicago Press (1931).

Allen, E. *Modeling with Itô Stochastic Differential Equations*. New York: Springer (2010).

Arnold, L. *Stochastic Differential Equations: Theory and Applications*. New York: John Wiley & Sons, Inc. (1974).

Baudoin, F. *Diffusion Processes and Stochastic Calculus*. Zurich: European Mathematical Society (2010).

Billingsley, P. *Probability and Measure*, Anniversary edition. Hoboken, NJ: John Wiley & Sons, Inc. (2012).

Bucy, R. S. Stability and Positive Martingales. *J. Differ. Equ.* **1**: 151–155 (1965).

Chan, K. C., Karolyi, G. A., Longstaff, F. A., Sanders, A. B. An Empirical Investigation of Alternative Models of the Short-Term Interest Rate. *J. Financ.* **40**: 1209–1227 (1992).

Ciesielski, Z. Hölder Condition for Realizations of Gaussian Processes. *Trans. Am. Math. Soc.* **99**: 403–413 (1961).

Cox, J. C., Ingersoll, J. E., Ross, S. A. A Theory of the Term Structure of Interest Rates. *Econometrica* **53**: 385–408 (1985).

Cysne, R. P. On the Statistical Estimation of Diffusion Processes: A Partial Survey. *Braz. Rev. Econ.* **24**: 273–301 (2004).

Daniell, P. J. Integrals in an Infinite Number of Dimensions. *Ann. Math.* **20**: 281–288 (1918).

Dineen, S. *Probability Theory in Finance*, 2nd ed. Providence, RI: American Mathematical Society (2013).

Donsker, M. A. *An Invariance Principle for Certain Probability Limit Theorems*. Mem. AMS, No. 6. New York: AMS (1951).

Doob, J. L. *Stochastic Processes*. New York: John Wiley & Sons, Inc. (1953).

Durham, G., Gallant, A. Numerical Techniques for Maximum Likelihood Estimation of Continuous-Time Diffusion Processes. *J. Bus. Econ. Stat.* **20**: 297–316 (2002).

*Stochastic Differential Equations: An Introduction with Applications in Population Dynamics Modeling*, First Edition. Michael J. Panik.
© 2017 John Wiley & Sons, Inc. Published 2017 by John Wiley & Sons, Inc.

Evans, L. C. *An Introduction to Stochastic Differential Equations.* Providence, RI: American Mathematical Society (2013).

Feller, W. Two Singular Diffusion Problems. *Ann. Math.* **54** (1): 173–182 (1951a).

Feller, W. Diffusion Processes in Genetics, in *Proceedings of the 2nd Berkeley Symposium on Mathematical Statistics and Probability*, Neyman, J. ed. Berkeley, CA: University of California Press: 227–246 (1951b).

Friedman, A. *Stochastic Differential Equations and Applications.* New York: Dover Publications, Inc. (2006).

Gard, T. C. *Introduction to Stochastic Differential Equations.* New York: Marcel Dekker, Inc. (1988).

Gihman, I., Skorokhod, A. *Stochastic Differential Equations.* New York: Springer-Verlag (1972).

Gompertz, B. On the Nature of the Function Expressive of Human Mortality, and on a New Mode to Determining the Value of Life Contingencies. *Philos. Trans. R. Soc. Lond.* **115**: 513–585 (1825).

Hahn, W. *Stability of Motion.* New York: Springer (1967).

Has'minskiy. *Stochastic Stability of Differential Equations.* Alphen: Sijtjoff and Noordhoff (1980).

Hurn, A., Jeisman, J., Lindsay, K. Seeing the Wood for the Trees: A Critical Evaluation of Methods to Estimate the Parameters of Stochastic Differential Equations. *J. Financ. Econ.* **5**: 390–455 (2006).

Iacus, S. M. *Simulation and Inference for Stochastic Differential Equations: With R Examples.* New York: Springer (2008).

Itô, K. On a Stochastic Integral Equation. *Proc. Imp. Acad. Tokyo* **22**: 32–35 (1946).

Jensen, B., Poulsen, R. Transition Densities of Diffusion Processes: Numerical Comparison of Approximation Techniques. *J. Deriv.* **9**: 18–32 (2002).

Jiang, D., Shi, N., Zhao, Y. Existence, Uniqueness, and Global Stability of Positive Solutions to the Food-Limited Population Model with Random Perturbation. *Math. Comput. Model.* **42**: 651–658 (2005).

Kloeden, P. E., Platen, E. *Numerical Solution of Stochastic Differential Equations.* New York: Springer (1999).

Kolmogorov, A. Grundbegriffe der Wahrscheinlich keitsrechnung. *Erg. Mat.* **2** (3): 78–81 (1933).

Krstić, M., Jovanović, M. On Stochastic Population Model with the Allee Effect. *Math. Comput. Model.* **52**: 370–379 (2010).

Lamperti, J. A Simple Construction of Certain Diffusion Processes. *J. Math. Kyoto Univ.* **4**: 161–170 (1964).

Lévy, P. *Processus Stochastiques et Mouvement Brownien.* Paris: Gauthiar-Villars (1948).

Lyapunov, A. M. Probleme General de la Stabilite du Mouvement. *Comm. Soc. Math. Kharkov* **2**: 265–272 (1892).

Mao, X. *Stochastic Differential Equations and Applications*, 2nd ed. Chichester: Horwood Publishing Ltd. (2007).

Maruyama, G. Continuous Markov Processes and Stochastic Equations. *Rend. Circ. Mat. Palermo* **4**: 48–90 (1955).

McKean, H. P. *Stochastic Integrals.* New York: Academic Press (1969).

Milstein, G. N. A Method of Second-Order Accuracy Integration of Stochastic Differential Equations. *Theory Probab. Appl.* **23**: 396–401 (1978).

Nicolau, J. Introduction to the Estimation of Stochastic Differential Equations Based on Discrete Observations. *Autumn School and International Conference, Stochastic Finance* (2004).

Øksendal, B. *Stochastic Differential Equations,* 6th ed. New York: Springer (2013).

Ozaki, T. Non-Linear Time Series Models and Dynamic Systems. *Handbook of Statistics,* Vol. **5**, Hannan, E. ed. North-Holland/Amsterdam: Elsevier: 25–83 (1985).

Ozaki, T. A Bridge Between Nonlinear Time Series Models and Nonlinear Stochastic Dynamical Systems: A Local Linearization Approach. *Stat. Sin.* **2**: 25–28 (1992).

Ozaki, T. A Local Linearization Approach to Nonlinear Filtering. *Int. J. Control.* **57**: 75–96 (1993).

Platen, E. Weak Convergence of Approximations to Itô Integral Equations. *Z. Angew. Math. Mech.* **60**: 609–614 (1980).

Platen, E. An Approximation Method for a Class of Itô Processes. *Liet. Mat. Rink.* **21** (1): 121–133 (1981).

Richards, F. J. A Flexible Growth Function for Empirical Use. *J. Exp. Bot.* **10**: 290–300 (1959).

Shoji, L. A Comparative Study of Maximum Likelihood Estimators for Nonlinear Dynamical Systems Models. *Int. J. Control.* **71** (3): 391–404 (1998).

Shoji, I., Ozaki, T. *Estimation for a Continuous Time Stochastic Process: A New Local Linearization Approach.* Res. Memo., No. 524. Tokyo: The Institute of Statistical Mathematics (1994).

Shoji, L., Ozaki, T. Comparative Study of Estimation Methods for Continuous Time Stochastic Processes. *J. Time Ser. Anal.* **18**: 485–506 (1997).

Shoji, L., Ozaki, T. Estimation for Nonlinear Stochastic Differential Equations by Local Linearization Method. *Stoch. Anal. Appl.* **16**: 733–752 (1998).

Simmons, G. F. *Introduction to Topology and Modern Analysis.* New York: McGraw-Hill Book Co., Inc. (1963).

Sørensen, H. Parametric Inference for Diffusion Processes Observed at Discrete Points in Time: A Survey. *Int. Stat. Rev.* **72** (3): 337–354 (2004).

Steele, J. M. *Stochastic Calculus and Financial Applications.* New York: Springer (2010).

Taylor, S. J. *Introduction to Measure and Integration.* New York: Cambridge University Press (1973).

Tvedt, J. *Market Structure, Freight Rates and Assets in Bulk Shipping.* Bergen: Norwegian School of Economics and Business Administration (1995).

Verhulst, P. F. Notice Sur la loi que la Population Suit Dans Son Accroissement. Correspondence Mathematique et Physique. *Ghent* **10**: 113–121 (1838).

# Index

*Stochastic Differential Equations: An Introduction with Applications in Population Dynamics Modeling*, First Edition. Michael J. Panik.
© 2017 John Wiley & Sons, Inc. Published 2017 by John Wiley & Sons, Inc.